FUNDAMENTALS OF PRESSWORKING

Other SME Books Authored or Edited by
David Alkire Smith

Die Design Handbook

Quick Die Change

also

Quick Die Change (video series)

FUNDAMENTALS OF PRESSWORKING

David Alkire Smith

Society of Manufacturing Engineers
Dearborn, Michigan

987654321

Library of Congress Catalog Card Number: 94-065574
International Standard Book Number: 0-87263-449-3

Additional copies may be obtained by contacting:

Society of Manufacturing Engineers
Customer Service
One SME Drive
Dearborn, Michigan 48121
1-800-733-4763

SME staff who participated in producing this book:

Larry Binstock: Senior Editor
Rosemary Csizmadia: Operations Administrator
Dorothy Wylo: Production Secretary
Frances Kania: Production Secretary
Judy Munro: Manager, Graphic Services

Printed in the United States of America

ABOUT THE SOCIETY OF MANUFACTURING ENGINEERS (SME) and FTA/SME

Founded in 1932, SME is a nonprofit professional society dedicated to the advancement of scientific knowledge in the field of manufacturing engineering. By providing direction for the evolution of manufacturing, SME has earned an international reputation for leadership in manufacturing technology.

Among its activities, SME produces and distributes a vast array of literature focusing on state-of-the-art, as well as traditional, manufacturing techniques. *Fundamentals of Pressworking* is a product of this ambitious publishing effort, which includes books, periodicals, videotapes, and an on-line data service.

Within SME are 10 associations and technical groups, each addressing the needs of a specific manufacturing area. Among them is the Forming Technologies Association (FTA/SME), whose mission is to provide the metal forming and fabrication discipline with information for developing competitive strategies. FTA/SME endorses *Fundamentals of* Pressworking as an effective vehicle for helping it accomplish its mission.

FTA/SME has more than 2,500 professional and student members in 37 nations around the globe.

CONTENTS

PREFACE

Fundamentals of Pressworking is the first book in many years to focus on the basics of pressworking. In it we examine, step by step, the complex press, and die and product interaction. The book's goal is to help the reader make the stamping process work without delays and expensive errors.

This book is truly a large group effort. The author is indebted to a host of individual experts and firms who supplied information, photographs, and drawings, and who reviewed the text for accuracy. Without their cooperation, this work would not have been possible. Because these contributors — who are listed on the following eight pages — donated time, talent, and design examples, the reader will benefit from a wealth of practical process information.

Some people at Ferris State University, Big Rapids, Michigan, played a large role in this book's development. Invaluable input was provided by Engineering Department Head Stephen J. Hickel, along with Associate Professors James A. Rumpf and Bill Thomas. Rumpf and Thomas also field tested a pre-publication version of *Fundamentals of Pressworking* when teaching college credit courses at General Motors Corporation, Ford Motor Company, and Steelcase Corporation during the 1993 fall semester.

Special thanks are due my wife, Marlyn, who spent countless hours proofreading the work and preparing the pre-publication version. The book's final format reflects the fine work of Editor Larry Binstock and the SME Book Publishing staff.

David Alkire Smith

CONTRIBUTORS

James J. Albrecht
Dayton Progress Corporation
Dayton, Ohio

Zain Ali, Design Engineer
The Bradbury Co., Inc.
Moundridge, Kansas

Lewis Alspaugh, President
IDC Corporation
Diamondale, Michigan

Taylan Altan, PhD.
Professor and Director
ERC/NSM
Ohio State University
Columbus, Ohio

Bob Anderson, President
Ready Tools
Dayton, Ohio

Ronald C. Anderson
Human Resource Manager
MascoTech Stamping Technologies, Inc.
Oxford, Michigan

John Andres, Chief Engineer
Specialty Market
Monroe Auto Equipment
Monroe, Michigan

Rich Armerling
Machine Repairman
John Deere Horicon Works
Horicon Wisconsin

Behnam Bahr, Ph.D.
Assistant Professor
Wichita State University
Wichita, Kansas

Jim Ballard
Blanking Supervisor
Worthington Steel Company
Monroe, Ohio

James Barrett, Jr., Ph.D.
President
Link Systems
Nashville, Tennessee

John Bates
President and CEO
Heidtman Steel Products, Inc.
Toledo, Ohio

Tom Beck, President
AuditAir, Inc.
Markham, Ontario, Canada

Mike Beuthin, Supervisor
Die Design
John Deere Horicon Works
Horicon, Wisconsin

Ashok Bhide
Data Instruments, Inc.
Acton Massachusetts

Grant Bibby, President
Orchid Automation Group, Inc.
Cambridge, Ontario, Canada

Fred Blackburn
Vice President of Manufacturing
Zippo Manufacturing Company
Bradford, Pennsylvania

Rollin Bondar, President
MPD Welding, Inc.
Troy, Michigan

Theodore A. Boop
Manager of Engineering Operations
H. P. Products
Louisville, Ohio

John A. Borns, President
JB Management Services
Sylvania, Ohio

David Bradbury
President and CEO
The Bradbury Co., Inc.
Moundridge, Kansas

Anthony M. Bratkovich, PE
Engineering Director
Association for Manufacturing Technology
McLean, Virginia

Roscoe Brumback
Director of Corporate Quality
JAC Products
Ann Arbor, Michigan

Scott Bulych, PE
Tool Engineer
A. G. Simpson Co., Limited
Scarborough, Ontario, Canada

Chuck Carlsson
Safety Director
Association for Manufacturing Technology
McLean, Virginia

Gregory A. Castignola
VP and General Manager
Worthington Steel Company
Monroe, Ohio

Ken Coco, President and CEO
GS Metals
Pinkneyville, Illinois

Troy Coffman, Vice President
Starr Welding
Walbridge, Ohio

Edward M. Colbert, Chairman
Data Instruments, Inc.
Acton, Massachusetts

Douglas A. Cope
Maintenance Supervisor
Worthington Steel Company
Monroe, Ohio

Todd W. Deal
Plant Superintendent
Car-Tec Depot, Inc.
Holland, Michigan

Marvin Deer
Manufacturing Engineer
The John Deere Company
Ankeny, Iowa

Rod Denton, President
Sun Steel Treating, Inc.
South Lyon, Michigan

Jim Derby
TD Marketing Services
Arvin TD Center
Columbus, Indiana

Chris Dittenbir
Plant Engineering Manager
GS Metals
Pinkneyville, Illinois

Goran Djuric, Tool Engineer
A. G. Simpson Co., Limited
Scarborough, Ontario, Canada

Joseph Donofrio, President
D & H Machinery, Inc.
Toledo, Ohio

Peter Dusina, Project Manager
Auto Steel Partnership Program
Southfield, Michigan

Hal Easley, Plant Manager
Chore Time
Milford, Indiana

Dennis R. Ebens, President
Rockford Systems, Inc.
Rockford, Illinois

Joseph M. Engerski, Sr.
General Sales Manager
Verson Corporation
Chicago, Illinois

Ron Erbe, President
R. D. Erbe and Associates
Okemos, Michigan

Daniel N. Falcone
Field Application Manager
Toledo Transducers, Inc.
Holland, Ohio

Luis Ferreira
Vice President of Tooling
and Product Development
A. G. Simpson Co., Limited
Scarborough, Ontario, Canada

Jim Finnerty
Product Specialist
Data Instruments, Inc.
Acton, Massachusetts

Wayne Fish, Facilities Manager
Zippo Manufacturing Company
Bradford, Pennsylvania

Nicholas Fisher, President
Metalworking Machinery Systems, Inc.
Elkhart, Indiana

Jeff Fredline, President
Fredline and Associates
Okemos, Michigan

Michael G. Gaines, Director
Manufacturing Systems
Walker Manufacturing Company
Grass Lake, Michigan

Daniel J, Gargrave
Vice President
Ready Tools
Dayton, Ohio

Robert E. Gauthier, Automotive Manager
Lamana, Inc.
Oak Park, Michigan

Henry H. Gehlmann
Blanking Area Manager (retired)
Ford Motor Company
Woodhaven, Michigan

Phillip A. Gibson
Advanced Energy Products, International
Atlanta, Georgia

Horst M. Glaser, Product Manager
Arvin TD Center
Columbus, Indiana

Donald L. Gleckler, Product Manager
H. P. Products
Louisville, Ohio

Leo Goepfrich
Tool and Die Manager
Ford Motor Company
Chicago Heights, Illinois

Todd Gonzales, Manager
Customer Manufacturing Support
National Steel Corporation
Livonia, Michigan

Jeffrey Gordish, Manager
Product Development
Management Technologies, Inc.
Troy, Michigan

Bob Green, Field Service Manager
E. W. Bliss Company
Hastings, Michigan

Charles A. Gregoire, Manager
Technical Development
National Steel Corporation
Livonia, Michigan

Paul Griglio, President
Pinnacle Associates
Lake Orion, Michigan

Michael J, Guthrie
Executive Vice President
Truemark, Inc.
Lansing, Michigan

Rick Haase, Chief Engineer
Unisorb Machinery Installation Systems
Jackson, Michigan

A. L. Hall, Stamping Engineering
Ford Motor Company
Dearborn, Michigan

Kevin Harding, Ph.D., Team Leader
Industrial Technology Institute
Ann Arbor, Michigan

Roger P. Harrison
Director of Training
Rockford Systems, Inc.
Rockford, Illinois

Donald J. Hemmelgarn
Applications Manager
The Minster Machine Company
Minster, Ohio

Michael R. Herderich, Senior Engineer
The Budd Company Technical Center
Auburn Hills, Michigan

Stephen J. Hickel, Head
Manufacturing Engineering
  Technology Department
Ferris State University
Big Rapids, Michigan

Brooke Hindle, Ph.D.
American Museum of American History
Smithsonian Institution
Washington, D.C.

William H. Hinterman, President
Hinterman Integrated Systems Company
Flint, Michigan

Joseph Hladik, Tool Designer
Western Electric Co.
Columbus, Ohio

Gary Hockin, Plant Manager
Lake Park Industries
Greenwich, Ohio

David M. Holley, Engineering Manager
Darnell and Diebolt Company
Detroit, Michigan

Joseph Ivaska
Vice President of Engineering
Tower Oil and Technology Company
Chicago, Illinois

Norbert Izworski
Product Development Engineer
Body and Assembly Division
Ford Motor Company
Dearborn, Michigan

Ray Jewell, Chief Engineer
Walker Manufacturing Company
Grass Lake, Michigan

Robert I. Johnson
Manufacturing Engineering Manager
General Motors BOC Division
Lansing, Michigan

Michael Jurich, Project Engineer
Monroe Auto Equipment
Monroe, Michigan

Hal Juster
Juster & Juster
Maple Grove, Minnesota

Shale Juster, President
Juster & Juster
Maple Grove, Minnesota

Eric Kaltenbacher
Electrical Engineer
Industrial Technology Institute
Ann Arbor, Michigan

Roger Karenan
Training and Development Director
Braun Engineering
Detroit, Michigan

Stuart P. Keeler, Ph.D.,Manager
Metallurgy and Sheet Metal Technology
The Budd Company Technical Center
Auburn Hills, Michigan

Mark Kenyon, Manager
EC Marketing
Industrial Controls Division
Eaton Corporation
Kenosha, Wisconsin

Karl A. Keyes, President
Feinblanking Ltd.
Fairfield, Ohio

Randy A. Knors
Manufacturing Manager
Walker Manufacturing Company
Culver, Indiana

Mike Koberstein
Director of Engineering
GS Metals
Pinkneyville, Illinois

Paul Kosaian, Area Manager
Ford Motor Company
Chicago Heights, Illinois

Gregory A. Kreps, Tooling Engineer
Capital Die, Tool and Machine Company
Columbus, Ohio

Rick A. Krieger, General Manager
Heidtman Steel Products, Inc.
Baltimore, Maryland

Roman J. Krygier, General Manager
Body and Assembly Operations
Ford Motor Company
Dearborn, Michigan

David Laing, Plant Manager
Parkview Metal Products
San Marcos, Texas

Mike Langford
Human Resources Manager
Olson Metal Products
Seguin, Texas

Mark Lantz
Vice President of Operations
Chore Time Brock
Milford, Indiana

Peter Latkovich
Manufacturing Engineering Manager
Weldotron Corporation
Piscataway, New Jersey

Phillip E. Laven
Director of Manufacturing
Truemark, Inc.
Lansing, Michigan

James L. Lehner
Applications Manager
HMS Products Company
Troy, Michigan

Dan Leighton
Forward Industries
Dearborn, Michigan

Ernie Levine, President and CEO
Admiral Tool and Manufacturing
Chicago, Illinois

Cecil Lewis, Director of Engineering
Midway Products Group, Inc.
Monroe, Michigan

John H. Ling, President and CEO
Bettcher Manufacturing Corporation
Cleveland, Ohio

John Litgen, Dieroom Foreman
Admiral Tool and Manufacturing
Chicago, Illinois

Jerry Looney
Advanced Manufacturing Engineer
Inter-City Products
Lewisburg, Tennessee

Robert G. Lown
Vice President (retired)
Greenard Press and Machine Company
Nashua, New Hampshire

Tracy Lowrey
Production Superintendent
Lakepark Industries
Greenwich, Ohio

Steven Lubar, Ph.D.
American Museum of American History
Smithsonian Institution
Washington, D.C.

Albert A. Manduzzi
Supervisor (retired)
Die Design and Standards
Ford Motor Company
Dearborn, Michigan

Michael R. Martin
Application Specialist
Darnell and Diebolt Company
Detroit, Michigan

John McCurdy, President (retired)
W. C. McCurdy Company
Oxford, Michigan

John McElroy, Editor
Automotive Industries Magazine
Detroit, Michigan

John Mertler
Mechanical Engineering Manager
Worthington Corporation
Columbus, Ohio

Bradley K. Mettert
Field Engineering Manager
Toledo Transducers, Inc.
Holland, Ohio

Rich Metzger
Director of Sales and Marketing
Verson Corporation
Chicago, Illinois

Ralph Meyers, General Foreman
Maintenance & Machine
Commercial Intertech Corp.
Youngstown, Ohio

Harry J. Micka
Manufacturing Engineering Manager
LA-Z-BOY Chair Company
Monroe, Michigan

Richard Micka
Vice President of Administration
LA-Z-BOY Chair Company
Monroe, Michigan

James Miller
Doctor of Osteopathy
Monroe, Michigan

Patrick Murley, Plant Superintendent
Worthington Steel Company
Porter, Indiana

Eugene J. Narbut
Die Room Unit Leader
Auto Alliance International
Flat Rock, Michigan

Paul Nawrocki, General Manager
Capital Die, Tool and Machine Company
Columbus, Ohio

Nancy Negohosian, Marketing Manager
HMS Products Company
Troy, Michigan

David J. Nelson
Manager, Applications and Estimating
Verson Corporation
Chicago, Illinois

Al Nichols
Area Leader, Stamping Engineering
Auto Alliance International
Flat Rock, Michigan

Gil Novak
Manager, Operations Support
Body and Assembly Division
Ford Motor Company
Dearborn, Michigan

Ken O'Brien
Vice President of Manufacturing
Walker Manufacturing Company
Grass Lake, Michigan

Steve Odum, Manager
Tooling and Equipment Engineering
Inter-City Products
Lewisburg, Tennessee

O. (Sam) Oishi, President
JIT Automation, Inc.
Scarborough, Ontario, Canada

Paul Pace, President
American Aerostar
Valencia, California

Dick Peoples, Metallurgist
Heidtman Steel Products, Inc.
Toledo, Ohio

Angelo Piccinini
Die Design and Die Standards Supervisor
Chrysler Corporation
Detroit, Michigan

Gerald A. Pool
Senior Manufacturing Engineer
Cadillac Motor Car Division
General Motors
Troy, Michigan

Robert W. Prucka
Director of Manufacturing
Midway Products Group, Inc.
Monroe, Michigan

Dan Quickel
Assistant General Manager
Marada Industries
Westminster, Maryland

Frank Randall
Engineering Manager
Zippo Manufacturing Company
Bradford, Pennsylvania

Anthony Rante
Manager, Mechanical Engineering
Danly - Komatsu LP
Chicago, Illinois

Tom Ready, Attorney
Ready, Sullivan & Ready
Monroe, Michigan

James A. Rumpf
Associate Professor
Ferris State University
Big Rapids, Michigan

Jerry Rush, Vice President
RACE, Inc.
Anaheim, California

Tom Schafer
Office of General Counsel
Danly - Komatsu LP
Chicago, Illinois

Jack Schron, Jr., President
Jergens, Inc.
Cleveland, Ohio

Tom Schippert
Training Director, MPD Welding
MPD Welding Inc.
Troy, Michigan

Phil Schlachter
Engineering Manager
E. W. Bliss Company
Hastings, Michigan

Daniel A. Schoch, PE
The Minster Machine Company
Minster, Ohio

Rick Schwartz
Application Specialist
The Minster Machine Company
Minster, Ohio

Aniese Seed, President (retired)
Toledo Transducers, Inc.
Holland, Ohio

Dennis Shirk, Vice President
Hydra-Fab, Inc.
East Detroit, Michigan

Richard Shirk, President
Hydra-Fab, Inc.
East Detroit, Michigan

Edwin Shoemaker
Vice President of Engineering
LA-Z-BOY Chair Company
Monroe, Michigan

David Skinner, Vice President
Toledo Transducers, Inc.
Holland, Ohio

Kenneth L. Smedberg, President
Smedberg Machine Corporation
Chicago, Illinois

Andrew D. Smith, Staff Member
Parts Evaluation SQA
Auto Alliance International
Flat Rock, Michigan

Gail A. Smith, Principal
Inmedia, Inc.
Dallas, Texas

Jeffrey L. Smolinski
Director of Operations
Rose Johnson
Grand Rapids, Michigan

Gary D. Smotherman
International Representative
U.A.W.
Detroit, Michigan

Scott A. Snyder, Facilities Engineer
Midway Products Group, Inc.
Monroe, Michigan

Janet Sofy
HMS Products Company
Troy, Michigan

Dave Sparks
Pressroom General Foreman
Chore Time
Milford, Indiana

Fredric Spurck, President
Webster Industries, Inc.
Tiffin, Ohio

Philip D. Stang
Senior Consultant
Gemini Consulting, Inc.
Morriston, New Jersey

John N. Steel
Metal Stamping Consultant
Lawrenceburg, Tennessee

Robert Storer, President
Toledo Transducers, Inc.
Holland, Ohio

James Story, Ph.D.
Senior Technical Specialist
Alcoa Technical Center
Alcoa Center, Pennsylvania

Fred Strouse
Senior Applications Engineer
Industrial Controls Division
Eaton Corporation
Kenosha, Wisconsin

Dick Studer, President
HPM Corporation
Mount Gilead, Ohio

Tim Stults, Manufacturing Engineer
Olson Metal Products
Seguin, Texas

Scott J. Sullivan
Industrial Engineer & Founder
TMS Time Management Systems
Galesville, Wisconsin

S. E. Swanson, M.D.
Neurological Associates
Ypsilanti, Michigan

Irby Tallant, Engineer in Charge
FCC Detroit Field Office
Farmington Hills, Michigan

Nick Tarkney
Director of Research and Education
Dayton Progress Corporation
Dayton, Ohio

Marty Taugner
Facility Engineering Supervisor
Twinsburg Stamping Plant
Chrysler Corporation
Twinsburg, Ohio

John Taylor, Field Applications Engineer
Syron Engineering & Manufacturing Company
Saline, Michigan

Mark R. Tharrett, Systems Engineer
General Motors Corporation
Warren, Michigan

Eric Theis, Vice President
Herr-Voss Corporation
Callery, Pennsylvania

Bill Thomas
Associate Professor
Ferris State University
Big Rapids, Michigan

Jack Thompson, President
Monroe Auto Equipment Company
Monroe, Michigan

George K. Tolford
Executive Vice President
Webster Industries, Inc.
Tiffin, Ohio

Mr. Ed. Tremblay, Vice President
Engineering and Manufacturing
P/A Industries
Bloomfield, Connecticut

Ray Vanderbok
Metalforming Program Manager
Industrial Technology Institute
Ann Arbor, Michigan

Allen J. Vanderzee
General Engineering Manager
Verson Corporation
Chicago, Illinois

Jeffrey L. Varner
Blanking Applications Manager
Worthington Steel Company
Monroe, Ohio

Ron Votava, Engineering Manager
Danly - Komatsu LP
Chicago, Illinois

William Lee Walker III
Vice President
Manufacturing Operations
Car-Tec Depot, Inc.
Holland, Michigan

Gordon Wall
Occupational Safety Consultant
Waterford, Michigan

Bernard J. Wallis
Chairman of the Board
Livernois Engineering Corporation
Dearborn, Michigan

Elmer Ward
Vice President of Manufacturing (retired)
Olson Metal Products Company
Seguin, Texas

Keith Weaver, Press Engineer
Capital Die, Tool and Machine Company
Columbus, Ohio

Dave Whyte
Stamping Engineering and Tooling Manager
Ford Motor Company
Dearborn, Michigan

Dale R. Williams
Field Application Specialist
Motorola Corporation
Schaumburg, Illinois

Kimball Williams
Senior Engineering Specialist
Eaton Corporation
Southfield, Michigan

Ron Wilson, Plant Manager
Fontaine Fifth Wheel Company
Rocky Mountain, North Carolina

Bill Winchell
Associate Professor
Ferris State University
Big Rapids, Michigan

Ken Windes
Customer Service Manager
E. W. Bliss Company
Hastings, Michigan

Joseph L. Wise
Maintenance Manager
Webster Manufacturing Company
Tiffin, Ohio

Terry Wissman
Managing Director
Granutech Environmental Systems
Division of The Minster Machine Company
Minster, Ohio

Wayne Wittaker, Vice President
Unisorb Machinery Installation Systems
Jackson, Michigan

James H. Woodard
Executive Vice President
Tridan Tool and Machine, Inc.
Danville, Illinois

Mike Young, President
Vibro/Dynamics
Broadview, Illinois

Sheldon E. Young, Chairman
Vibro/Dynamics
Broadview, Illinois

Timothy Zemaitis
Metallurgical Engineer
Sun Steel Treating, Inc.
South Lyon, Michigan

CHAPTER 1

HISTORICAL PRESS DEVELOPMENT

ANCIENT PRESSES AND PRESSING

Presses have been used since ancient times, with some types still unchanged today. One type still used by primitive cultures is the simple, uncomplicated lever press (Figure 1-1). Here a pole or timber is placed under a rock ledge, which serves as a fixed pivot. The pressure is supplied by stones placed on the end of the timber.

Figure 1-1. *A simple lever press used to extract or press substances from vegetable matter.*

Like the lever press, the hydraulic press provides a controllable force that can be maintained as long as needed. It is widely employed in food processing to press oil and juices from vegetable matter.

Another simple press that originated in antiquity — in China — used the wedge and ramp principle. In this case, a rectangular timber framework contained the wooden pressing platens. Two or more wedges were driven together to apply pressure to the pressing blocks. A press like this was used for lacquer production in 19th-century Japan.

Screw Presses

The screw press, which was well developed by the 16th century, was used mainly for printing and minting coins.

Figure 1-2. A tool room screw press employs the centuries-old design principle of early printing and coinage presses.

Screw presses of metal construction, fitted with a heavy, large-diameter, screw-actuating hand-wheel, develop great force upon contact with the work. The amount of energy stored, and hence the work performed, is controlled by how fast the hand-wheel is rotating before contacting the work.

Manually operated screw presses still find limited use in small die construction and tryout work. A tool room screw press is illustrated in Figure 1-2. Typical applications include die testing, die assembly, and disassembly involving pressing operations.

Limited use is still made of such presses for forming mold cavities of very ductile steel with a hardened steel male master called a *hub*. The process is known as *hubbing*. After the hubbing, steel receives the impression of the hub to the correct depth. The formed mold cavity is hardened for wear resistance, usually by carburizing. Hand-operated screw presses also remain useful for shear-fitting irregularly shaped punches into tool steel die blocks and strippers. Hand presswork operations, such as hubbing mold cavities and shear-fitting dies, have been largely replaced with either conventional or wire-burn electrical discharge machining (EDM).

Current production of screw presses is limited mainly to power-actuated types, which are very useful for precision forging and coining operations.

Trip Hammers

To increase forging productivity, a hammer head of substantial weight was attached to the end of a lever. Early trip hammers were operated by foot power, leaving the user's hands free to position the work. Later, water or steam power actuated presses were arranged to deliver blows continuously. They could also deliver blows at designated moments when the operator tripped a foot treadle or a hand-operated lever. Such hammers were useful for a variety of operations, especially hot forging rifle barrels and wrought iron articles.

Simple Drop Hammers

Early drop hammer presses for pressworking sheet metal consisted of a wooden framework or housing in which a fixed lower die and a movable upper die were placed. For forming simple metal parts, such as embossed decorative metal panels, the lower die was made of zinc and the upper die was of lead, cast directly onto the zinc lower die. This zinc lower die served as part of the mold.

The lead upper die was raised in several ways, depending on production requirements and available power source. One simple lifting arrangement used a rotating horizontal drum at the top of the machine, driven by water or steam power. To lift the upper die, a rope was attached to the die and made several turns around the drum. The drum was pulled taut by the operator. Tightening the rope greatly increased friction of the rope on the drum, thus raising the upper die. Releasing the rope permitted the heavy upper die to fall onto the lower die with the workpiece in place. Several blows might be required to completely form the work.

The lead upper die, dulling much more rapidly than the harder zinc lower die, was sharpened by simply being dropped onto the lower die several times without a workpiece in place. The zinc lower die was either manually redressed or recast when dull. Similar methods are still used to a limited extent for forming low-volume stampings such as aircraft sheet metal parts. Concern over the toxicity of lead is one factor that may cause this forming method to be replaced by such techniques as hydroforming.

Drop Hammer Development

Both flat and closed die hot forging were natural applications for the drop hammer. Increasing the weight by providing a heavy ram guided by the machine housing resulted in more powerful blows. The upper die was attached to the ram by a dovetail slot and/or bolts.

Other methods for raising the weight of the ram and upper die included lifting a long flat board attached to the ram by power-driven frictional rollers or by steam or pneumatic cylinders.

The steam-lifting arrangement was improved by employing a double acting cylinder to lift the ram and drive it downward. In this form, the drop hammer evolved into the more powerful and productive steam hammer.

C-frame Presses

In the 19th century, fabricating wrought iron into structural components and steam boilers required punching many rivet holes. Since wrought iron contained abrasive impurities, drilling the holes was slow and the tool, made of plain high-carbon steel, required frequent resharpening. This activity led to the rivet punch, which in turn evolved into the C-frame press (Figure 1-3). Its screw and attached punch were actuated by manually turning a wrought-iron bar inserted through the cross-drilled holes in the screw head.

C-frame punches were used extensively in boiler and bridge construction as well as general steel fabrication employing rivets. The next step in the evolution of the boilermakers' rivet punch replaced the screw mechanism with a hand-operated hydraulic pressure system. The mechanism employed was essentially the same as the familiar hydraulic bottle jack. Examples of this improvement were used from 1895 and perhaps earlier. Portable hand-actuated C-frame structural steel punches are still produced, with the essential design features unchanged for approximately a century.

Historically, punching holes was much faster than drilling. Today, the punching process remains more economical for materials ranging from foil to thick high-alloy steel plate.

POWER PRESS DEVELOPMENT

An important step in power press development was replacing the actuating screw in both the screw press and C-frame

Figure 1-3. *Example of an early screw-actuated rivet hole punch, which evolved into the C-frame press.*

boilermaker's punch, with a horizontal crankshaft fitted with a flywheel.

In 1855, Augustus Alfred, a New England farmer, machinist, and clockmaker, built the small straightside press illustrated in Figure 1-4. The press, like many screw presses of the period, was equipped with a hand-actuated flywheel and slide guiding system.

Figure 1-4. *The 1855 straightside clockmaker's press built by Augustus Alfred. (Smithsonian Institution)*

The combination of flywheel momentum and the mechanical advantage of the crankshaft provided sufficient force to punch out clock parts by hand. The machine in the figure shows wear on the flywheel, indicating it was also power driven with a flat leather belt.

Power C-frame Presses

Machine deflection, a problem common to C-frame presses, results in angular misalignment under load. Tie rods installed on the front of the press reduce, but do not eliminate, angular deflection.

Current C-frame presses seldom incorporate the *open-back inclinable* (OBI) feature originally developed to facilitate gravity discharge of parts and scrap through the rear of the machine. Today, air blowoff and conveying devices serve this purpose. Angular deflection is restricted to acceptable limits by heavy frame construction. The frame structure rests directly on the foundation or shop floor. This type of press is known as an *open-back stationary* (OBS) machine.

HYDRAULIC PRESSES

Like the mechanical press, the hydraulic press has an interesting evolutionary history. For example, one pioneering firm in the field, the Hydraulic Press Manufacturing Company (HPM) of Mount Gilead, Ohio, was founded by Augustus Q. Tucker, a distinguished student of mechanical engineering who owned extensive apple orchards. In 1867 he started research that — 10 years later — resulted in the first practical hydraulic cider press, a model illustrated in Figure 1-5.

The original design featured dual moving carts, which carried apples in cage-like containers. The two carts and bed arrangement permitted the pressing duty cycle to be maximized. Pulp could be removed and new apples loaded during the pressing cycle. This ingenious dual cart cider press predated the invention of the dual moving bolster system for stamping presses by more than three quarters of a century. The latter was invented by the engineer Vasil Georgeff of Chicago press builder Danly Machine Corporation in 1956.

Many parts comprising the framework, and even the vertical ram, which was actuated from the bottom of the press, were made of wood. The original working fluid was water.

Tucker's press worked so well that enthusiastic stockholders provided expansion capital in 1877. Under Tucker and his successors the firm prospered. Improved HPM fruit presses

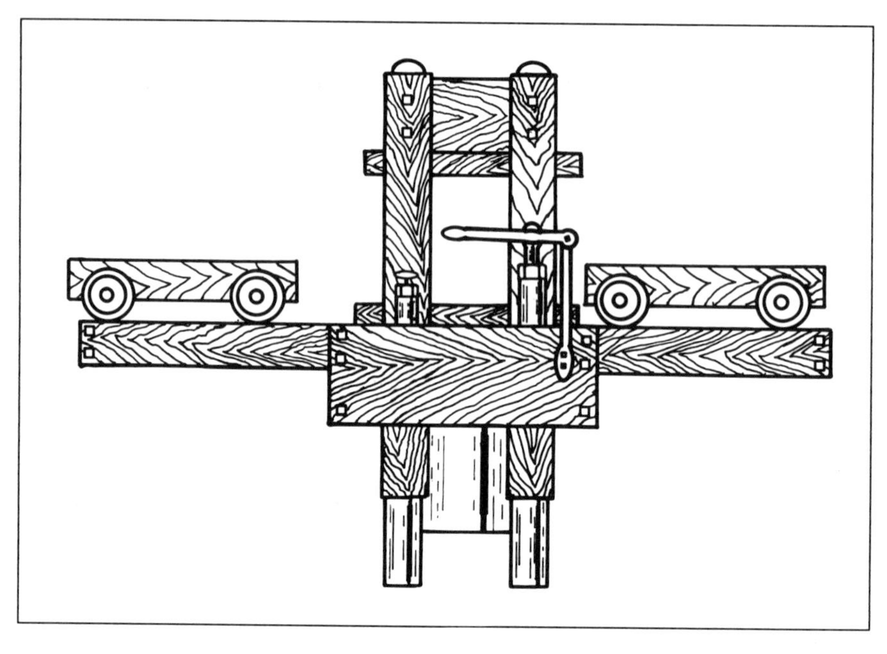

Figure 1-5. *A model of the first practical hydraulic cider press. Note the dual moving carts and the pumping mechanism. (The Hydraulic Press Manufacturing Company)*

Figure 1-6. *A historic photograph showing compression molding of thermoplastic material utilizing a hydraulic press operation. (The Hydraulic Press Manufacturing Company)*

employed mainly metal parts. The later production models, most of which are still in service, functioned essentially in the same manner as the compression molding press for thermoplastics (Figure 1-6).

HPM did not invent the hydraulic press. Earlier uses in Europe included the application of both screw and hydraulic presses for compressing black powder military rocket propellant. The hydraulic press proved superior in uniformity and safety of rocket production, especially considering the former method used a drop hammer. However, HPM, a company responsible for many developments, adapted the hydraulic press to high-production pressworking applications.

Hydraulic Presses for Metalworking

Applications where prolonged clamping pressure is required, such as die casting and plastic molding operations, almost exclusively use hydraulic presses. Increasingly, hydraulic presses are applied in metal stamping operations.

A major advantage of hydraulic presses for deep drawing is the availability of full tonnage anywhere in the press stroke. Figure 1-7 is a historic HPM photograph of a deep drawing hydraulic press application. Increasingly, hydraulic presses are built with double actions or controllable hydraulic die

Figure 1-7. A historic photograph of a deep drawing hydraulic press operation. (The Hydraulic Press Manufacturing Company)

cushions, permitting precise drawing speed and blankholder force control.

Mechanical stamping presses currently have capacities up to 6000 tons (53.376 MN) or more. Such machines are generally very large. Higher tonnages or more compact construction is practical in modern hydraulic presses. Hydraulic presses for cold forging can have up to 50,000 tons (445 MN) capacity. Some fluid cell presses have force capacities over 150,000 tons (1334 MN).

BIBLIOGRAPHY

1. Roy Ellen, *A Vertical Wedge Press from the Banda Islands, Technology and Culture,* Volume 33, Number 1, January, 1992.
2. Brooke Hindle and Steven Lubar, *Engines of Change, The American Industrial Revolution, 1790-1860,* Washington: Smithsonian Institution Press, 1985.
3. *It Started with an Apple,* 75th Commemorative Publication of The Hydraulic Press Manufacturing Company, Mount Gilead, Ohio: Hydraulic Press Manufacturing.
4. David A. Smith, *Quick Die Change,* Dearborn, Michigan: Society of Manufacturing Engineers, 1991.
5. Taylan Altan and others, "Improvement of Part Quality in Stamping by Controlling Blankholder Force and Pressure," Columbus, Ohio: The Ohio State University Engineering Research Center for Net Shape Manufacturing, May 1992.

CHAPTER 2

TYPES OF PRESSES

GAP-FRAME AND STRAIGHTSIDE

There are over 300,000 presses in use in the United States, and many more worldwide. The housing or structure of presses falls into two predominant categories, *gap-frame* and *straightside*. These frame types are found in both mechanically driven and hydraulic presses. Mechanical press components, such as the clutch, brake, counterbalance system, die cushions, electrical, and pneumatic control systems, are of similar design in both the gap-frame and straightside types.

Selection of type and size is determined by factors such as:

- Force requirements;
- Size of the press bed area and required stroke;
- Acceptable angular and total deflection under load;
- Press stroking rates; and
- Feeding device requirements.

In North America, the mechanical drive system is more popular than its hydraulic counterpart. Thus, while this chapter covers many structural designs common to both mechanical and hydraulic types, it focuses mainly on mechanical drive and control systems.

Gap-Frame

The principal feature of gap-frame machines is the C-shaped opening. For this reason, gap- frame presses are also referred to as C-frame presses. Figure 2-1 illustrates a typical gap-frame press.

In press force capacities up to approximately 250 tons (2,224 kN) and larger, gap-frame presses are less costly than straightside presses having the same force capacity and control features. In the 35 to 60 ton (311 to 534 kN) force range, they may cost approximately half as much as a straightside press.

The C-shaped throat opening has the advantage of permitting access to the die from three sides, allowing pressworking operations to be carried out on the corners and sides of large sheets of

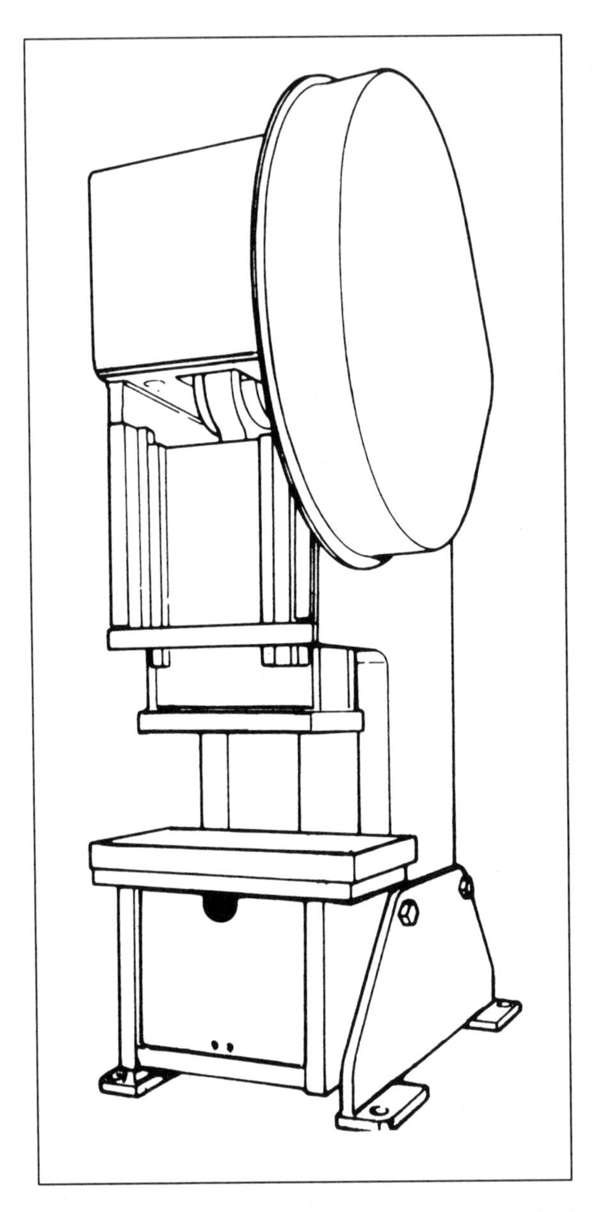

Figure 2-1. A typical gap-frame open back inclinable (OBI) mechanical press. (Verson Corporation)

material. The open back is also accessible for discharging finished parts and scrap, as well as feeding stock. The open accessibility feature is also useful for trying out and repairing dies in the press, changing new dies, and making adjustments to produce a different style part.

The main disadvantage of gap-frame presses is an unavoidable angular misalignment that occurs under load. Limiting the amount of angular misalignment requires very robust construction, which adds to the weight and cost of the machine.

Straightside

Straightside presses derive their name from the vertical columns or uprights on either side of the machine. The columns, together with the bed and crown, form a strong housing for the crankshaft, slide, and other mechanical components.

The housing, or frame, of most straightside presses is held together in compressive preload by prestressed tie rods; some straightside presses have solid frames. Generally, a solid-frame straightside press is less expensive than one having tie rods. However, those having tie rods are easier to ship disassembled in large sizes, and generally better able to withstand overloads.

Freedom from angular deflection under load is one reason for choosing a straightside rather than a gap-frame press for work involving close tolerance dies. The part dimensional accuracy and number of hits between necessary die maintenance often improve by a factor of three or more.

Figure 2-2 illustrates some of the principal mechanical components of a straightside press having double-end drive gears and two connections. The *bed* is the base the machine must rest on to ensure proper machine functioning.

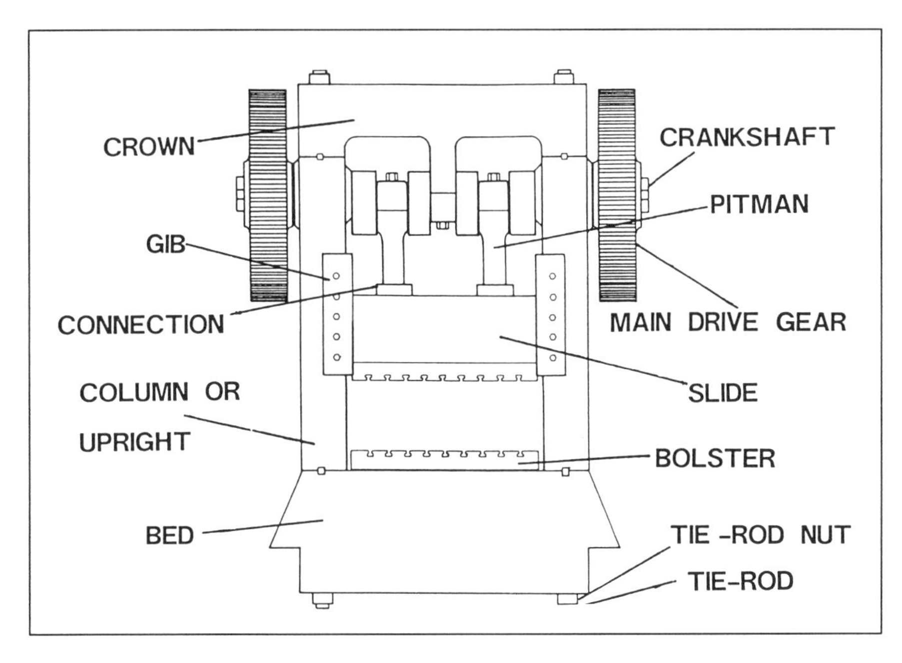

Figure 2-2. *A straightside mechanical press having double-end drive gears and two connections. (Smith & Associates)*

The *columns* support the crown and have gibs attached that guide the slide. The crankshaft end bearings may be contained in the columns or crown.

The *crown* serves many functions, depending upon machine design. Typically, the clutch, brake, motor, and flywheel mount on the crown of the press. The *gears* shown in Figure 2-2 may be open, with only a safety guard to contain the gear if it should fall off due to a failure such as a broken crankshaft. In modern designs, the gears are fully enclosed and run in a bath of lubricant. This serves to lessen noise and ensure long gear life.

The *tie rods* hold the housing assembly in compression. The pitman, connection, bolster, and other parts have similar functions in both gap-frame and straightside presses.

TYPES OF MECHANICAL PRESS DRIVES

In nongeared, or direct drive presses (Figure 2-3) as they are also known, the flywheel is mounted on the end of the crankshaft. The flywheel is motor-driven by means of a belt drive. Directly driven presses are capable of much higher operating speeds than geared types. Speeds range from under 100 strokes per minute to over 1,800 for short-stroke high-speed operation.

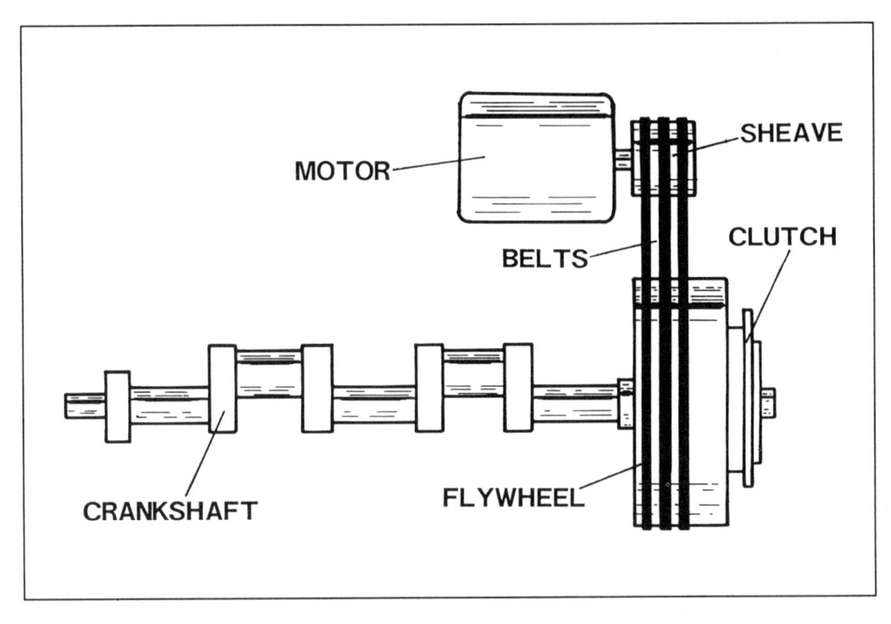

Figure 2-3. *A nongeared drive illustrating a double-throw crankshaft directly driven by the flywheel through a frictional clutch. Energy is transferred from the motor to the flywheel by several V-belts.*

Nongeared presses find widespread application in blanking and shallow forming operations. They have several major advantages over all other press types. First, the design is simple. There are few bearings and no gears to wear out. Also, frictional losses are lower than those of mechanical geared and hydraulic presses.

The high operating speeds provide much greater productivity than geared presses. The direct drive press is very popular for precision progressive die and high-speed perforating operations.

Two main factors limit application of the direct driven press. First, the full rated force of the machine is only available very close to the bottom of the stroke, typically 0.060-inch (1.524 mm) from bottom dead center. A second disadvantage is that the ability to deliver rated forces is substantially reduced if the press is operated at less than full speed.

In single gear reduction presses (Figure 2-4), the flywheel is mounted on the backshaft and the power is transmitted through a pinion to a main gear mounted on the crankshaft. Some single gear reduction presses have main gears mounted on both ends of the crankshaft, which is mounted on a pinion. Single geared presses typically operate in the speed range of 35 to 100 strokes per minute (SPM), with some smaller machines operating at speeds of up to 150 SPM or more.

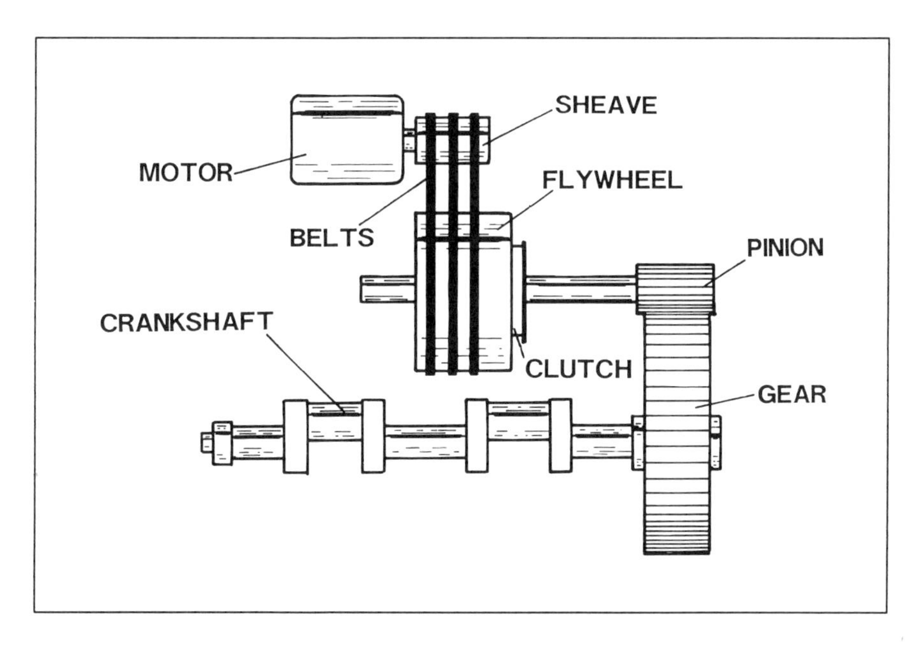

Figure 2-4. A single-geared drive illustrating a double-throw crankshaft driven by a single end gear and frictional clutch. The flywheel drives the crankshaft end gear by means of a smaller pinion gear.

Because these presses utilize a gear reduction with the flywheel on the high speed backshaft, more flywheel energy can be provided than in a nongeared press, making the single-geared press better suited for drawing and heavy forming operations.

Figure 2-5 illustrates how an angular misalignment, proportional to the torque transmitted through a crankshaft with two throws, occurs in single-end drive presses. This causes the side of the ram nearest the driven end of the crankshaft to reach bottom dead center before the other side. The amount of ram tipping tends to be proportional to the force delivered by the machine, provided the ram is evenly loaded.

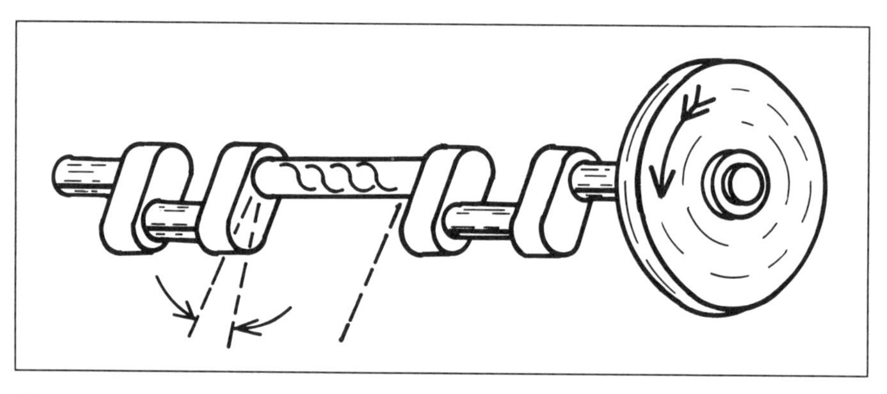

Figure 2-5. *Angular misalignment proportional to the torque transmitted through a crankshaft with two throws occurs in single-end drive presses. This factor results in a ram tipping alignment error.*

The error will be made worse if the largest load is placed on the side of the press opposite the driven end of the crankshaft. Presses having driving gears on each end of the crankshaft are often specified for heavy presswork. Single-end drive presses can be used for precision high-speed presswork. The twist problem is limited by using large diameter eccentric-type crankshafts.

Figure 2-6 illustrates a single-gear reduction twin end drive on a press having a crankshaft with two throws. Driving the crankshaft equally on both ends provides much more accurate left-to-right, ram-to-bed alignment under load than is the case with the single-end drive system. It is important that the machining and

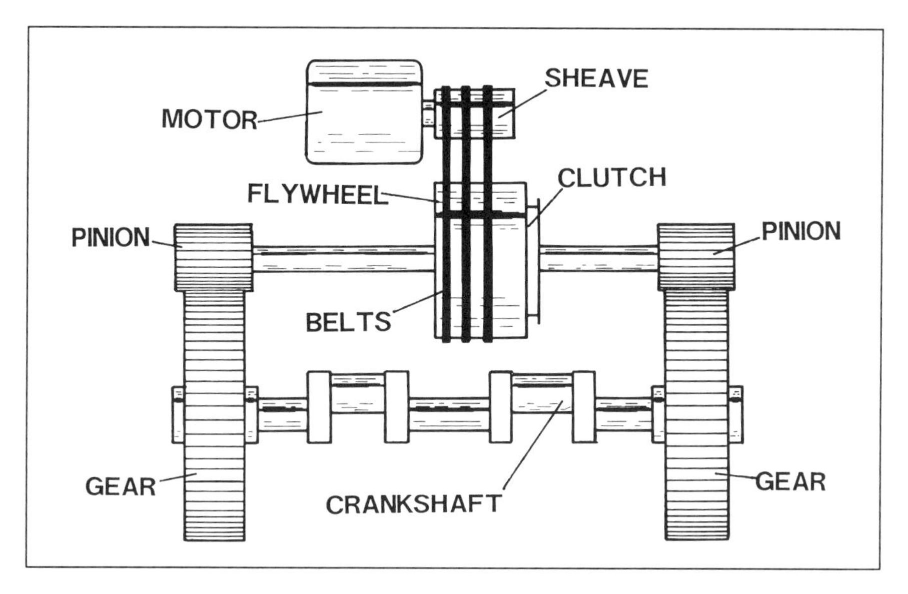

Figure 2-6. Providing a driving gear on either end of the crankshaft avoids the angular misalignment under load illustrated in Figure 2-5.

timing of the gears, keyways, and crankshaft be precise to avoid binding and ensure smooth operation.

Presses having two gear reductions from the flywheel to the crankshaft are termed double-gear reduction presses. These machines, which normally achieve a speed range from 8 to 20 SPM, are used for difficult applications such as heavy deep drawing, cold forging, and flanging large parts such as truck frame rails. Large transfer presses also frequently employ double-gear reduction. Figure 2-7 illustrates one type of clutch and gearing arrangement for a double-gear reduction press.

TERMINOLOGY AND COMPONENT IDENTIFICATION

Press Terminology

The following terms are used to describe some principal characteristics and specifications of power presses. Many of the terms apply to both mechanical and hydraulic gap-frame and straightside presses. Terms that describe the maximum size of die and force capacity are very important.

Shut Height. The space available between press bed or bolster and slide or ram is called the *shut height* and is always measured

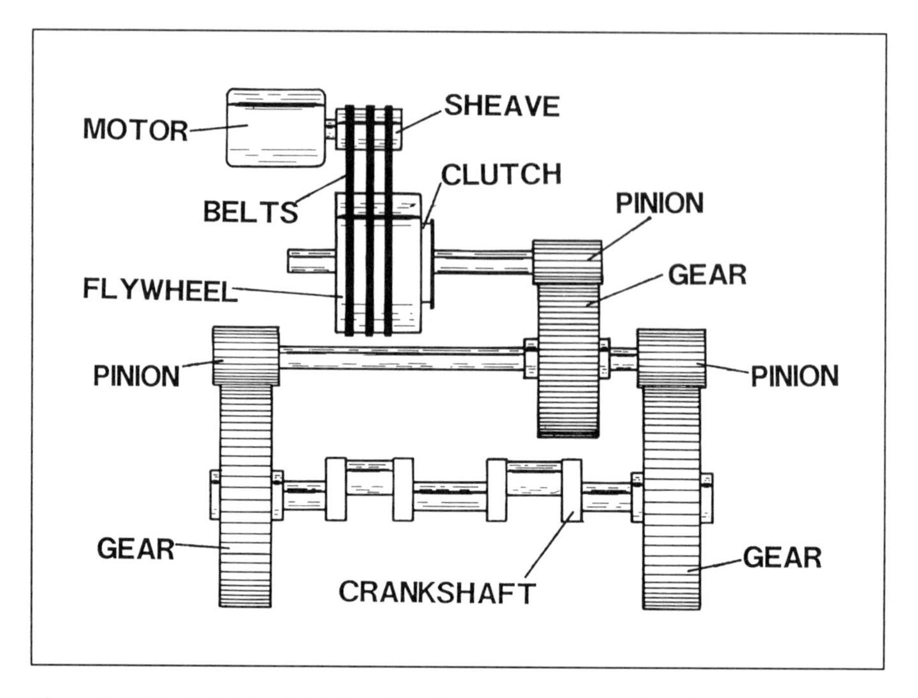

Figure 2-7. *A typical style of clutch and gearing arrangement used on double-gear reduction presses.*

with the press shut or at bottom dead center. It may be specified as the vertical space between the ram and either the top of the bed or the bolster, as illustrated in Figure 2-8.

When a die must be put in an existing press, the distance from the top of the bolster to the bottom of the ram is the figure that should be used. This distance, specified with the screw adjustment at maximum and minimum values, determines the range of closed heights of dies that will fit into the press.

At times, more shut height is needed than can be accommodated with the press bolster in place. Some shops remove the bolster and fasten the die directly to the press bed. This is poor practice. The bolster's thickness is needed to stiffen and assist in spreading the load evenly over the bed's structural members.

Bed or Bolster. The bolster adds stiffness to the press bed and has tapped holes or, preferably, T-slots to permit the die to be fastened securely. The most important bolster measurements are the left-to-right and front-to-back dimensions. These determine the width and length of die that can be accommodated.

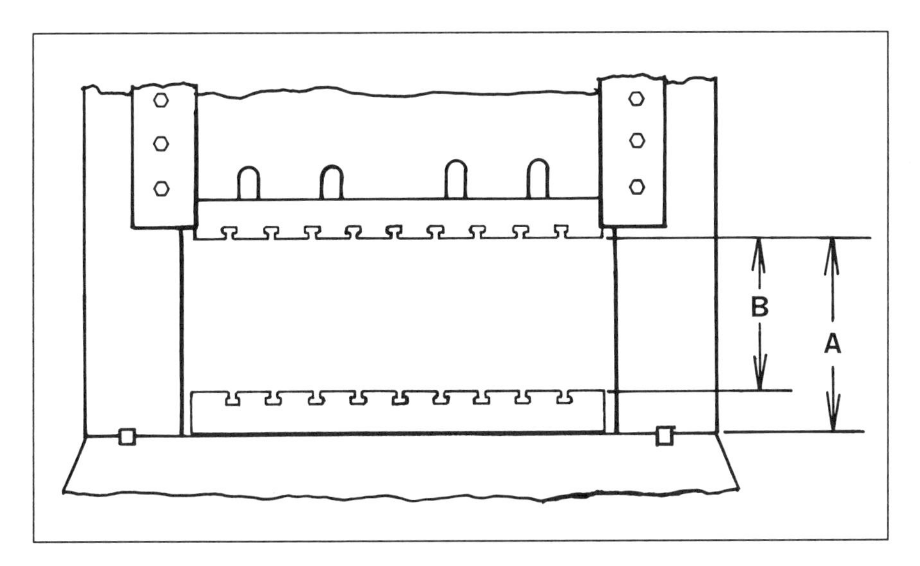

Figure 2-8. Example of shut height measurement taken (A) from the ram to the bed, and (B) from the ram to the bolster.

Occasionally, a shop will place in a press a die that overhangs the edges of the bolster. This is a very poor practice; the die is inadequately supported. Also, establishing proper safety distance and other safeguarding measures to protect the operator may not be possible.

Press Frame Members. The members that make up the framework or housing of presses determine the force capacity and amount of deflection that occur in the machine.

MOVING PRESS PARTS

The moving press parts store, control, and transmit energy supplied by the motor to the die and workpiece, as the force required to accomplish the intended pressworking operation. The principal components are:

- The clutch, which transmits energy from the flywheel to the crankshaft (or eccentric drive in some cases), through reduction gearing.
- The brake, which is used to stop the press and hold the slide and attached mechanism in place.
- The flywheel, which stores the energy supplied by the motor.

- The motor, which furnishes energy to the flywheel.
- Gears to reduce the speed and increase the torque delivered by the flywheel through the clutch.
- The pitman or eccentric strap, which transmits the motion of the crankshaft or eccentric drive to the slide by means of the connection.
- The connection, which connects the pitman or eccentric strap to the slide through a bearing.
- The slide or ram to which the upper die is fastened; it is guided by gibs attached to the machine frame or housing.
- The adjusting screw, located at each slide connection point.
- The counterbalance, to counterbalance the weight of the slide, upper die, and attached linkage.

Mechanical Press Clutches and Brakes

Virtually all mechanical presses transmit the energy stored in the flywheel to the press by means of a clutch mechanism. Otherwise, the press would cycle continuously whenever power was applied to the flywheel. When the clutch is not engaged, the press is stopped and maintained in a stationary position by a brake.

The clutch is a mechanism used to control the coupling of the flywheel (or gear on a geared press) with the press crankshaft. Many older presses utilize a mechanical clutch, which is a full revolution type. When a full revolution clutch is activated, it cannot be disengaged until the crankshaft has made one complete revolution.

Most modern presses are equipped with an air friction clutch and brake arrangement commonly called a partial revolution clutch. The partial revolution clutch can be disengaged at any point in the stroke before the crankshaft has completed a full revolution. The air friction clutch permits versatility of press use such as:

- Rapid dependable engagement;
- Increased press speed;
- Emergency stopping capability in mid-stroke; and
- Long-term reliability.

Correct clutch and brake action is essential to safe press operation. Historically, advances in control reliability have been achieved by adding redundancy to control systems such as dual solenoid clutch actuating valves, brake monitors, and electrical control systems having anti-repeat features. Today, these features are required by law for many pressworking applications.

Full Revolution Clutches

The simplest type of clutch maintains engagement for the full revolution of the press once actuated. Such clutches are simple, low in cost, and reliable, provided they are not abused by overloading, misadjustment, or lack of maintenance. Most full revolution clutches are of a simple mechanical design that operates by means of engaging one or more rolling keys or pins. The engaging mechanism may be actuated manually, or by means of pneumatic, hydraulic, or electrical solenoid devices.

Since the clutch cannot be disengaged before a full revolution is completed, a positive means, such as a physical barrier guard, is required to protect personnel in the area. Few new presses are built with such clutches. Figure 2-9 illustrates an example of a full revolution clutch installed on the crankshaft of a small press.

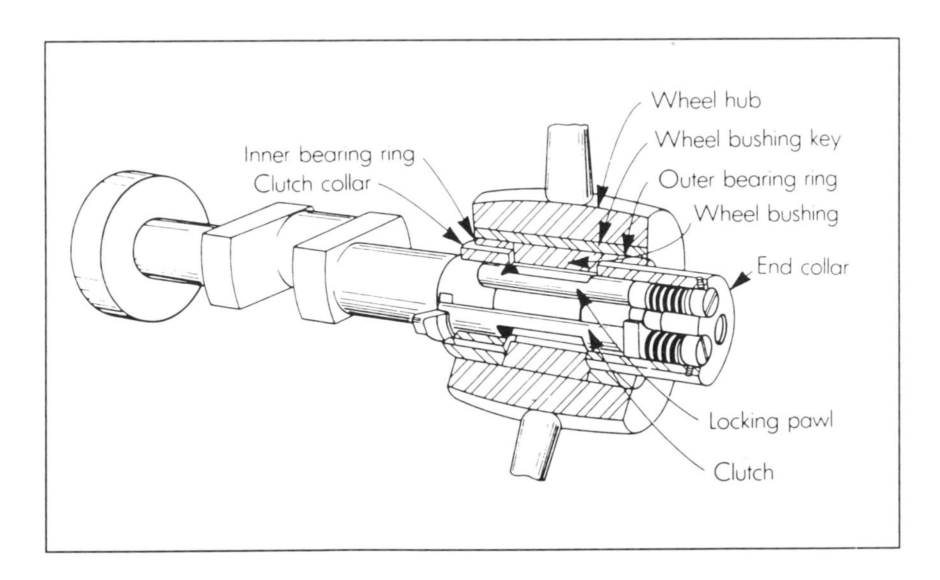

Figure 2-9. A full revolution clutch mounted on the press crankshaft. (E.W. Bliss Company)

An obvious disadvantage is the inability to stop the press before top of stroke when a misfeed is detected. Many full revolution presses cannot transmit power in the reverse direction. This should be considered if a motor needs to be run in the reverse direction to free a press that has become stuck on bottom of stroke.

Older presses can often be upgraded for smoother, more reliable operation by retrofitting improved clutch and brake systems. Also, the electrical controls, which may no longer meet current safety requirements, can be replaced. Usually, the most satisfactory way to retrofit the press is to install a completely new control package especially designed for the application. Such systems are available from several suppliers.

Brakes

A brake serves to stop and hold the press crankshaft or eccentric drive stationary when the clutch is not engaged. In its simplest form, the brake is a drum attached to the crankshaft. A metal band lined with frictional material is attached to the press frame. A spring with screw adjustment is used to apply constant braking action to the drum. Such brakes are commonly used only on small presses having full revolution clutches. Since the inertia of the moving parts is low, required braking action is minimal.

Most mechanical power press brakes are applied by spring pressure. The brake is released by applying air pressure to an air cylinder that depresses the springs holding the frictional surfaces in contact.

Piston-actuated brakes are often combined with a friction clutch in an integral package. Figure 2-10 illustrates such a system. Its main advantages are simplicity and lower cost compared to that of a separate clutch and brake. The design ensures that the clutch must release before the brake can be applied. The combined unit is simple for a trained technician to service.

One design uses a fabric-reinforced circular rubber air tube rather than a piston to accomplish the air actuation. The tube does not require lubrication and is leak-free throughout its useful life. However, the safe reliable functioning of the air valve still requires clean filtered air and lubrication.

Very large presses, or machines designed for rapid stopping, may employ a brake that is a separate unit. Presses having eddy-current drives, such as the Eaton Dynamatic®, require a separate air-released spring-actuated brake. If the Dynamatic drive unit also incorporates an electromagnetic brake, the mechanical brake

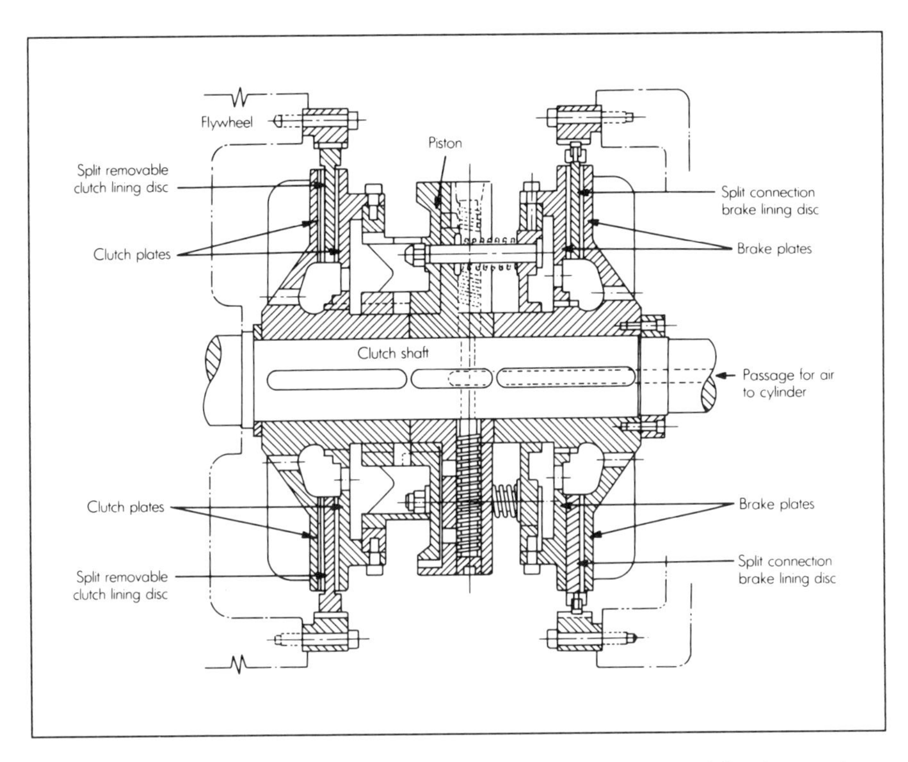

Figure 2-10. *Single piston-actuated combined clutch and brake system. (Verson Corporation)*

is called a holding brake. Here, the majority of the stopping action is accomplished by electromagnetic attraction between stationary and moving parts of the brake in the eddy-current drive.

Eddy-current Clutches and Brakes

Clutches actuated by electromagnetic attraction are termed *eddy-current clutches.* Electrical eddy currents are induced in conducting materials subjected to a changing electromagnetic field. In the case of large eddy current drives such as some Eaton Dynamatic® units, electromagnetic coils are contained within the flywheel assembly.

The clutch is engaged by applying a DC electrical current to the rotating coils through slip rings. The eddy currents, set up in the clutch member attached to the press rotary driven shaft, produce a magnetic field that permits torque to be transmitted from the flywheel to the driven shaft with very little slip.

The amount of slip, or difference in speed, is controlled by the amount of DC current supplied to the coils. This provides a useful way to slow down the press speed during the stroke while the flywheel runs at essentially full speed. This type of drive is called a *constant energy system* (CES), since full flywheel energy is available at any point in the stroke. This feature finds application in large presses where the slowdown function is useful to assist in deep drawing operations or to control moving die member impact problems.

The same principle may also be used to arrest the motion of the press with a set of brake coils using eddy-current action. If everything is functioning properly, the frictional holding-brake dissipates very little energy.

The eddy-current drive system is extensively used on large, single-stroked presses. Many applications are found in nonsynchronous automotive tandem press lines.

Clutch and Brake Specifications

Power press clutches and brakes are rated in several ways. To ensure correct application, long life, and safe operation, it is important to match the clutch and brake to the press.

The amount of torque a clutch can transmit before slipping is often given in inch-pounds or Newton-meters. Air-actuated clutches range in torque capacity from under 25,000 inch-pounds (2,825 Nm) to well over 500,000 inch-pounds (56,500 Nm).

The torque capacity is rated at a stated air pressure, typically from 60 to 80 psi (4.08 to 5.44 bar). Increasing the amount of air pressure results in greater torque-transmitting capacity. Therefore, it is very important not to exceed the manufacturer's recommended maximum values for production applications. The resulting increase in torque may overload the clutch and other press parts, resulting in accelerated wear and possible catastrophic failure.

The torque the brake will withstand without slipping is also rated in inch-pounds or Newton-meters. It is normally less than the clutch torque capacity. The function of the brake is to arrest press movement and hold the slide at any point in the stroke. In its usual configuration, it is applied by mechanical springs and released by air pressure. Brake torque capacities typically range from under 15,000 inch-pounds (1,695 Nm) to over 250,000 inch-pounds (28,250 Nm).

Both the clutch and brake must dissipate heat, since the application and release are not institutions. The capacity to dissipate heat is often given in horsepower or kilowatts. These units are readily converted into other engineering units such as British Thermal Units (BTU) or calories.

The heat dissipation capacity of a clutch or brake increases with rotational speed, since this also increases air movement. Clutches and brakes on single-stroked machines often have fans to increase the heat dissipation capacity.

Some Requirements for Safe Clutch-Brake Operation

Exceeding the torque capacity or heat dissipation capacity of a clutch and brake is dangerous. Such abuse will very likely result in rapid wear. Should the clutch fail, there is danger of operator injury. Following good practice is essential.

It is important that the clutch regulator air setting be no higher than that required to provide rapid engagement and transmit the full torque value required to develop rated press tonnage. Higher settings should not be used because:

- Air is wasted;
- Excessive torque can be transmitted;
- Press damage can result from excessive forces; and
- More air must be exhausted which increases stopping time.

The only reason for increasing the clutch air pressure temporarily is to assist in getting a stuck press off bottom. This procedure should be resorted to only after a careful evaluation of the problem, including the press and clutch manufacturer's recommendation. Should the press not have reached bottom dead center (BDC), increasing the air pressure and attempting to inch will make the problem worse. In such cases the motor will need to be reversed before inching the machine.

The clutch-engaging solenoid air valve admits air to release the brake and engage the clutch. There is always a danger that the clutch air valve may stick in the open position. Should this occur, the press will continue to cycle. A special double air valve is used to greatly lessen this possibility. In the event of a malfunction, the valve, which is self-checking, automatically locks in the off position until the fault is corrected.

In order to avoid erratic operation of the clutch/brake pneumatic system, a supply of clean filtered air is an important requirement. An in-line air filter is used to furnish clean dry air. Often the filter is an integral part of the clutch pressure air regulator. If needed, an air-line lubricator is also installed to keep the system working freely.

To ensure rapid actuation, the clutch air valve is located close to the clutch. The air supply piping must be large enough to ensure rapid actuation, especially in single-stroking applications.

A small surge tank is often located near the air valve to supply the required volume of air rapidly. One or more quick-release valves may be located on the clutch housing to reduce press stopping time. As the clutch wears, the actuating piston or air-tube travel increases. This results in a greater air capacity requirement to both actuate and release the clutch and brake. Increased stopping time is one result.

Wet Clutches and Brakes

The term *wet clutch and brake* refers to a unit operating in a housing filled with oil. A film of oil between the friction disks and intermediate contact plates transmits torque while absorbing most of the heat. Wet clutches fitted with oil coolers are excellent in severe single-stroking applications.

Clutch Adjustment and Maintenance

For safe operation, it is essential to follow the manufacturer's recommendations for frequency of maintenance. Should any significant increase in engagement or stopping time occur, the clutch and brake must be examined by a qualified maintenance technician to determine and correct the cause.

Some clutches and brakes have provisions for adjustment to compensate for wear. This is usually done with solid shims. Here, the manufacturer's recommendations for proper running clearances and wear limits before parts require adjustment or replacement must be strictly followed.

Repairs to clutches and brakes are usually limited to replacement of worn or damaged parts. Any attempt to modify a clutch, to increase the torque or braking capacity, should only be done upon a careful engineering evaluation, and in accordance with the manufacturer's instructions.

Welded repairs generally should not be attempted. The parts are subjected to cyclical loading during operation as well as during frequent heating and cooling cycles. The possibility of crack formation and propagation leading to sudden failure is unacceptable. Very serious pressroom accidents have occurred when unwise welded repairs have been attempted.

Press Capacity Factors

The press builder normally designs presses for specific applications, taking into account three main criteria:

- Component strength;
- Flywheel energy; and
- Torque capacity.

Component Strength. The physical strength of various parts of the press must be sufficient to withstand cyclical loading within the machine's rated force, and limit maximum deflection within design tolerances. Standard mechanical engineering formulas are used by press designers. The goal is to build a robust machine that provides decades of normal service, and can withstand occasional abuse without premature failure.

Flywheel Energy. The motor furnishes energy to the flywheel. Once the flywheel is up to speed and not being cycled, the motor need only supply enough energy to make up for frictional losses. The flywheel stores the energy, and certain amounts are removed by each working press stroke necessary to perform work.

Torque Capacity. The press must have the ability to take the energy of the flywheel and transmit it through the clutch, gears (if a geared press), crankshaft, connection, and slide to perform the required work without exceeding the safe working capacity of any component.

The energy stored in the flywheel increases as the square of the flywheel rotational speed. Thus, presses having variable speed drives vary greatly in the amount of flywheel energy available, depending upon the speed adjustment setting.

While reducing the flywheel speed lowers the total energy available, it does not reduce the press tonnage capacity. The tonnage capacity is based on the strength of the machine component parts.

When the flywheel speed goes below its standard rating, the flywheel energy is reduced by the square of the speed reduction. Likewise, if the speed is increased, the flywheel energy is increased by the square of the speed increase.

The following calculations apply to a 60-ton (534-kN) nongeared press with a variable speed drive adjustable from 50 to 200 strokes per minute. The 60 tons of force is available at a rated distance of 0.060 inch (1.524 mm) above bottom dead center at a rated speed of 100 strokes per minute.

Doubling the speed to 200 strokes per minute will result in four times the usable flywheel energy being stored. Because this press is designed with component strength and torque capacity of 60 tons, the additional stored flywheel energy is not usable. All press tonnage and torque capacity ratings remain unchanged.

If the speed is reduced by 50% to 50 strokes per minute, the flywheel energy is actually reduced by 75% because at one-half speed there is only one-fourth the energy left in the flywheel. The 60-ton (534-kN) press thus would supply 15 tons (133 kN) through 0.060 inch (1.524 mm) or 60 tons through approximately 0.015 inch (0.381 mm). At 50 strokes per minute the press is still a 60-ton press from a component strength and torque capacity standpoint, but the flywheel energy is reduced to 25% of the standard rating. Thus its ability to do work at the 50-stroke-per-minute speed is greatly reduced. Figure 2-11 illustrates the relationship of flywheel speed to usable force and energy.

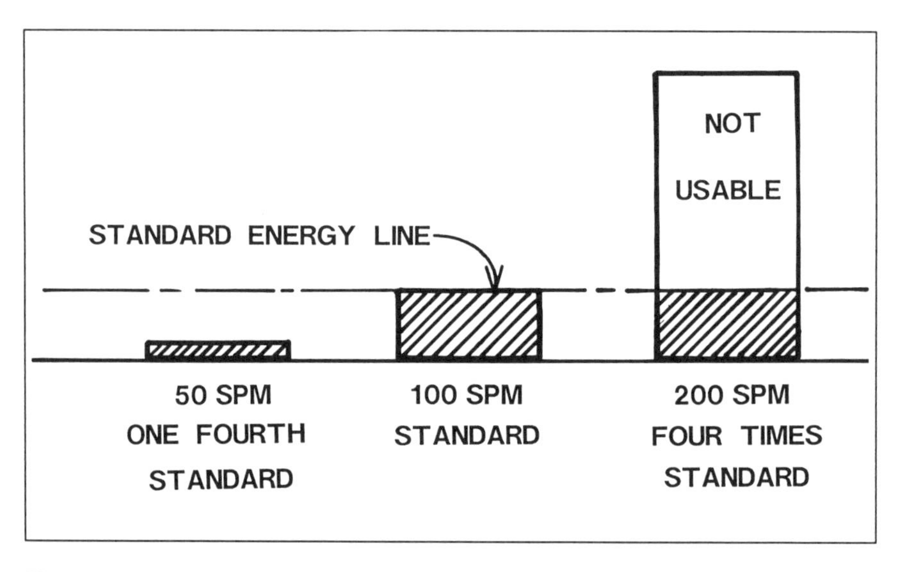

Figure 2-11. Relationship of flywheel speed to usable force and energy.

The speed range must be taken into account in selecting a press, so this loss of energy at slower speeds does not present a problem. Even if the press does not stall at slow speeds, excessive drivebelt wear, and other drive component wear, may result.

Press Motors and Drives

The most commonly used arrangement is to belt-drive the flywheel by means of a pulley on the output shaft of the motor. The periphery of the flywheel usually has grooves for multiple V-belts. The pulley, or sheave as it is commonly known, is sized to obtain the proper press speed. In some cases, the press speed can be varied within a limited range by changing the diameter of the sheave in accordance with the press manufacturer's recommendations. This arrangement is illustrated in Figures 2-3, 2-4, 2-6, and 2-7.

Presses driven by single-speed induction motors operate at a nearly constant speed. For example, a four-pole induction motor having a rated speed of 1,725 RPM at full load will approach the 1,800 RPM speed of the rotating magnetic field within a few RPM under low-load conditions. As the work done by the press is increased, the speed of the motor drops. Full horsepower will be delivered at the rated motor speed of 1,725 RPM.

Should more energy be used than the motor can restore to the flywheel during successive strokes, the motor speed will decrease to less than the rated full-load speed. This results in excessive current being drawn that can damage the motor unless properly protected with fuses and/or a circuit breaker.

Variable Speed Motor Drives

A variable speed drive permits the press speed to be adjusted for optimum performance, taking into consideration the die, press feeding system, part ejection method, and production requirements.

There are many motor drives used on mechanical presses, including:

- Direct current;
- Multiple speed induction;
- Variable frequency induction;
- Mechanical variable speed; and
- Eddy current.

In addition, the Eaton Dynamatic constant energy system (CES) is a type of variable speed drive. Also, various types of gear reduction and hydraulic drives are in limited use.

Direct-current motors historically have been the choice for handling large starting loads, while providing adjustable output speed. At one time, a mechanical motor-generator set was needed to convert the industrial three-phase power source to direct current. These units required expensive maintenance and were inefficient. In addition, power-wasting components such as resistor banks or rheostats were needed to achieve motor speed adjustment.

Modern DC motors use solid state components, such as large silicon controller rectifiers (SCRs) and high-power transistors, to replace both the motor-generator set and inefficient speed adjusting systems. A small and efficient combined power supply and speed control is now used. Typical sizes for press applications range from 5 through 500 horsepower (3.729 through 372.9 KW). The high maintenance items on DC drive motors are the brushes and the rotating commutator. Also, a fan is often required to cool the motor during low-speed operation.

An induction motor must have some slip, or loss of speed below that of the rotating magnetic field produced by the stator (stationary) windings, in order to develop torque. The no-slip or synchronous speed of the motor is determined by the frequency (in AC cycles per second [CPS] or hertz [HZ] times 60 seconds) divided by the number of poles in the stator.

The minimum number of induction motor poles (like the north and south poles of a magnet), is two. Thus a two-pole induction motor operated from a 60 HZ power source will run at 3,600 RPM minus the slip, a function of the developed torque. In the same way, a four-pole motor will run at nearly 1,800 RPM, a six-pole motor at 1,200 RPM, and an eight-pole motor at 900 RPM.

The number of poles is determined by how the coils in the stator are connected. By changing the connections with a switch or relay, the motor speed can be varied by two or three discrete steps. This drive speed control system is little used; the mechanical complexity of the switching required for a three-phase system is a disadvantage. Also, only a limited number of discrete speeds are available.

Since the speed of an induction motor varies in proportion to the frequency of the AC power source, changing the supply frequency can control the motor speed. This is practical with solid-state controllers.

One method is to convert the 60 HZ AC input to direct current. The DC is then efficiently converted to AC of the frequency required to drive the motor at the desired speed by solid-state switching devices. While the electronic controls needed to accomplish the conversion are complex, control designs employing modern solid-state integrated circuit components permit the control package to be relatively small, cost effective, and highly reliable.

Figures 2-12 and 2-13 illustrate a simplified concept of a mechanical variable-speed drive. The drive unit is available as a package unit containing a single-speed motor with a belt and movable sheave arrangement that changes the speed of the drives' output shaft. These drives are normally supplied in variable speed ratios ranging from two-to-one to four-to-one.

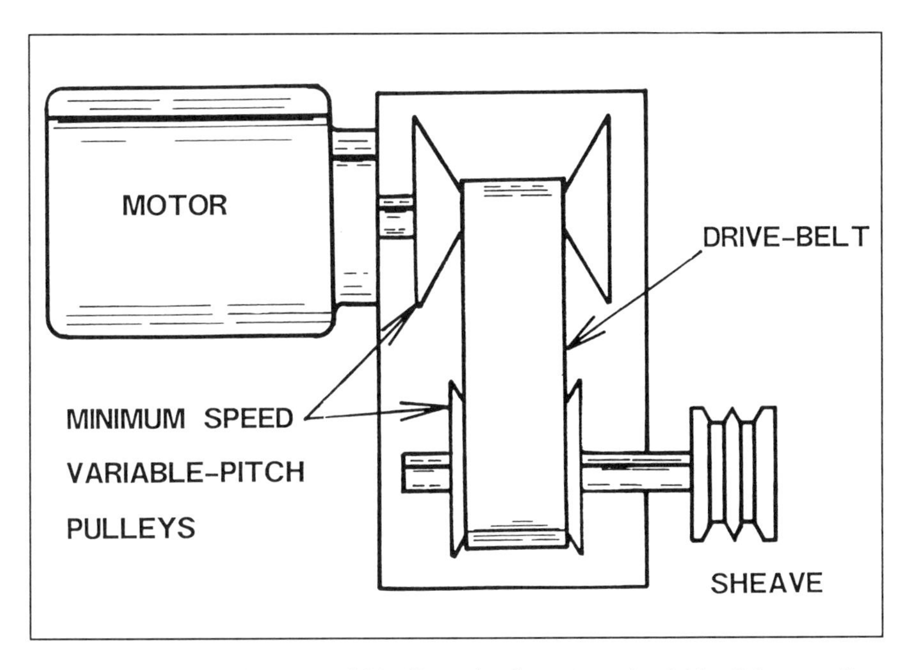

Figure 2-12. V-belt variable-speed drive illustrating the motor and variable-pitch cone pulleys set for minimum speed.

Some presses, especially those used in high-speed perforating work, may have a conventional induction motor that drives a V-belt variable speed drive built into the press. In this case, one of the cone pulleys is attached directly to the flywheel.

Variable speed eddy-current motor drives (Figure 2-14) are available as a packaged unit. They contain a single-speed motor and an eddy-current coupling that changes the drive's output shaft speed.

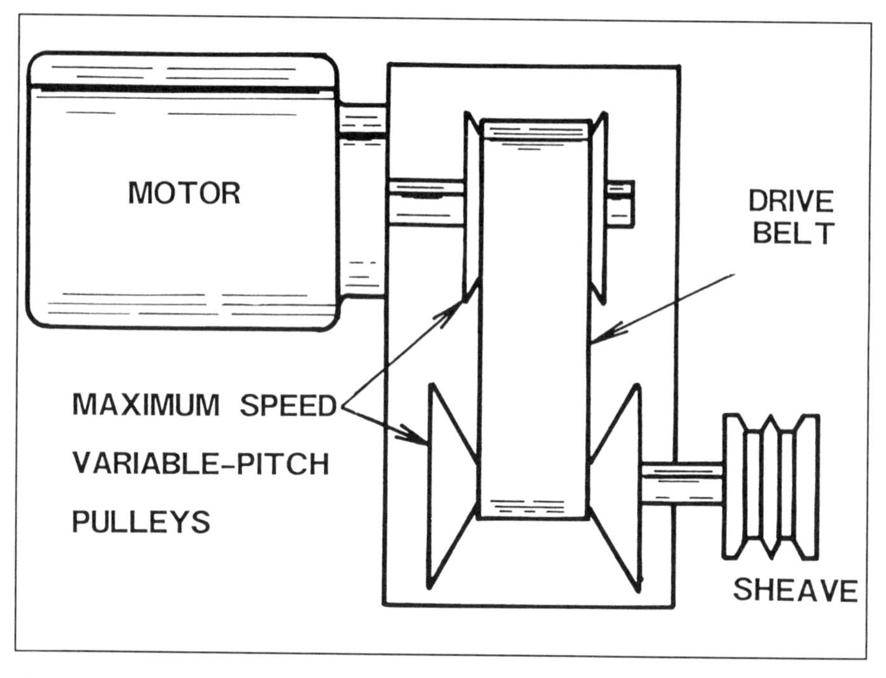

Figure 2-13. V-belt variable-speed drive illustrating the motor and variable-pitch cone pulleys set for maximum speed.

While the eddy-current drive speed ratio is theoretically infinite, its effective range is normally in the 2:1 range for best efficiency, although speed ranges through 5:1 are usable. Often, a problem with excessive loss of flywheel energy limits the speed range.

Output ratings are available from 5 through 300 horsepower (3.729 through 223.7 KW). An auxiliary blower is often included to assist with cooling the unit.

Reduced Voltage Starting Devices

In sizes of up to 300 horsepower (224 KW) or larger, fixed speed press motors are designed to be line-started. This is done by connecting them directly to the power source with a simple relay contactor having thermally-delayed overload protection. Current inrush to the motor is limited by the resistance of the windings and an increase in reactance due to the design of line-start motor rotors. However, the initial current inrush is from five to seven times the full-load current rating of the motor.

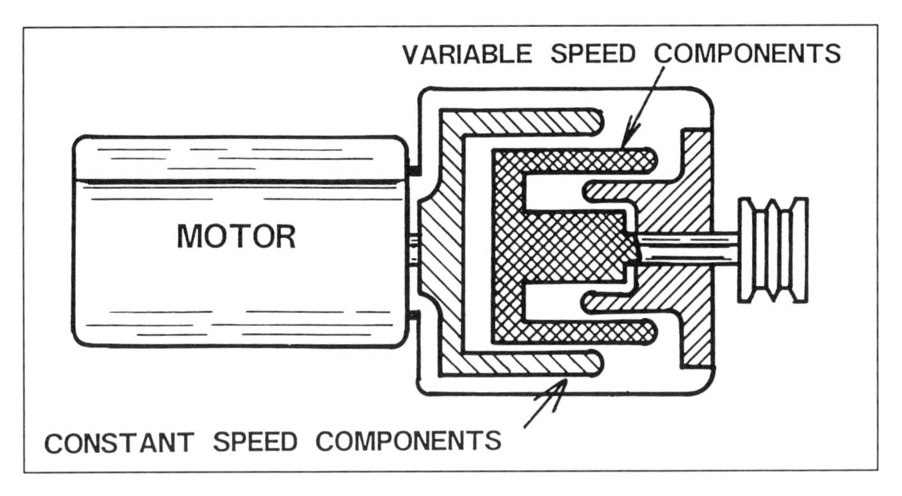

Figure 2-14. A variable speed eddy-current motor drive available as a packaged unit. It contains a single-speed motor and an eddy-current coupling that changes the drive's output shaft speed.

The high current drawn when starting large motors can result in a momentary drop in voltage that may disrupt other equipment on the same supply line. Also, many commercial users are charged a penalty by the utility company for the peak electrical demand of the plant. Even if only one or two large press motors are in use, the starting surge can result in substantial peak penalty charges. If many large press motors are involved, it is wise to establish a procedure to stagger starting them at the beginning of the day or shift.

If the starting voltage is reduced to half the running value, the starting current drops to half the full voltage value. Since the torque is the product of the input power, the starting torque is approximately one fourth the full voltage value.

Reduced starting torque substantially increases the amount of time required to bring the flywheel up to speed. However, an advantage is that there is much less drive belt wear and less stressful torque reaction in the motor stator windings. Energy savings may also be realized, because the resistive losses in the windings increase as the square of the current.

Some reduced-voltage starters operate by switching the winding connections with relays in the case of multiple voltage motors. Because of the circuit's complexity, this method is seldom used. A more common method makes use of a starter that reduces the starting voltage by means of solid-state switching devices. Also, these starters often have a provision to reduce the voltage supplied to the motor when reduced loads are experienced. This feature reduces power consumption.

An added benefit of solid-state starting systems is a provision for dynamic braking by applying DC current to the windings to stop the flywheel. Power factor correction can also be accomplished by switching the current timing to keep the current drawn in phase with the applied voltage.

If the peak current and voltage are out-of-phase, a higher current is required to supply the same amount of actual power. Utility companies usually apply a penalty to the billing of commercial customers with power factor problems. Solid-state reduced-voltage motor starters are available on many new presses as well as on retrofit packages. The savings in motor, belt, and flywheel brake wear and energy (including penalty factors) often can provide short payback times, especially for equipment that is started frequently.

Gears

Figures 2-2, 2-4, 2-6, and 2-7 illustrate the use of gears in press drive applications. In mechanical press drive applications, gears are essential to both reduce the flywheel speed and increase available torque.

Gears are expensive press components. Proper lubrication is an absolute necessity. Where pairs of gears must work together, such as in the double-end drive system illustrated in Figures 2-2, 2-6, and 2-7, it is essential that the load be shared equally. Figure 2-15 illustrates how the gears shown in Figure 2-2 are keyed to the press crankshaft.

Pitman

The English term *pitman* has long referred to any man who worked in a pit, such as a coal miner. Likewise, the man who worked in a saw pit, cutting boards from logs with a large two-man handsaw, was called a pitman. The upper end of the saw was pulled from above by the top man.

When water power was applied to the sawing operation, the connecting link between the crank driven by the water wheel and the lower end of the saw was called a pitman. This term was applied in emerging technologies employing such a mechanical linkage,

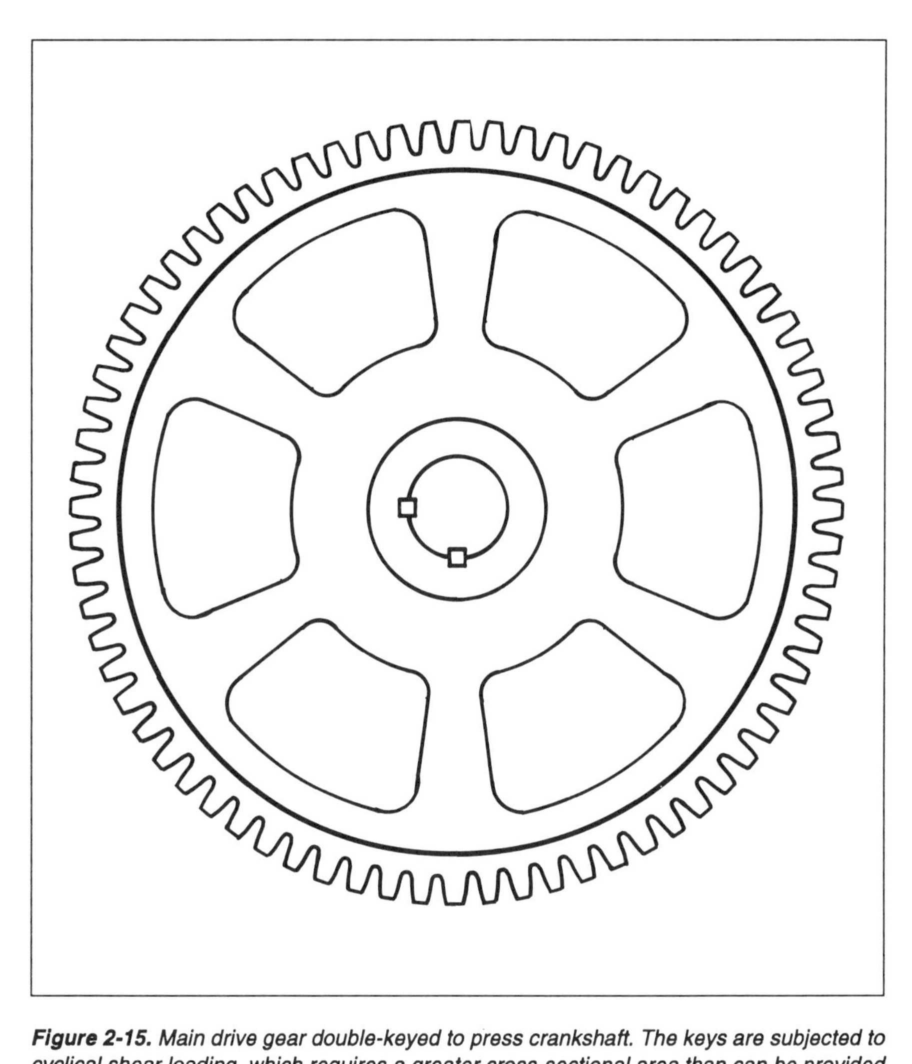

Figure 2-15. *Main drive gear double-keyed to press crankshaft. The keys are subjected to cyclical shear loading, which requires a greater cross-sectional area than can be provided by a single key.*

such as the steam engine and stamping press. Thus the link between the crank and slide is generally termed a pitman. When the press is driven by eccentrics, the link between the eccentric and connection is called an eccentric strap.

The flexible attachment of the pitman or eccentric strap to the slide is called the connection. A means to accomplish shut height adjustment by an adjustable screw mechanism is nearly always a part of the connection system.

AIR COUNTERBALANCES

The correct adjustment of the press counterbalance air pressure is required for the safe operation of a press. The purpose of the press counterbalance system is to offset or counterbalance the weight of the press slide and upper die. Excessive clutch wear results if the setting is too high. Low settings cause excessive brake wear and can cause the brake to overheat dangerously. Safe operation requires that the counterbalance have enough capacity to hold the slide and its attachments at any point in the stroke without the brake applied.

Air counterbalances are air cylinders mounted in the press frames and connected by the cylinder rod or rod brackets to the press slide. Air pressure is placed in the counterbalance cylinder, that tends to lift and slide to a neutral weight condition. As die weight is added to the slide, additional air pressure is added to the air counterbalance to compensate for this added weight. Air counterbalancing aids in the overall performance of a press by:

- Counterbalancing the weight in the slide and the die components attached to the slide;
- Taking up bearing clearances before the die closes;
- Assisting in stopping the press because less load is applied to the brake; and
- Helping maintain constant gear tooth contact by taking up the clearance necessary for proper gear functioning.

Counterbalance System Components

Figure 2-16 illustrates the main components that make up a mechanical press counterbalance system. The principal component is the pneumatic cylinder the piston rod of which attaches to and counterbalances the weight of the slide and upper die. Some presses have two or more cylinders. Usually, double-action presses have separate counterbalance systems for each slide.

A pressure gage and adjustable air regulator are provided on the press to permit accurate adjustment to the correct setting. Some presses may have regulators of the self-relieving type that automatically bleed excess air when the pressure setting is lowered. Self-relieving regulators should have the pressure adjustment raised slightly until air is heard being admitted to the system after bleeding the pressure to a lower value. Bleeding the system can be speeded up by opening the blowdown valve.

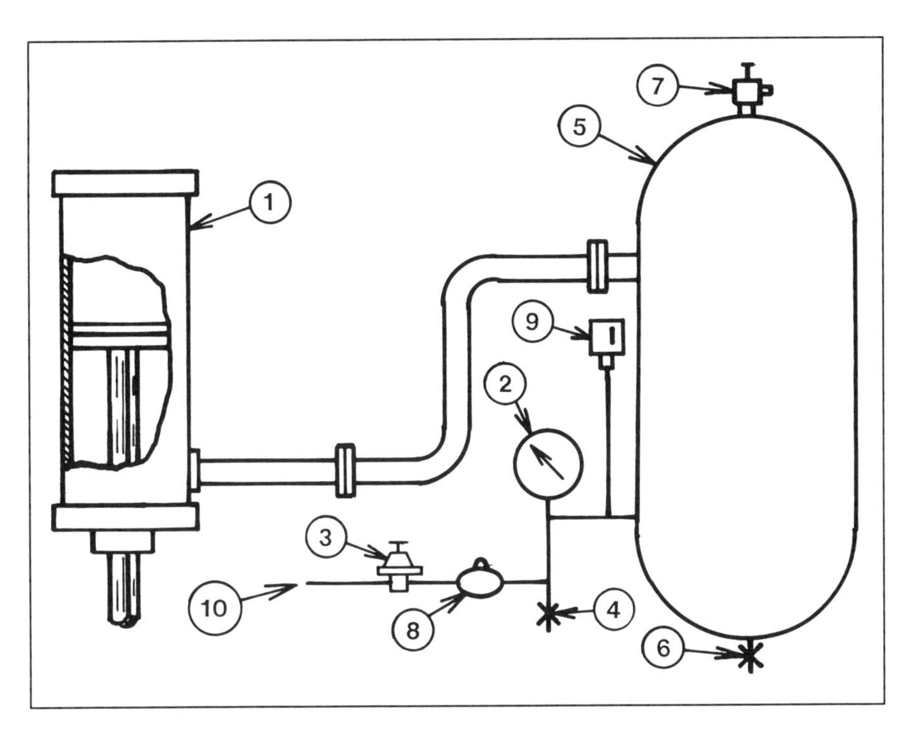

Figure 2-16. Diagram of a typical mechanical press counterbalance system. (1) counterbalance cylinder with a piston rod that attaches to the press slide (not shown); (2) pressure gage; (3) pressure regulator; (4) blow-down valve; (5) surge tank; (6) water drain valve; (7) safety pop-off valve; (8) check valve; (9) low air pressure switch; (10) shop air inlet.

A means is also required to prevent a sudden loss of pressure from the counterbalance system. This function may be served by a check valve. In addition, a pressure actuated switch is included to open the main motor run circuit in the event that the pressure falls below a minimum value.

Retaining Counterbalance Parts

Should a counterbalance piston rod become detached from the slide or piston, the piston and rod can easily blow the cover plate off the top of the cylinder on some older presses. Often, these press parts will be propelled through the plant roof, making a second hole on the way down. For this reason, safety rules require that presses incorporate a means to retain counterbalance parts in the event a component breaks or loosens. The press manufacturer should be consulted to find out if older presses meet this requirement. If not, the manufacturer should cooperate in supplying the correct design for the required modifications. Badly designed add-on retaining devices can become airborne, endangering workers.

Periodic maintenance inspections should include a check of the piston rod's attachment to the slide. During a press overhaul, the rod ends should be checked for stress cracks with die penetrant or magnetic particle inspection. Some older presses are especially prone to this type of failure.

Figure 2-17 illustrates an example of a counterbalance cylinder attached to the side of the press frame and slide. The surge tank, piping regulator, and other pneumatic components are also shown. This method has several advantages over mounting the cylinder on the top of large machines. First, the cylinder is readily accessible for inspection and maintenance. The design is robust and compact. The cylinder does not protrude from the top of the machine. The clevis and pin attachment for the cylinder, together with the cylinder rod attachment to the slide by means of a L-bracket, assures retention of parts in the unlikely event of detachment.

Setting Correct Counterbalance Pressure

Setting the counterbalance pressure to a value that will correctly counterbalance the weight of the slide and its attachments (upper die, risers, parallels, etc.) is generally the most certain means of achieving the right setting rapidly. To do this, three things are required:

- The weight of the upper die and attachments is known;
- An accurate table of pressure settings for the upper die weight is available to the diesetter; and
- The pressure gage used is accurate.

Many presses have a chart that gives the correct pressure setting for various upper die weights. If the information is missing, a chart of correct settings can be obtained from the manufacturer. If this is not possible, the information can be determined from measurements and engineering calculations.

Dies should have the upper, lower, and total die weight clearly identified to facilitate both safe die handling and correct counterbalance setting.

Safety laws require that dies be stamped to indicate upper die weight for proper air counterbalance pressure adjustment. Further, it is required that the total die weight be stamped when equipment handling may become overloaded. OSHA regulations

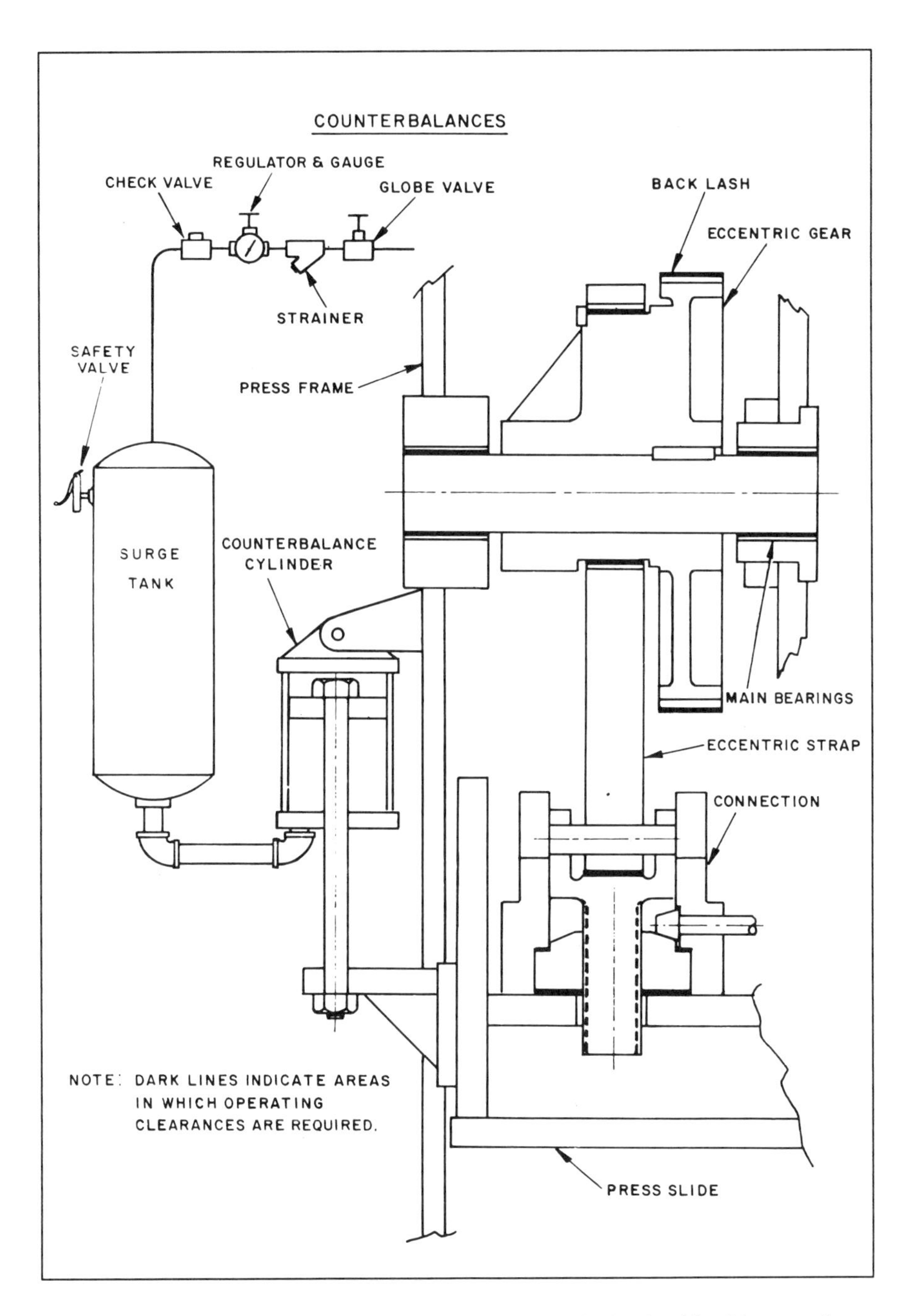

Figure 2-17. *An example of a counterbalance cylinder attached to the side of the press frame and slide. This method is preferred to top mounting the cylinders; less overhead clearance is required. (Verson Corporation)*

specifically state "stamped" rather than marked or painted. There is always a danger that an incorrect figure may be painted on the die should it be repainted.

Having the upper, total, and lower die weights accurately and clearly marked has many benefits that go beyond correct counterbalance settings. For example, if a die is to be sent out for work, the truck driver will need to know how heavy it is. If there is doubt as to the correct weight, the driver may insist on hauling one half at a time. A lot of problems can be avoided if the facts needed for safe correct decision making are available.

Shut heights should also be stamped on the dies. To be safe, the diesetter should always double check the shut height with a tape measure. It may have been altered during die repair or modification.

Common Errors in Counterbalance Adjustment

Large presses have big surge tanks that may take a long time to fill. A common mistake is to make a big change in the regulator setting when a small change, followed by a wait of several minutes to allow the system to stabilize, is all that is needed. The diesetter should check the final setting for correct adjustment and stability before the dieset can be considered complete.

Incorrect pressure settings are often caused by inaccurate or missing gages. Press vibration and pressure pulsations can ruin the accuracy of a low-grade gage in a short time. To avoid this problem, it is highly recommended that a high-quality liquid-filled gage with a built-in pulsation snubber should be used. Another good solution is to equip each press with a quick-disconnect fitting and use a portable gage of known accuracy. Special miniature diagnostic fittings are made for this purpose. The portable gages should be checked against a master gage periodically.

Some newer presses designed for quick die change feature automatic counterbalance adjustment based upon a computerized database of die numbers. In most cases, the correct pressure must still be determined and entered into the database. Failing to update the database and relying on manual adjustment after problems develop is a common error.

Establishing Correct Counterbalance Settings

If the information is missing, several procedures may be used to establish the correct air pressure adjustment. It is very important that a system be worked out to avoid trial and error adjustments at every dieset. This may include listing the correct counterbalance pressure for the press and die combination in the diesetting instructions.

Checking the drive-motor current while the press is cycled is an accurate way to find the right air pressure setting. A reading that increases as the slide descends and drops sharply on the upstroke indicates the pressure is too high. Amperage readings that are high on the upstroke and increase as the top of stroke is approached mean the setting is too low.

When checking ammeter readings, one must consider that the press motor must supply the energy lost by the flywheel as the press does work at the bottom of the stroke. A current surge is normal when that occurs.

Press-strokes-per-minute indicators that operate by measuring the RPM of the drive motor, flywheel, or crankshaft can be used in place of an ammeter. If the press speeds up on the downstroke, the setting is too low. A loss of speed means the air pressure is too high. SPM meters that measure the time of each stroke, rather than actual speed throughout the stroke, cannot be used for this purpose. SPM indicators that measure the time interval to complete each stroke are built into some press tonnage meters.

The dial indicator method can be used to establish correct counterbalance pressure if the press counterbalance adjustment information is not available. To use this method:

- Stop the press at 90 degrees on the downstroke;
- Follow proper lockout safety procedure;
- Place a dial indicator so the tip touches the slide;
- Exhaust the air from the counterbalance to a value below that required to counterbalance the slide — do not release the brake;
- Slowly raise the counterbalance air pressure until the dial indicator shows that the counterbalance has lifted the slide; and
- Make a record of this setting so this procedure will not need to be repeated.

This procedure usually establishes a setting very close to that of the ammeter method. Generally, it is the best method if a chart of correct counterbalance versus upper die weights is to be developed. This method is also useful for determining if the data on a newly developed chart is accurate.

Counterbalance Maintenance

Lubrication must be supplied to both the cylinder piston packing and the rod gland packing. Dry piston packing on large presses emits a characteristic sound best described as that of a cow mooing. If the problem is not corrected promptly, the packing will fail.

Lubrication is either supplied by a manual hand pump or metered automatically. In each case it is important to make sure that the correct lubricant is applied as needed. This must be part of the total preventive maintenance program.

If properly lubricated, the packing should give decades of service. If the rod packing gland or piston packing are permitted to run dry for very long, the packing will fail and the rod, piston, and cylinder body become scored. This results in downtime and expensive repairs.

Water can accumulate in the surge tank(s) and must be drained by a valve. Otherwise there will not be sufficient space in the surge tank, and excessive pressures will cause erratic press operation. In some cases, the safety valve will open at the bottom of each stroke, resulting in wasted compressed air.

Frequency of draining depends upon dryness of the air supply. Usually a weekly schedule is sufficient. Under conditions of high temperatures and humidity, the compressed-air dryer may be overloaded and a daily schedule may be required.

OSHA rules also require that a means be provided to prevent a sudden loss of pressure from the counterbalance system. This function is served by a check valve. In addition, a pressure-actuated switch is included to open the main motor run circuit should the pressure fall too low. Tests should be conducted periodically to assure that each shutoff device shuts off the main motor as intended.

Piping must be periodically inspected for leaks. Pipe unions and the large bolted joints in the surge-tank piping often work loose from press vibration, especially if shock loads are experienced. Air leaks may produce system malfunctions. A leak loud enough to be heard is costing a great deal of money every year. The shock and vibration associated with pressworking can cause flange fittings, pipe unions, and fittings in general to loosen and leak.

All press piping should be kept in good repair and inspected periodically for leaks. Often this is best done at night or on a weekend when production is not running. The main air valve to each press can be shut off and any pressure drop noted over a period of time.

Automatic Counterbalance Control

One newly developed system sets counterbalance pressure automatically by weighing the die. The microprocessor-based control unit weighs the die at the top of each stroke, by means of strain gages attached to the pitman or eccentric strap. Pressure adjustments are made based on data stored in nonvolatile computer memory. A desirable method is to operate the system in conjunction with the existing OSHA-required counterbalance controls installed by the press manufacturer. Any failure of the automatic adjuster must not result in an unsafe condition.

PRESS CONSTRUCTION MATERIALS

The most common material for construction of crankshafts and other highly stressed parts such as pitmans and gears, is plain medium carbon steel, typically AISI-SAE 1045. However, alloy steels are employed for demanding applications.

Gray cast iron and iron alloys are also used. Economy may be a factor in the choice of gray iron, especially in the case of older equipment. However, cast irons, especially iron alloys such as nodular iron, damp vibration better than steel. For this reason, irons are used in constructing many good-quality high-speed press frames.

Parts made of cast irons and medium carbon steel are found to fail frequently in service. The AISI-SAE 4140 chrome-molybdenum alloy is a good choice. To a lesser extent, AISI-SAE 6150, a chrome-vanadium alloy, is used for highly stressed machine parts.

Both medium carbon and the popular alloy steels are readily cast or forged to shape. Properly done, welded repairs are generally successful in these materials.

PRESS ELECTRICAL CONTROLS

Protecting personnel and having safely functioning pressroom equipment should be the user's paramount concern when specifying and installing press electrical controls. In many nations, government officials issue power press safety rules, which are subject to change. These rules, which usually have the force of law, should be considered consensus requirements for safe press operation. Manufacturers engaged in metal stamping should strive to meet or exceed these standards to reduce employee exposure to injury.

Control Reliability

Press electrical control circuits must be designed to be inherently reliable. For example, if critical components, such as the clutch/brake control valve, are actuated by two independent electrical solenoid control valves, self-checking should be built into such components to prevent operation if either valve malfunctions. After failure of any part of the engaging means, the controls must not permit operation until the control circuit is reset after the fault is corrected.

The control voltage should be isolated from the main electrical supply. It must not be possible to actuate the press if any part of the control circuitry is accidentally grounded. Another use of redundancy is in dual actuating button contacts and logic relays to greatly lessen the probability of any one component failure causing unwanted press activation.

Modern press controls are sometimes designed around solid-state microprocessor-based circuitry. It is important that solid-state controls have built-in redundant self-checking features that ensure control reliability. Standards organizations and government regulations usually set strict requirements for assured fail-safe operation of press controls. The manufacturer or appropriate certification testing agency should certify that the needed control reliability is built into all safety-related press devices.

Presses employ mechanical cams that operate electrical contacts at presettable degrees of rotation. These contact signals are used to initiate a number of functions ranging from automation control to positive top stop activation. The cams are contained in one or more boxes and rotate one time per full revolution of the crankshaft.

One critical function that may be employed is activation of a take-over circuit that mutes the operator's controls on the upstroke. Should the chain or drive coupling fail on the upstroke, the press will continue to cycle unexpectedly. This can occur because there is no cam signal that the top of stroke was reached. This failure mode alone has caused a great many tragic and altogether avoidable amputations over the years. Monitoring devices are required to stop the press in the event of a cam drive failure.

Several safety systems can be used to detect loss of the control cam drive function. One is to provide dual cam boxes driven separately by opposite ends of the crankshaft or eccentric drive. In this case a separate top stop contact is contained in each cam box. Some modern integrated systems make use of a combination of one or more rotary resolvers and, in some cases, a motion detector to determine the integrity of the chain drive.

The flywheel stores energy and is engaged by a clutch. The clutch operates in conjunction with a brake to stop and hold the slide in place when the clutch is not engaged. The flywheel, clutch, and brake operate as an integrated system to supply power to the press when required, and stop the machine safely. The clutch driving and driven members are spaced close together to permit rapid action. The brake maybe an integral part of a clutch-brake package, or located on the opposite side of the press.

There is always a danger that mechanical failure of the system will prevent the press from stopping properly, or permit it to cycle without actuation. A common cause of this problem is deterioration of one or more system components. Bearings, clutch disks, and brake components are subject to normal wear as well as deterioration due to shock loading. Deterioration of the clutch/brake system results in increased press stopping time, defined as the elapsed time between actuation of the brake and the press stopping.

Some electrical cams are equipped with an extra set of contacts, to stop the press if the top stop setting is overrun a fixed amount. Should this occur, the press controls prevent actuation of a successive stroke. Such systems are better than no protection but, in practice, the cams actuating the contacts are seldom set close enough to the normal stopping point to provide early warning of clutch/brake failure.

A common type of electronic brake-monitoring device measures elapsed time between the deactivation of the clutch/brake actuating valve and the stopping of the crankshaft. The crankshaft's stoppage is sensed by a rotary motion detector. This type of monitor is illustrated in Figure 2-18.

Such systems provide a digital readout of the actual stopping time in milliseconds. One millisecond is 0.001 second. An adjustable setpoint sets the actual stopping time, plus a slight added amount determined by the employer, to allow for slight variations in stopping time. Stopping time monitors are often built into modern electronic control systems as a standard feature.

The employer must ensure that this setting provides proper safety distance for the operator. Should a stopping time in excess of the preset limit occur, the press control circuitry must not permit a successive stroke. The root cause of the problem must be determined and corrections made.

An obvious first step is to examine the clutch and brake to determine if significant wear or damage has occurred. The em-

Figure 2-18. *Example of an electronic stopping time monitor installed on an existing press. (Link Systems)*

ployer must not permit an arbitrary increase in the alarm time setting, especially if insufficient operator safety distance results.

If resetting the alarm and repeated cycling of the machine are required for diagnostic purposes, the press must be removed from service. Repeatedly resetting the alarm during production, without taking corrective action, can result in serious accidents and must not be permitted. In such cases, the brake stopping time monitor signals that a dangerous condition exists. Continued production utilization, by means of resetting the control, defeats the warning function and very likely will result in serious operator injury, if hands-in-die operation is being practiced.

Figure 2-19 illustrates a simple motor starting circuit. The electrical schematic symbols differ in many ways from those used for electronic equipment. For example, parts of a single motor starting contactor are shown as four normally open contacts labeled M1 through M4. The contactor is essentially a large relay. The contactor starting coil is shown as a circle. Electrical control schematics are shown in this way because parts of a single device, such as a contactor, are shown at various points on the circuit drawing.

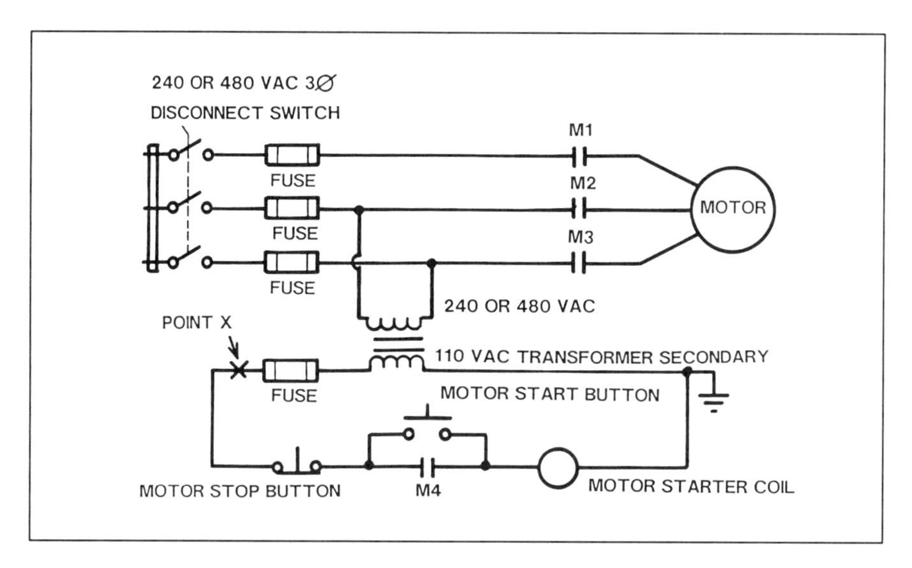

Figure 2-19. *A simple three-phase motor starting and stopping circuit.*

An important safety feature is that the motor starter coil disconnects the motor from the incoming three-phase power if power is interrupted momentarily. To restart the motor, the start button must be manually depressed. When the motor contactor is energized, contact M4 seals the contactor coil circuit in the *on* state, until a power interruption occurs or the circuit is unsealed when the stop button is depressed.

Figure 2-19 illustrates point "X" where additional electrical controls can be wired into the motor stop circuit. Figure 2-20 illustrates a series circuit of two pressure switches, an emergency stop button, and a safety block receptacle. The circuit wiring may be interrupted at point "X" in Figure 2-19, and the series circuit illustrated in Figure 2-20 connected at that point.

Often, emergency stop buttons and safety block interlock plugs are wired into the series electrical circuit with the low air and counterbalance pressure switches. An example of a correctly installed safety block interlock plug is illustrated in Figure 2-21. It is important that all press wiring be done according to the manufacturer's recommendations. Tests should be conducted periodically to assure that each shutoff device will shut off the main motor as intended.

A typical mechanical press has one or more pressure switches for safe operation. Air counterbalance systems have a pressure-

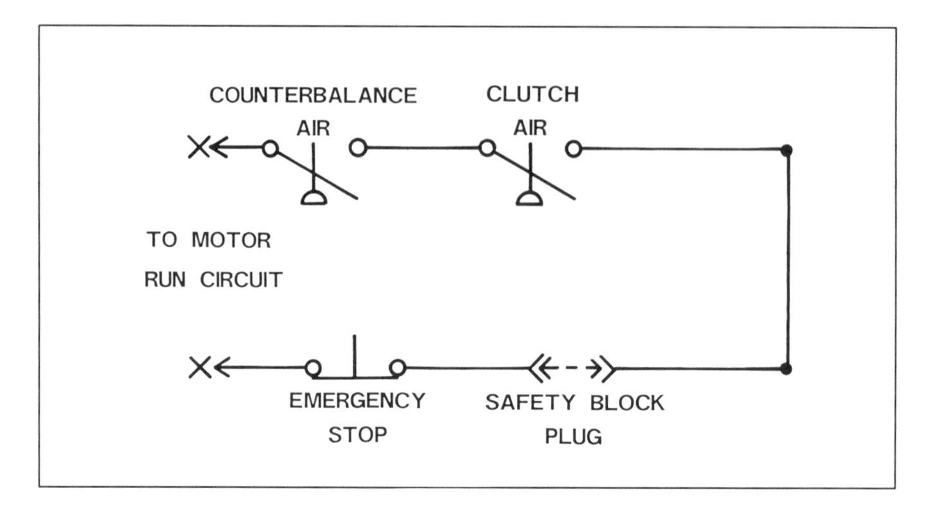

Figure 2-20. *A press electrical series circuit featuring two pressure switches, an emergency stop button, and a safety block receptacle. Interrupting the current flow in this circuit opens the main drive motor contactor illustrated in Figure 2-19.*

Figure 2-21. *A correctly installed safety block with an attached interlock plug and receptacle.*

actuated switch that opens the main motor run circuit in the event of a drop below a minimum value.

A similar switch, wired in series with the counterbalance switch, is used to detect sufficient air pressure to correctly activate the clutch. Hydraulically actuated die clamps may also require pressure switches to ensure that correct clamping pressure is maintained.

The pressure switches, which have adjustable setpoints, are of the normally open type. The correct pressure setting must be in accordance with the press manufacturer's recommendation. A minimum amount of clutch actuating and counterbalance air pressure must be present before the motor can be started. Should a dangerous drop in pressure occur, either switch must open at the correct minimum setting, shutting off the motor.

To avoid press damage in a loss of recirculating lubricant pressure, a pressure switch is usually provided to stop the press. Other switches are often wired into this safety circuit to stop the press for various potentially harmful conditions, such as a lack of sufficient hydraulic overload precharge pressure and over-travel of the slide adjustment motor.

Pressing the emergency stop button also opens the motor run circuit, shutting off power to the motor. Usually, several large emergency stop buttons are conspicuously located on the press. All buttons are of the normally closed type and wired into the series circuit.

One or more safety blocks, plugs, and receptacles of the type illustrated in Figure 2-21 are provided. It is extremely important that the safety block be interlocked with the press electrical controls so the press cannot be cycled with the block in the press.

The block is designed to withstand only the static weight of the slide, attached linkage, and upper die including a safety factor. Accidentally cycling the press under power may cause the block to fail, often resulting in serious press and die damage. Should blocks of different lengths or multiple blocks be needed, each block must have an interlock plug and designed receptacle.

Electrical Controls of Used Presses

Anytime a used press is purchased or moved from storage, one should automatically assume that serious wiring errors are present. Again, an approved print should be used to verify that the entire circuit is wired correctly. All wiring should be placed neatly in the raceways with proper wire markers used in all cases. Any damage to the electrical enclosures should be repaired, or the entire enclosure replaced. All external safety block interlock receptacles and pressure, cam, and emergency stop switches should be checked for jumpers.

Once the press is running, all emergency shutoff functions should be carefully checked for correct operation. Complete verification must be made to assure that all safety devices required by the press manufacturer and applicable law are present and functioning correctly.

The electrical controls of older presses often do not meet functional, control reliability, or current safety standards. The controls may have to be modified or replaced to meet such requirements.

Changing safety requirements may mandate improvements such as double solenoid clutch/brake air valves and stopping-time monitors. In some cases, such needs can be accommodated by modifying existing wiring, including adding needed logic circuitry in accordance with the manufacturer's recommendations. When other add-on improvements, such as tonnage meters, light curtains, electronic shut height indicators, and electronic cam switches, are installed on older presses, the same safety considerations apply. Figure 2-22 illustrates the operator terminal of a modern integrated press control system.

If all changes are properly made and carefully documented to conform to applicable safety regulations, older equipment can provide additional years of service. In cases where fire or mechanical damage requires extensive repair of the control panel, it is often more cost effective to completely replace the old system than recondition it. The same economic cost justification should be made when extensive upgrading of the control system is needed to comply with current safety standards.

INTEGRATING PRESS CONTROL SYSTEMS

Modern integrated press controls may cost much less than rewiring the press panel and adding on features such as programmable cam switches and tonnage meters as discrete items. Such work can be cost justified and budgeted as an ongoing equipment upgrading program.

Many new presses are equipped for total integrated process control through advanced microprocessor-based technology. Existing presses often can easily be retrofitted with such systems.

Properly designed, installed and used, modern integrated controls can increase productivity, safety, and parts quality. In addi-

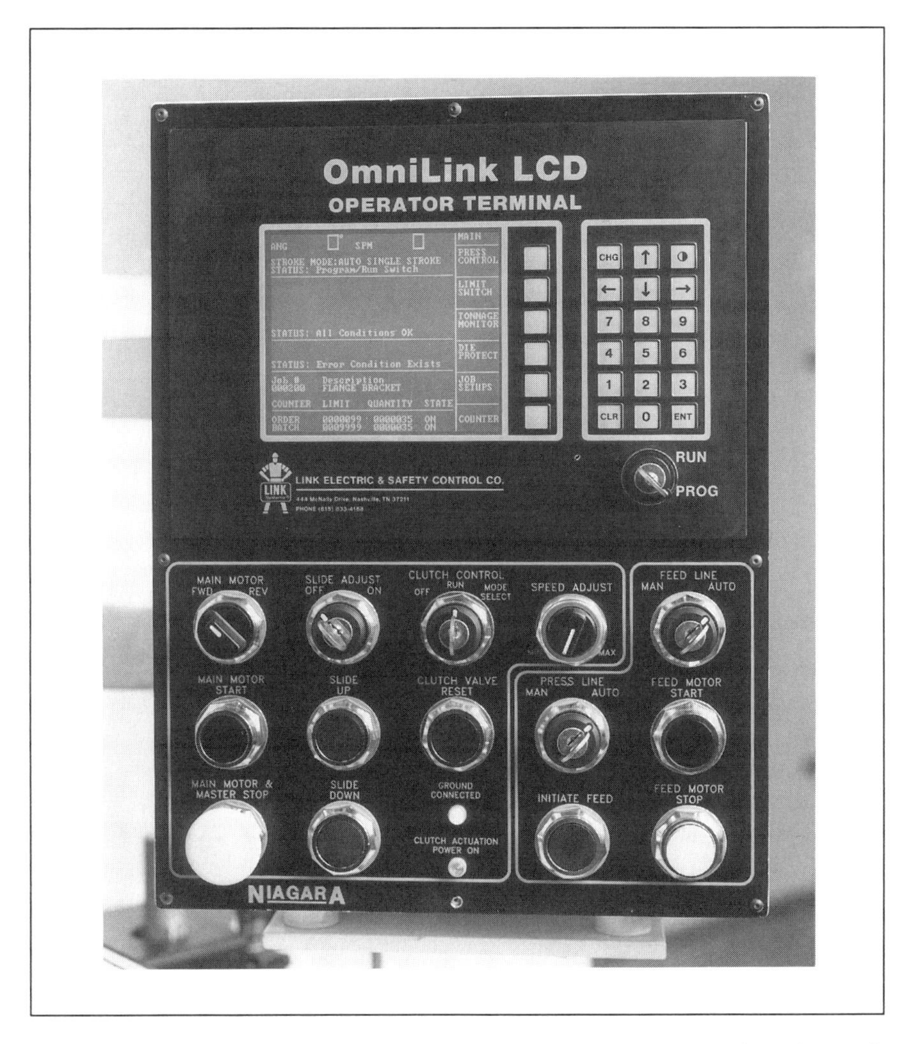

Figure 2-22. *The operator terminal of an integrated press control system. (Link Systems)*

tion, some systems can record or transmit production information such as part counts and machine uptime.

Figure 2-23 illustrates a block diagram of an integrated press control system with a high level of subsystem integration. It can be applied to manually fed presses, automatically fed press systems, and presses integrated into production cells.

The following functions can be provided by such systems:

- Programming and display of all functions by means of one easy-to-use terminal;

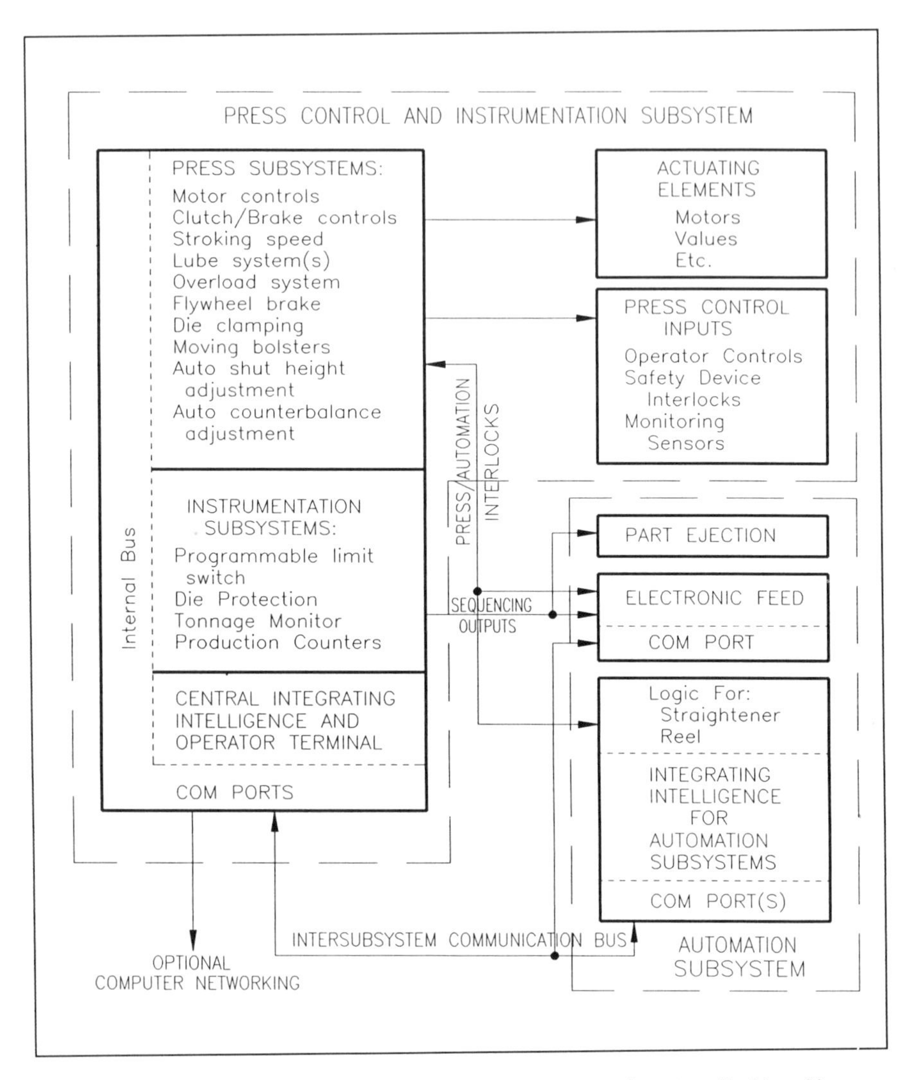

Figure 2-23. *A block diagram of an integrated press control system tied in with many subsystems. (Link Systems)*

- Programmable multi-output limit switch;
- Clutch/brake control;
- Automatic setup of shut height and counterbalance pressure;
- Control of automation and ancillary equipment;
- Die protection and process monitoring;

- Setup parameter storage for multiple jobs;
- Multiple modes of production counting;
- Tonnage monitoring and waveform signature analysis; and
- Two-way data communications capability.

Operator Terminal

The terminal may either employ a conventional cathode ray tube (CRT) or liquid crystal display (LCD) to provide system information. User selectable display screens provide information on system configuration and operation. For example, display screens might indicate stroking mode selection, press crankshaft angle, shut height, actual slide position, or strokes per minute, as well as provide complete system diagnostics.

A desirable feature is provision for a number of discrete inputs for diagnostic messages to be sent from auxiliary automation equipment. In this way, a wide range of diagnostic messages from automation and other ancillary equipment can be displayed at one convenient location.

Display and programming functions are accomplished through a tactile keyboard resistant to shop grime. This display terminal links to all other system microprocessors in the equipment controller through a serial communication interface. Good systems have high noise immunity, allowing the display unit to be located some distance from the main controller rack and permitting integration into remote operator stations.

The control of auxiliary press and automation systems is accomplished by means of programmable limit switch outputs, which can also be used to provide control and interfacing logic for auxiliary systems. Typically, outputs are available as solid-state AC, solid-state DC, or mechanical relay contacts.

Desirable features of the main controller rack are compactness and modular construction. If modular construction is employed, outputs can be installed as needed. Often, they can be mounted in existing enclosures with motor and other electrical controls, to provide a consolidated control system. Plug-in modules are added or changed to add new features or to facilitate servicing.

The latest integrated system designs feature a comprehensive press control that provides clutch/brake logic. In addition, the logic for such auxiliary press systems as hydraulic overloads, flywheel brakes, main drive motor dynamic braking, eddy-current drive engagement timing, lubrication systems, feeders, die change systems, and other functions can be provided. Automatic setup features, such as counterbalance and shut height adjustment, can be accomplished at the press by the diesetter and double-checked as a means of avoiding both machine and human errors.

The rotary resolvers and encoders used with press control systems should be of rugged design and not subject to picking up false signals due to the high levels of electrical noise present in many pressrooms. The shaft must connect in a 1:1 ratio with the press crankshaft to offer crankshaft position information.

Setup of Control Information

Some functions, such as the stroking mode selector, must provide for supervisory control. This is also used when selecting features to mute safety devices such as light curtains on the upstroke. Usually this is accomplished by means of a password or a mechanical key selector. In addition to the usual inch, single-stroke and continuous modes, setup/stop time measurement, and timed inch modes are some features that may enhance quick die change and setup.

Programmable limit switch, feed length, counterbalance, and shut height adjustment; tonnage monitor settings, and die protection programs can be recalled with a few key strokes. This provides a paperless record-keeping system and quick, consistent job changeovers.

Some system options are hard-wired at the time of system installation. Typically, safety-critical logic, such as the clutch/brake function, is stored in read-only memory programmed by the equipment supplier.

User programmable functions not safety-critical are stored in nonvolatile memory, meaning that battery backup is not needed to prevent loss of stored data.

Brake monitoring is usually a standard feature supplied with integrated press control systems. This function is cost effective and simple to accomplish, since the system requires a rotary position resolver or encoder to provide signals for a number of other functions. A rotary transducer is a major part of the cost of an add-on brake monitor.

For purposes of checking stopping time to determine operator safety distance for placement of light curtain safety devices and two-hand controls, the stopping time should be measured at approximately 90 degrees on the downstroke. This test is usually made with portable test equipment. However, it can be performed whenever needed, as a programmable feature of an integrated control system.

An important indicator of press performance is clutch engagement time. A clutch engagement monitor can be provided to indicate excessive time from the energization of the air valve(s) until the clutch engages. This feature compares a programmable limit on clutch engagement time with measured time and displays the reading on the operator terminal. The clutch/brake valve can be de-energized if the measured time exceeds the limit.

Automatically adjusted top stop is a good feature for variable speed presses. As stroking speed is increased or decreased, the top stop command signals are advanced or retarded so that the slide stops at the top of the stroke. In the same way, preprogrammed changes in automated die protection and pneumatic control signals can be linked to press speed.

Integrated press controls normally provide a number of inputs and outputs to interface with automation and auxiliary equipment. Inputs initiate top stop, cycle stop, and master stop of the press when signaled to do so by auxiliary equipment. Inputs are also provided to tell the press control whether feed line equipment is in manual or automatic mode and when an automatic feeder has completed its feeding action.

The press must not stroke in the continuous or automatic single-stroke modes unless the feeder is in the automatic mode. When stroking in the continuous mode, the press should be stopped if the feed-complete input does not agree with the correct programmable downstroke crankshaft angle for the job.

Auxiliary stop outputs are also required to stop independently driven automation when a cycle stop, master stop, or light curtain stop signal is received by the press control. Discrete inputs are also provided by the operator terminal, so that diagnostic messages can be displayed to explain faults or stop conditions for automation equipment.

Independently driven feeds, destackers, transfer systems, scrap choppers, and parts extraction systems are often used on automated power presses. Other devices requiring control include air blowoff devices, air cylinder ejection of parts or scrap, and solenoid valve operated stock lubrication systems. This automation equipment must be controlled, timed, or sequenced in relation to the stroking action of the press.

Electronically programmable limit switch outputs are provided to coordinate the action of automation equipment with the press. These outputs can be programmed to turn on or off at given press crankshaft angles. In addition, the outputs that can be programmed turn on at a given angle and off at a specified time later. One feature of some operator terminals is a data communication link to electronically roll feed controllers to program the feed length for each job.

To keep track of production and provide for scheduled maintenance, the press strokes are counted. The parts counter may be preset to signal the press to stop when the required production is completed. A batch and quality check counter may be used to signal when to perform quality checks and change part containers.

Many maintenance efforts, such as lubricant filter changes, can be accomplished more economically if based on the number of press strokes rather than elapsed time. The total stroke counter can signal when such servicing is needed.

Inputs for die protection sensors that can be monitored from the operator display terminal are a very useful feature. A number of functions can usually be monitored. These may include part in place, part ejection, double blank detection, stock buckle detection, and the correct operation of an in-die transfer mechanism.

Including the tonnage monitoring function in an integrated system is usually a less costly option than stand-alone equipment. Some features that add cost to conventional monitors are job memory capability and a means to obtain a reliable signal to rezero the monitor. Integrated systems normally have the capability to recall multiple job setup parameters. This can easily include tonnage setup data and the optimum crankshaft angle at which to rezero the DC balance of the monitor.

The operator terminal CRT or LCD display can also serve to monitor waveform signature data. This information can be used to determine if harmful snap-through energy is present and to diagnose various press and die problems.

Two-way communications capability is a common feature of integrated control systems. If the feature is not built in, it often can be added by installing a plug-in communications module. In simple systems, production counts and press downtime status codes may be all the information relayed to a central data collection center. Other press information transmitted may include tonnage monitor data including waveform signatures, bearing temperatures, and stock status. Two-way communications can permit setup parameters for each job to be downloaded from a central control point.

Electronic data collection has progressed to the point that virtually any form of information gathering is possible. However, not everything that is possible is cost effective. Communication to a central data collection station is recommended only if it is shown to be the most cost effective method of accurately reporting production status.

Increasingly, stamping producers are relying on cross-functional work teams to set up and operate work cells having one or more presses. Team members are given the opportunity to manage the business of the stamping work cell. These functions may include issuing orders for stock to vendors, running jobs in the correct order, identifying lots of finished parts, and accurately reporting completed work.

Avoiding Wire Errors

In the author's experience, serious accidents and a great many potentially disastrous conditions have resulted from unqualified persons making randomly conceived wiring changes in press panels. An up-to-date electrical wiring diagram, meeting all safety requirements, should be kept inside the control panel of each press. Any changes, such as those required to install tonnage monitors, light curtains, electronic cam switches, and other auxiliary equipment, should be approved by the press manufacturer or control supplier and marked on that print.

All wires should have numbered wire markers on both ends and these numbers should be indicated on the print. This will help enable a qualified individual to troubleshoot and maintain the integrity of the circuit. Anyone who opens the press control panel to do any wiring of any nature whatsoever must be thoroughly trained and possess a complete understanding of the functions of all control circuitry.

BIBLIOGRAPHY

1. N. Fisher, *Principles of Mechanical Power Presses*, SME Technical Paper MS76-285, Society of Manufacturing Engineers, Dearborn, Michigan, 1976.
2. C. Wick, J. T. Benedict, R. F. Veilleux, *Tool and Manufacturing Engineers Handbook*, Volume 2, Fourth Edition — Forming, pp. 5-35 to 5-37, Society of Manufacturing Engineers, Dearborn, Michigan, 1984.
3. Oliver Evens, *The Young Mill-Wright & Miller's Guide*, Philadelphia, 1795.
4. William H. Hinterman, *Automatic Counterbalance Control*, SME Technical Paper EM93-152, Society of Manufacturing Engineers, Dearborn, Michigan, 1993.
5. Many standard electrical symbols used on press wiring diagrams are based on JIC Electrical Standards for Mass Production Equipment EMP-1-67 (pp. 24-27) and General Purpose Machine Tools EGP-1-67 (pp. 53-57), available from The Association for Manufacturing Technology, McLean, Virginia. Since AMT no longer endorses the JIC standards, the dated material is provided as a service only. Users may want to consider the use of the recently revised standard entitled Electrical Standards for Industrial Machinery ANSI/NFPA 79-1991, published by the National Fire Protection Association, Quincy, Massachusetts.
6. J. Barrett, *Electronic System Integration in Metalforming*, presented at Precision Metalforming Association (PMA) Metalform Symposium, 1992.
7. A. Bhide, "A Programmable System for Press Automation," *Stamping Quarterly*, Winter 1989.
8. H. Juster, "How One Metal Stamping Company Improved Quality and Delivery Time Using Computers," *Stamping Quarterly*, Spring 1992.
9. Dr. Behnam Bahr and Zain Ali, *Microcomputers and Flexible Manufacturing Systems*, SME Technical Paper MS91-338, Society of Manufacturing Engineers, Dearborn, Michigan, 1991.
10. D. Smith, "Using Visual Indicators in World-class Pressworking," *The Fabricator*, January/February 1993.
11. *The What, Why and How of OSHA Requirements for Mechanical Power Presses and Press Brakes*, Link Electric & Safety Control Company, Nashville, Tennessee, 1981.

CHAPTER 3

GAP-FRAME PRESSES

Gap-, or C-frame presses, which derive their name from their C-shaped throat opening, have many desirable features. These include excellent accessibility from the front and sides for diesetting and operation. The open back is available for feeding stock as well as ejecting parts and scrap.

Gap-frame presses are easier to set up than straightside presses, and give diesetters greater access to locate and bolt the die in place. They also generally have less overall height than a straightside press of comparable tonnage — a valuable asset when overhead clearance is limited. These machines also cost much less than straightside presses.

Figure 3-1 illustrates an older-style gap-frame press with all guarding removed. In this style, known as open back inclinable (OBI), the press frame is secured in a cradle, permitting the machine to be inclined backward. This facilitates gravity loading, as well as part and scrap discharge through the open back of the press. The frames of most older OBI presses are of cast construction, most commonly gray cast iron.

In Figure 3-1, two pretensioned tie rods appear across the open front of the machine. Lugs are cast into the frame of the machine to accept the tie rods, which are installed as an option to limit machine deflection under load.

The development of timed air blowoff devices and a variety of small conveyors has lessened the demand for the OBI style of presses. Today, many are operated in the vertical position. While the style is not obsolete in the inclined configuration, few press builders supply OBI presses, except on special order.

OPEN BACK STATIONARY GAP-FRAME PRESSES

The open back stationary (OBS) gap-frame press illustrated back in Figure 2-1 is more compact and often a more robust machine than the older OBI style that it has largely replaced. The OBS press has a box-like structure. Nearly all OBS presses are fabricated of

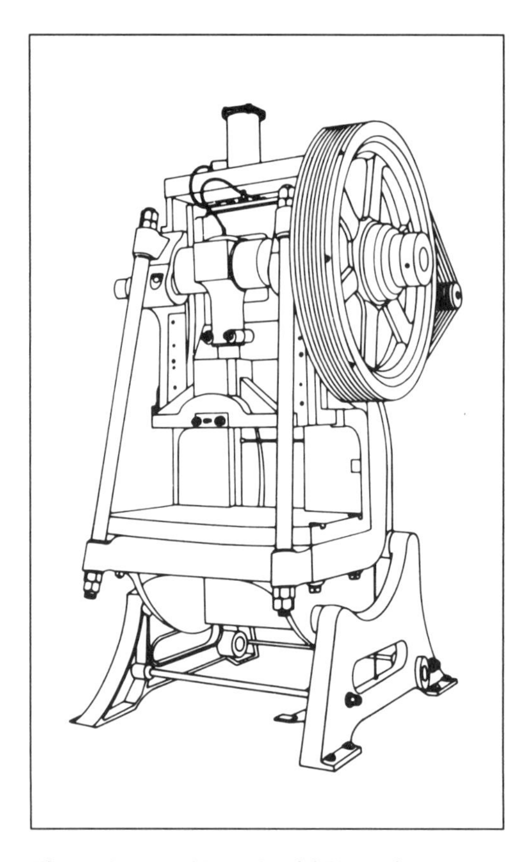

Figure 3-1. An older style of OBI gap-frame press with all guarding removed.

heavy steel plate and assembled by welding. Figure 3-2 illustrates a machine having two slide connection points.

One advantage of steel plate over cast iron is that steel has a substantially higher modulus of elasticity and, hence, proportionally greater stiffness. The modulus of elasticity for steel is approximately 30,000 ksi (206 GPa) compared to as low as 9,600 ksi (66 GPa) for ASTM class 20 gray cast iron and approximately 24,500 ksi (169 GPa) for ductile (nodular) irons.

Gap-frame presses are popular for short-run work where high accuracy of die alignment or close part tolerances are not necessarily controlling factors. Precision progressive dies can be successfully operated in gap-frame presses in spite of the angular deflection limitation. Success factors include:

- Maintaining excellent machine fit and alignment;
- Derating the press tonnage, which reduces angular deflection;
- Operating at high speeds so that much of the punching work is done by the kinetic energy of the slide, which does not produce machine deflection, and
- Installing properly prestressed tie rods with tubular spacers across the front opening to reduce deflection.

The chief limiting factor of gap-frame presses is that they have more deflection than straightside presses for a given load. The deflection has both a vertical and an angular component. The angular deflection or misalignment is due to the throat opening being spread as tonnage is developed. In many applications, this angular misalignment under load may not be objectionable.

Measuring Gap-Frame Press Stiffness

An older accepted American standard used by builders of gap-frame presses for measuring gap-frame stiffness is 0.0015 inch per inch (0.038 mm per mm) of throat depth, measured from the center line of the connection to the back of the throat opening. This measurement includes both vertical and angular deflection.

Angular deflection is by far the greatest concern because it results in misalignment between the punch and die. ANSI standard B5.52 M specifies the allowable vertical and angular deflections in machines built to metric standards.

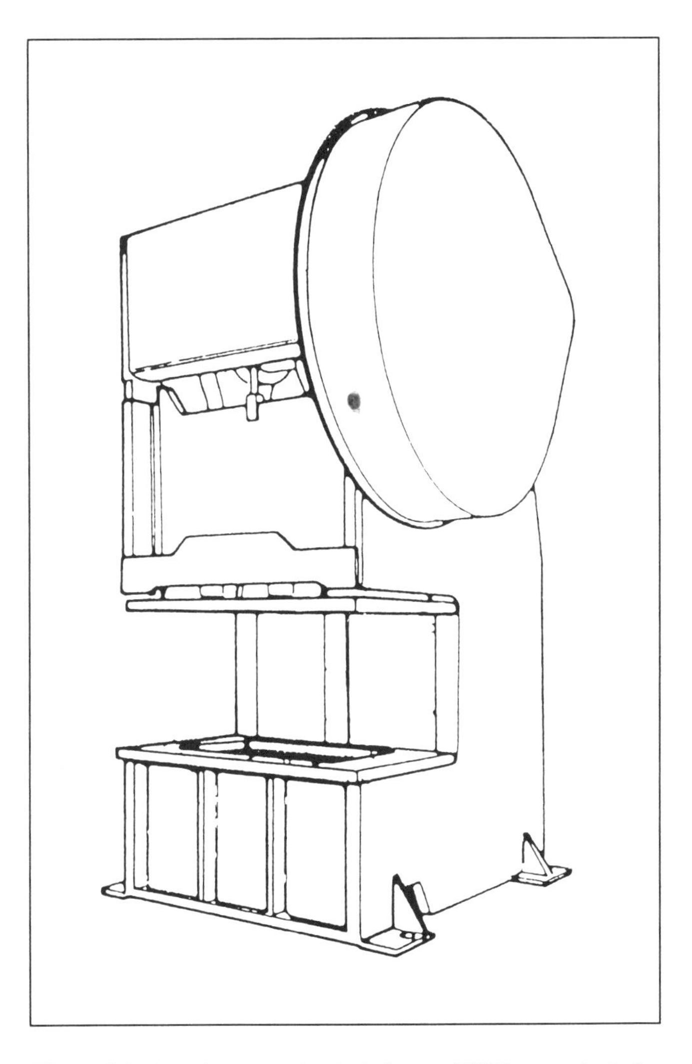

Figure 3-2. A modern open back stationary (OBS) press, featuring heavy welded plate construction to limit deflection and two slide connections. (Verson Corporation)

In the ANSI metric total deflection test method, a jack capable of full rated tonnage is placed under the slide at the center line of the slide and bed. The vertical deflection is measured between the center line of the slide and bed. It should not exceed 0.002 inch per inch (0.05 mm per mm) of distance from the center line to the throat opening. Some press builders design for a lower value of deflection. For example, Danly Machine allows only 0.0015 inch per inch (mm per mm) for standard frames.

In the ANSI metric angular deflection test method, the angular misalignment is measured with a test bar and two dial indicators. The procedure is:

1. A rectangular test bar is attached to the slide with an angle plate or other means to assure that it is at a 90-degree angle to the slide face.
2. Two dial indicators spaced 3.937 inches (100 mm) apart are placed on a single supporting rod attached to the press bolster.
3. Both dial indicators are carefully adjusted to zero.
4. A large hydraulic jack is placed directly under the connection and full press tonnage applied.
5. The dial indicators should show a difference of no more than 0.0047 inch (0.12 mm). This corresponds to an angular deflection of 0.0012 inch per inch, or 0.0144 inch per foot.

Since this measurement is intended to pick up an angular value, measuring the difference from front to back across the slide face will give similar results.

For many applications, angular misalignment under load is harmful. If the job cannot be run in a straightside press, some reduction in angular deflection can be achieved by installing tie rods across the open front of the press.

Adding Tie Rods

Some older gap-frame presses have existing lugs for tie rod installation in the front of the machine. In other cases, the manufacturer may supply lugs, that can be welded in place. Provided the rods are properly prestressed, a significant reduction in deflection results.

The best method of installing tie rods is to use them in conjunction with tubular steel spacers around the rods. The spacers help ensure that the tie rods are not overstressed, and serve to further stiffen the machine.

In a case study of a 200-ton (1779-kN) gap-frame press, installing properly prestressed tie rods with correctly machined spacers reduced the total deflection from 0.026 inches (0.66 mm) to 0.0134 inch (0.34 mm). The tie-rod diameter is 3.000 inches (76.2

mm). Each spacer cross-sectional area is 10.5 square inches (6774 square mm).

Figure 3-1 illustrates tie rods that are installed on the front of the press to reduce the angular deflection under load. The tie rods are prestressed a fixed amount. Care must be exercised not to overstress them, which can cause die misalignment. The use of carefully fitted tubular steel spacers around the tie rods is advised. No attempt should be made to increase press force capacity by this method. At best, adding prestressed tie rods to a gap-frame press will result in a reduced total angular misalignment. The physically limiting factor is that the cross-sectional area of the tie rod and spacer, if used, is small compared to that of the press frame. Also, adding tie rods to the front of the press limits access to the die opening, making the point of operation more difficult to guard properly.

The following design criteria are recommended if tie rods are to be added. The spacer should have 1.5 times the area of the tie rod. In addition to adding stiffness, the spacers will reduce the alternating load in the tie-rod thread.

The tie-rod area should be sufficient to support half the force capacity of the machine. A conservative nominal prestress in the tie rod is approximately 14,000 psi (96,516 kPa).

BULLDOZERS

Figure 3-3 illustrates a hydraulic bulldozer. Mechanical bulldozers originated in the middle of the 19th century. Metal working

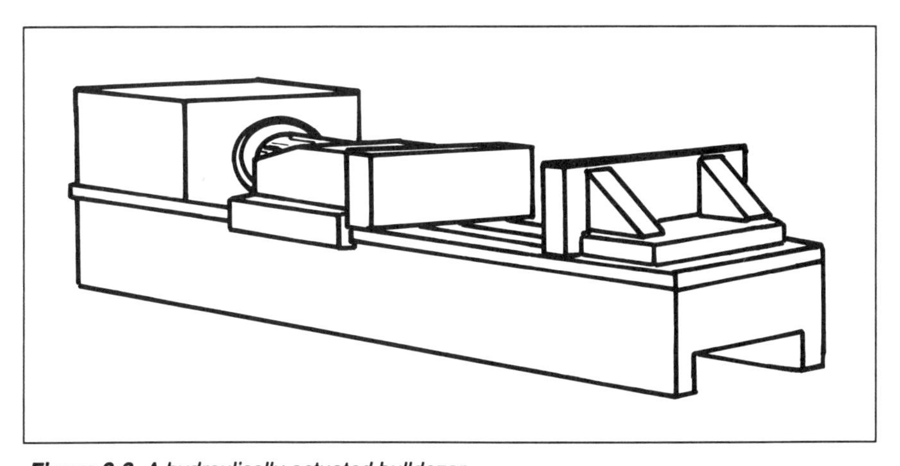

Figure 3-3. A hydraulically actuated bulldozer.

bulldozers were in use before the development of the familiar earth-moving bulldozer.

Early mechanically driven machines were typically double or triple back-geared, capable of exerting great force. The design found widespread application in all sorts of heavy-plate fabrication, such as punching rivet holes and bending operations.

Essentially, a bulldozer is an especially strong C-frame press laid on its back. They are still used for heavy fabrication work, built in capacities from 25 to over 1,000 tons (222 to 8896 kN). Modern machines are hydraulically powered, providing excellent control of the forming speed and force desired. They are excellent for straightening applications as well as punching plate and structural steel shapes.

The dies required are simple and inexpensive. The ability to handle the workpiece with an overhead crane and the unrestricted work area makes them suitable for forming hoops, rings, and large irregular structural shapes.

BIBLIOGRAPHY

1. D. Reid, *Fundamentals of Tool Design*, Third Edition, Chapter Seven: "Design of Pressworking Tools," Society of Manufacturing Engineers, Dearborn, Michigan, 1991.
2. Danly Machine, O.B.I. frame deflection standards.
3. C. Wick, J. T. Benedict, R. F. Veilleux, *Tool and Manufacturing Engineers Handbook*, Volume 2, Fourth Edition — Forming, pp 5-12 to 5-13, Society of Manufacturing Engineers, Dearborn, Michigan, 1984.

CHAPTER 4

STRAIGHTSIDE PRESSES

Straightside presses are named for the vertical columns or uprights on either side of the machine. The columns, together with the bed and crown, form a rigid box-like housing.

A major advantage of the straightside press compared to the gap-frame machine is freedom from angular misalignment under load. Maintaining true vertical motion throughout the press stroke is critical in minimizing tool wear while maintaining accurate part tolerances.

Examples of high-volume close-tolerance stampings include electrical connectors, snap-top beverage cans, spin-on oil filter cartridge bases, and refrigeration compressor housings. Tiny computer connectors are stamped at press speeds up to 1,800 strokes per minute (SPM) or more. Often two to eight or more parts are completed per hit.

Precision stampings are also produced at low speeds. For example, large refrigeration compressor housings may be stamped at press speeds less than 12 SPM. The housing consists of two mating halves, which must fit together precisely to properly align the internal parts.

All these stampings have several common factors, including:

- Zero defects is a goal that must be approached or attained.
- All variables must be minimized.
- Defective stampings may result in product failures.

Avoiding angular deflection under load is an important reason for choosing a straightside press for these and other close tolerance stampings, wherever possible.

Straightside presses are available in very high force capacities. Mechanical presses are built with force capacities through 6,000 tons (53.4 mN) or more.

One of the largest single-slide straightside mechanical presses in the world is a USI-Clearing 6,000-ton machine. The bed is 66 inches (1.68 m) wide by 494-inches (over 41 feet or 12.3 m) long. It was built in Chicago in 1968, and transported to Cleveland, Ohio disassembled. It is in daily use at Midland Steel Products, producing truck frame rails from high-strength steel. In some cases, the left and right rails are formed in a single hit from two blanks. The frames are used for semi-tractors, medium trucks, and recreational vehicles.

Force capacities of 50,000 tons (445 mN) or more are available in hydraulic straightside presses. The very large hydraulic machines are used in warm and cold forging applications as well as various rubber-pad and fluid-cell forming processes.

MECHANICAL VS HYDRAULIC STRAIGHTSIDE PRESSES

Increasingly, both single and double-action hydraulic presses are used for forming large, irregularly shaped parts for the automotive and appliance industries. An advantage for deeply formed or drawn parts is that full force is available anywhere in the press stroke. Mechanical presses have the full rated force available only very near the bottom of the stroke. A chart showing distance from bottom of stroke versus available force is called a force curve. The force curves for six different mechanical presses are illustrated in Figure 4-1.

Repair costs due to abuse and/or poor maintenance practices are high for both types. Mechanical presses have an advantage of lower mechanical losses or lower energy consumption in many applications.

Overload Protection

Hydraulic presses limit overloading by restricting the maximum pressure supplied to the actuating cylinder(s). Overload protection for mechanical presses can be provided by placing a hydraulic overload cylinder in series, with the force delivered to each connection (Figure 4-2).

When a preset maximum limit is exceeded, an overload valve dumps the precharged oil from the overload cylinders and trips a limit switch, stopping the press. The cause of the overload condition is first corrected, and the overload system recharged by actuating a key-locked switch. Such systems can accommodate maximum shut height errors of approximately 0.75 inch (19 mm).

Other types of press overload devices include shear collars or washers, shear pins, and stretch links. These devices are simple and low in cost. However, their failure point is uncertain and may

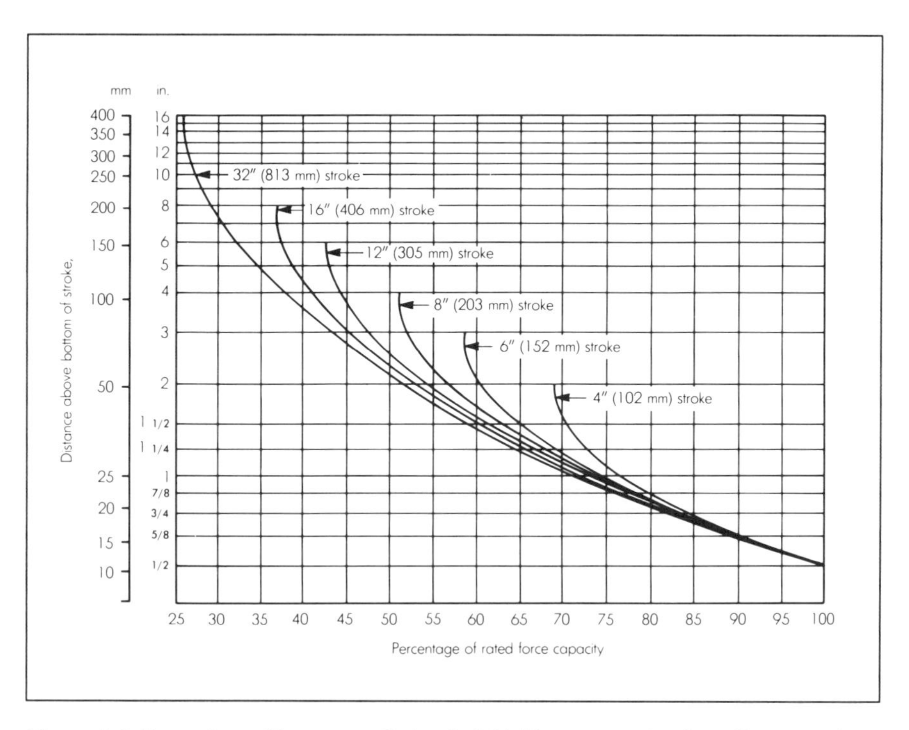

***Figure 4-1.** Percentage of force capacity for straightside presses at various distances above bottom of stroke. (Danly Machine Corp.)*

not be immediately detected, resulting in possible press damage and poor quality work.

Most, but not all, straightside presses employ tie-rod construction. The rods hold the press housing in compression. They provide a means to move large presses in sections. Should the press become stuck on bottom, the rods can be heated to relieve the prestress. The rods also limit press overloading.

As long as the press columns are maintained in a preloaded condition by the tie rods, the deflection in the die space occurs at a linear rate as a function of increasing tonnage. However, once an overload condition exceeds the tie-rod preload, the crown lifts off the press column. When crown lifting occurs, press stiffness is greatly decreased, limiting overloading as illustrated in Figure 4-3.

Some press builders supply straightside machines of non-tie-rod construction. When purchasing such a press, it is wise to specify some form of overload protection, such as a hydraulic overload system.

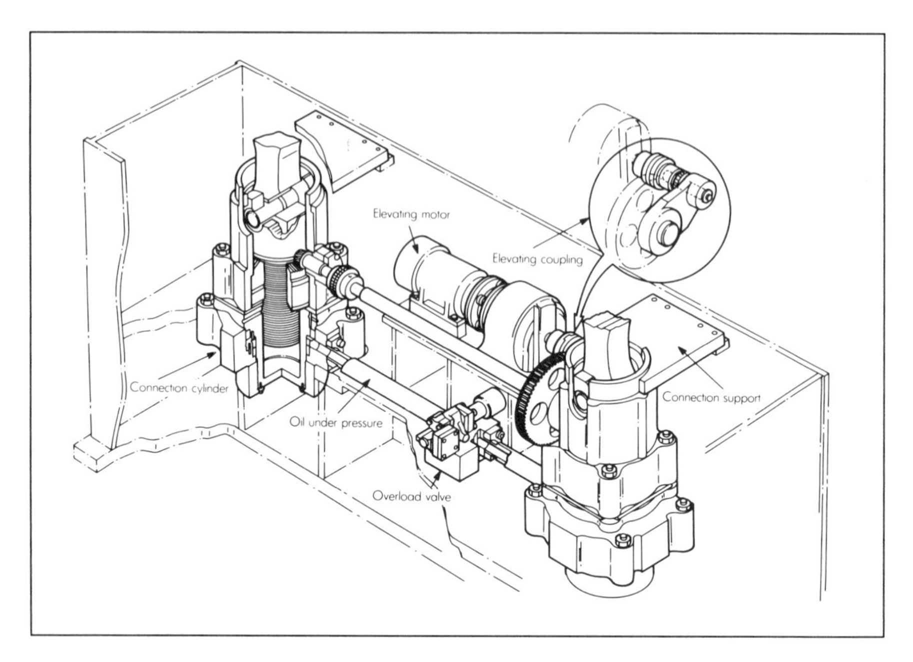

Figure 4-2. Precharged hydraulic cylinders under each connection provide fast-acting resettable overload protection. (Verson Corporation)

Press Slide Connections

A mechanical press connection is the point of attachment of the pitman or eccentric strap to the slide. Ball and socket type bearings are frequently used in smaller machines. A connection bearing of the type shown in Figure 4-2 has both a bronze-lined saddle-type bearing and a wrist pin to transmit force to the slide.

Connection Strength

The connection is designed to transmit large compressive forces to the slide. If subjected to an extreme overload, the ball and socket type may be damaged by a crack or deformation of the socket. In the event of a large overload, the bronze bearing material may be extruded out of a saddle-type bearing, and bend or break the wrist pin.

Connection damage is most likely to occur due to excessive stripping or reverse loads. Generally, the connection is designed to withstand reverse loads of only 10% of machine force capacity. Reverse loads are concentrated on the wrist pin or ball connection

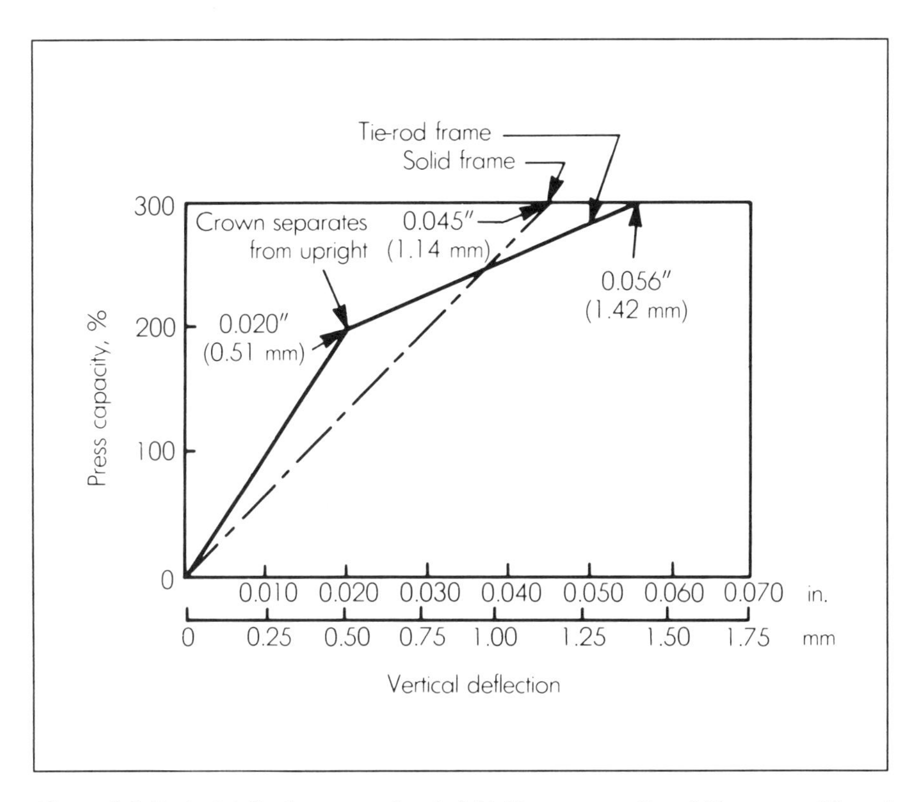

Figure 4-3. *Typical deflection curves for straightside presses with solid frames and tie-rod frames.*

retaining ring screws. Failure of these parts frequently results from excessive snap-through forces in blanking operations.

The connection may incorporate a screw adjustment mechanism. Larger machines have an electrically or pneumatically powered adjustment screw drive. Normally a mechanical brake in the motor automatically engages to hold the adjustment in place. In the case of presses having multiple connections, a single motor is used in conjunction with shafts, bevel gears, and flexible couplings to drive all adjustment screws in synchronism (Figure 4-2).

SINGLE-CONNECTION PRESSES

Straightside presses with single connections often are built to provide very high force capacities in a machine having a relatively

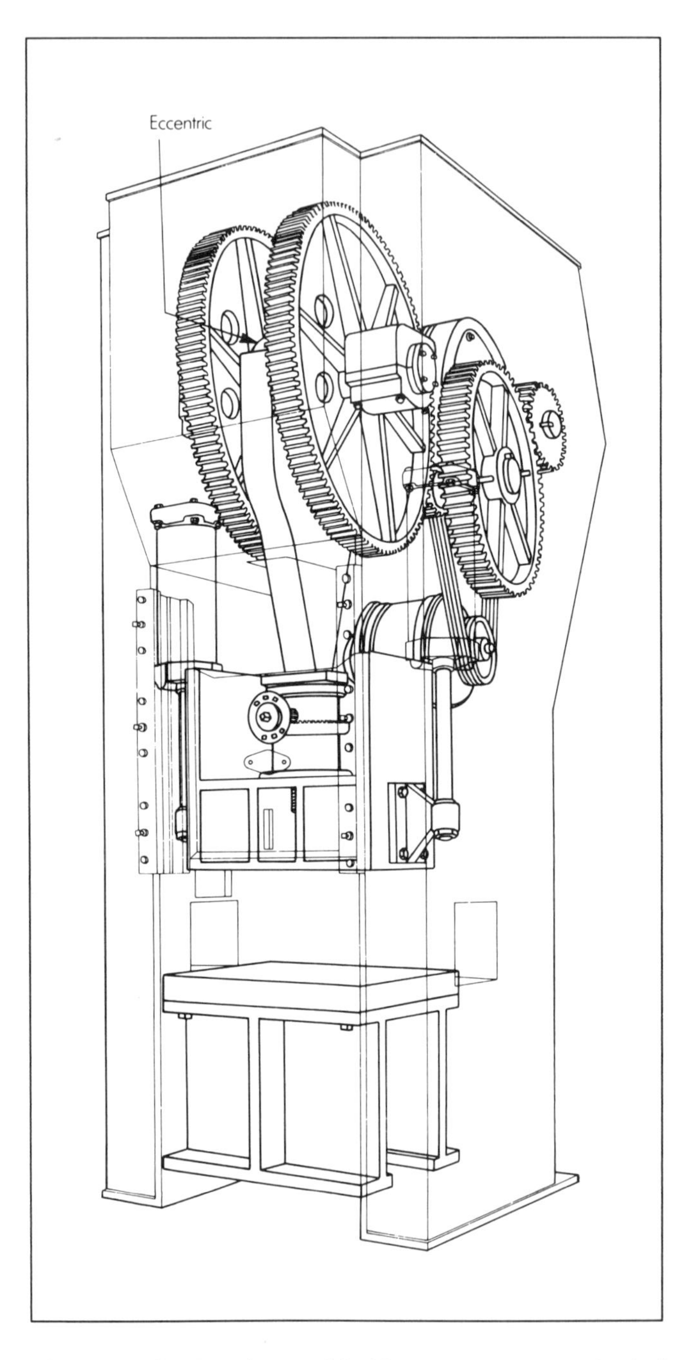

Figure 4-4. Single-action, straightside, eccentric-type mechanical press. (Verson Corporation)

small bed size. Figure 4-4 illustrates a high tonnage single-action, straightside, eccentric-type mechanical press.

The gear train is of the double reduction type. A large gear on either side drives the eccentric. This type of machine is very useful for heavy forming, as well as both warm and cold forging work.

The heavy construction, together with the narrow bed and slide, results in low deflection under load. A press of this type is ideal for blanking work involving thick high-strength materials. Very stiff machines, with bed sizes no larger than necessary, are subject to much less snap-through energy release than presses with bed sizes that are much too wide for the application.

In order to avoid ram tipping problems, the load must be carefully centered under the connection. While this is always very important, it is especially necessary in single-point presses. Keeping the load centered minimizes the pressure on the gibbing and reduces die wear.

TWO-POINT PRESSES

Straightside presses with two connection points are in very widespread use. Figure 2-2 illustrates a crankshaft-driven press with two connections. The slide shown in Figure 4-2 has two connections that are driven by gear-actuated eccentrics.

Two-point presses are guided by the correct adjustment of the pitman straps from left to right. Some front-to-back guiding of the slide is provided by the saddle bearing and wrist pin type connection. However, the majority of the front-to-back guiding is provided by the gibbing. Again it is important to center the load, especially from front to back.

Figure 4-5 illustrates a two-point eccentric-driven straightside press. Typical electrical control components needed to power and control the machine are shown. Note that the rotary cam switches are separately driven. This adds an extra measure of safety in addition to chain breakage monitors.

Eccentric Drive for Two- and Four-point Presses

Double or quadruple gear driven eccentrics normally rotate in opposite directions to aid in slide guiding and avoid lateral thrust. This feature is found in most two- and four-point presses that are top-driven by eccentrics.

Underdrive Straightside Presses

Figure 4-6 illustrates a four-point underdrive single-action straightside press. The slide is actuated by four eccentric straps that act in tension to pull the slide down. Double-gear reduction is employed. The final reduction is accomplished by involute-tooth spur gears. The first reduction gears are of the herringbone type. This style of gear, while more costly than the spur type, runs more smoothly and has a long service life, if not abused.

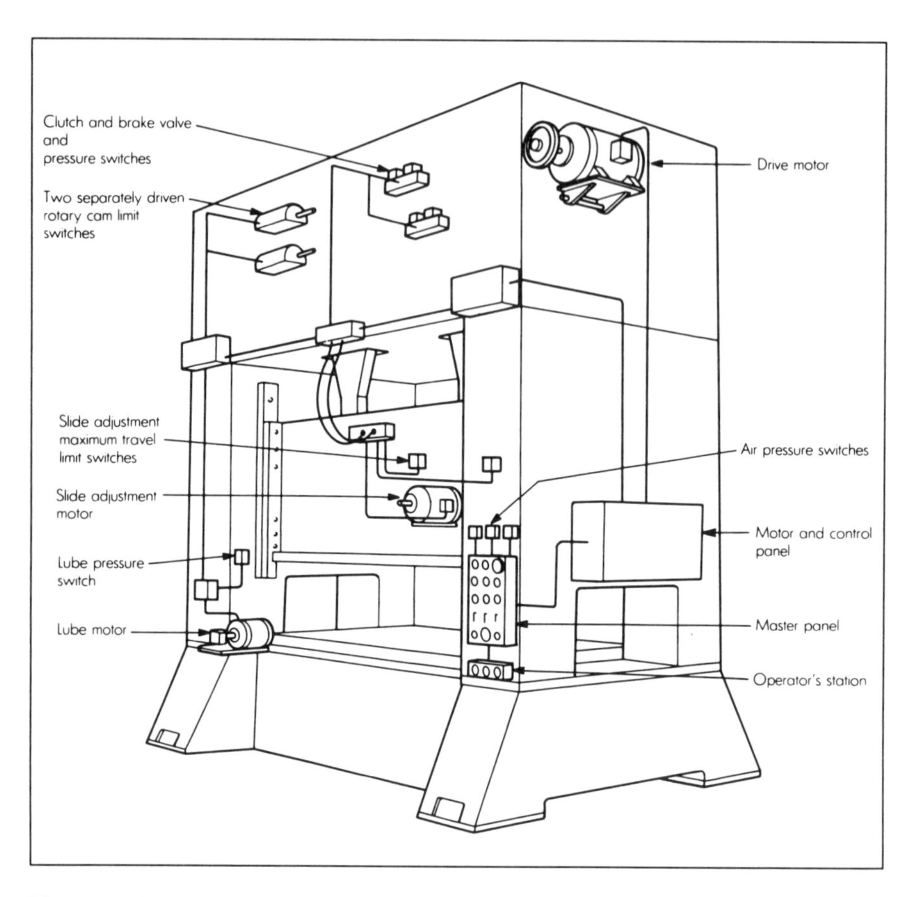

***Figure 4-5.** Two-point eccentric-driven straightside press illustrating typical electrical control components. (Verson Corporation)*

Timing of the gearing from the left to right side of the press is critical. Driven by the clutch, couplings are provided on the shaft that either are adjustable or may be fitted with offset keys.

Advantages of Underdrive Presses

Most underdrive presses in North America are installed in large automotive stamping plants built in the post World War II industrial expansion period. The main feature, compared to a top-drive machine, is that a lighter housing is required and there are no tie rods. The housing serves to guide the slides and contain control equipment and the counterbalance system.

Less ceiling height is required since clearance is not needed for the drive components. However, a deep basement is needed. A basement or press pit is often required for the efficient removal of scrap, so this may not be a disadvantage.

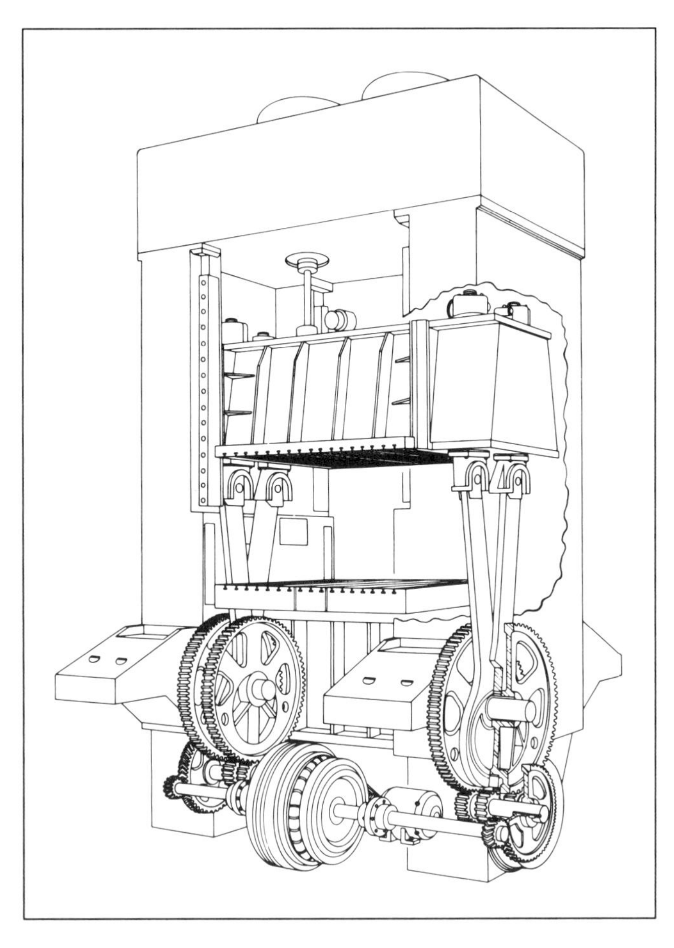

Figure 4-6. Four-point underdrive straightside press (USI-Clearing).

Underdrive presses are built with single, double, and triple actions. The triple-action machine is not advised for new construction. Good product design practices have eliminated the need for the third action in nearly all current die designs. If a third action is required, the press is much more complex because a toggle dwell action is needed for the inner slide. This adds greatly to the cost of the machine. Also the dwell time adds approximately an additional second to the machine cycle time.

Figure 4-7 illustrates the placement of pneumatic piping, tanks, and controls for a typical straightside press. The system is typical of a good pneumatic system for a press equipped with an air-actuated friction clutch and die cushions.

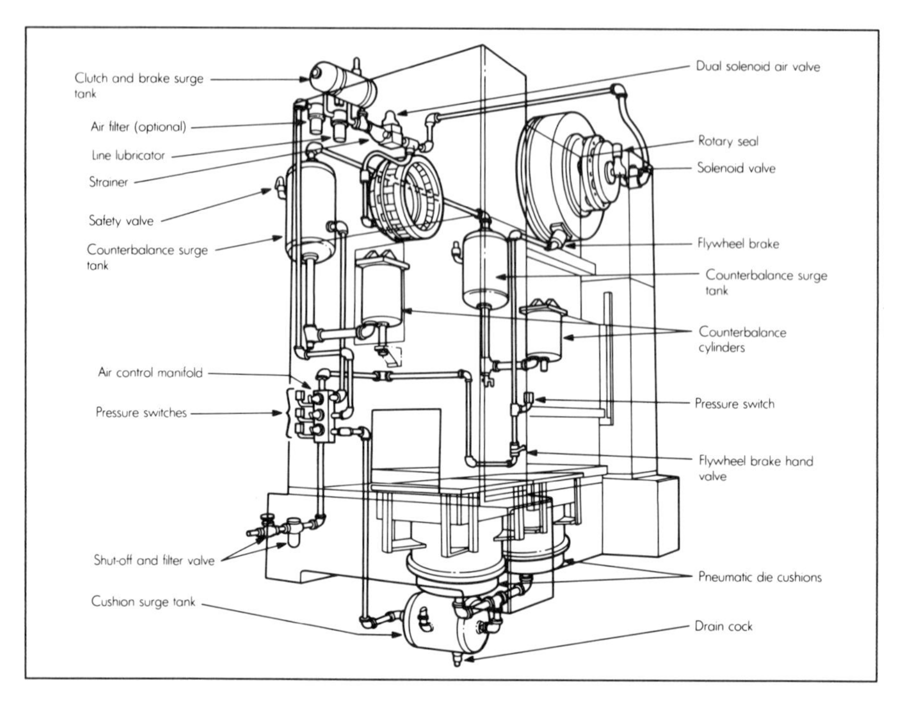

Figure 4-7. *Pneumatic piping, tanks, and controls for a straightside press. (Verson Corporation)*

BIBLIOGRAPHY

1. C. Wick, J. T. Benedict, R. F. Veilleux, *Tool and Manufacturing Engineers Handbook*, Volume 2, Fourth Edition — Forming, Chapter 5, Society of Manufacturing Engineers, Dearborn, Michigan, 1984.
2. D. Smith, *Die Design Handbook*, Section 4, "Shear action in Metal Cutting," Society of Manufacturing Engineers, Dearborn, Michigan, 1990.
3. D. Smith, *Die Design Handbook*, Section 13, "Dies for Large and Irregular Shapes," Section 27, "Press Data," Society of Manufacturing Engineers, Dearborn, Michigan, 1990.

CHAPTER 5

HYDRAULIC PRESSES

The hydraulic press is one of the oldest basic machine tools. In its modern form, it is well adapted to presswork ranging from coining jewelry to forging aircraft parts. Modern hydraulic presses are, in some cases, better suited to applications where the mechanical press has been traditionally more popular.

ADVANTAGES OF HYDRAULIC PRESSES

The mechanical press has been the first choice of many press users for years. The training of tool and die makers and manufacturing engineers in North America has been oriented toward applying mechanical presses to sheet metal pressworking.

Modern hydraulic presses offer good performance and reliability. Widespread application of other types of hydraulic power equipment in manufacturing requires maintenance technicians who know how to service hydraulic components. New fast-acting valves, electrical components, and more efficient hydraulic circuits have enhanced the performance capability of hydraulic presses.

Factors that may favor the use of hydraulic presses over their mechanical counterparts may include:

- Depending on the application, a hydraulic press may cost less than an equivalent mechanical press.
- In small lot production, where hand feeding and single-stroking are practiced, additional press wear is not incurred.
- Die shut height variations do not change the force applied.
- There is no tonnage curve derating factor.
- Forming and drawing speed can be accurately controlled throughout the stroke.
- Hydraulic presses with double actions and/or hydraulic die cushions are capable of forming and drawing operations that would not be possible in a mechanical press.

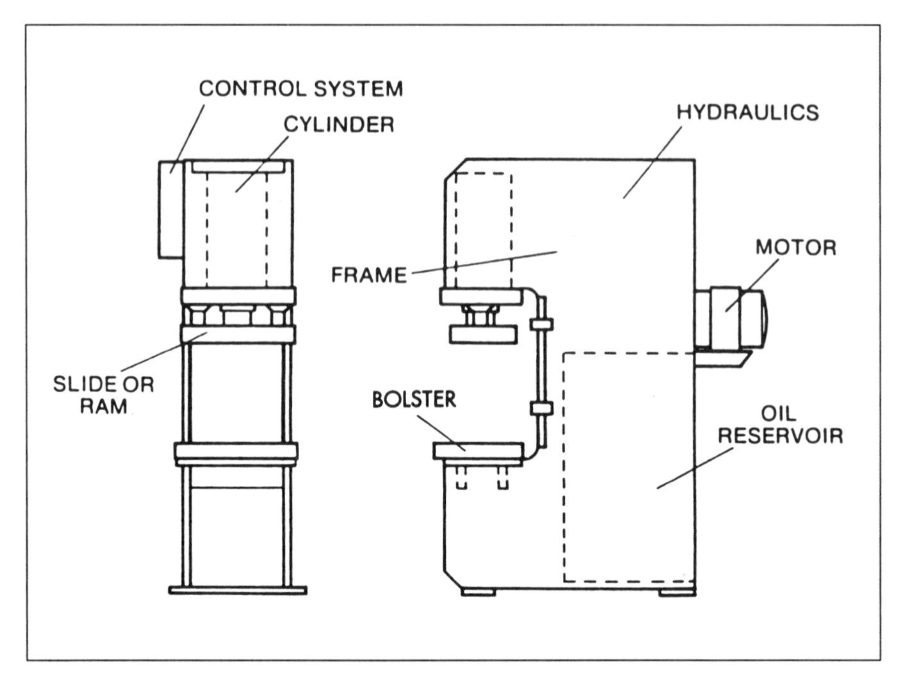

Figure 5-1. A typical hydraulic press featuring gap-frame construction. (Greenerd Press & Machine Company)

Example of a Gap-frame Hydraulic Press

Like the mechanical press, hydraulic presses deliver a controlled force to accomplish work. The style of the press frame and the hydraulic components vary depending on the intended use. Figure 5-1 illustrates a gap-frame or C-frame hydraulic press.

The press shown in Figure 5-1 has a frame and bolster similar to the construction used for open back stationary (OBS) mechanical presses. The frame is of robust construction to limit both angular and total deflection. The bolster and ram provide a surface for tool mounting. The ram is actuated by a large hydraulic cylinder in the center of the upper part of the frame. Additional alignment is provided by two round guide rods.

The motor drives a rotary pump, which draws oil out of the reservoir housed in the machine frame. The control system has electrically actuated valves that respond to commands to advance and retract the slide or ram. A pressure regulator is either manually or automatically adjusted to apply the desired amount of force.

UNIQUE FEATURES OF HYDRAULIC PRESSES

Figure 5-2 (A) illustrates why the rated force capacity of a mechanical press is available only near the bottom of the stroke. The full force of a hydraulic press can be delivered at any point in the stroke. This feature, illustrated in Figure 5-2 (B) is very useful.

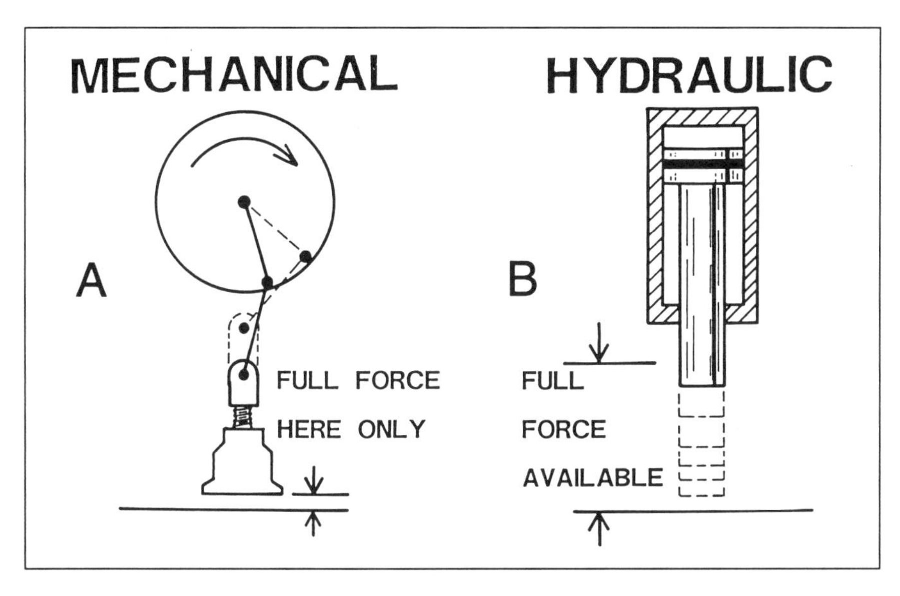

Figure 5-2. *The rated capacity of a mechanical press is available only at the bottom of the stroke (A); the full force of a hydraulic press can be delivered at any point in the stroke (B).*

Deep drawing and forming applications often require large forces very high in the press stroke. Few mechanical presses have tonnage curves that will permit such severe applications.

Another advantage is that the stroke may be adjusted by the user to match the requirements of the job. Only enough stroke length is required to provide part clearance. Limiting the actual stroke will permit faster cycling rates, while reducing energy consumption.

The desired preset hydraulic pressure provides a fixed working force. When changing dies, different shut heights do not require fine shut height adjustment. Different tool heights or varying thicknesses of material have no effect on the proper application of force.

Availability of full machine force at any point in the stroke is very useful in deep drawing applications. High force and energy requirements usually are needed throughout the stroke. The ram speed can also be adjusted to a constant value that is best for the material requirements.

Built-in Overload Protection

The force that a hydraulic press can exert is limited to the pressure applied to the total piston area. The applied pressure is generally limited by one or more hydraulic relief valves.

A mechanical press usually can exert several times the rated maximum force in the event of an accidental overload. This extreme overload often results in severe press and die damage. Mechanical presses can become stuck on bottom due to a large overloads, such as part ejection failures or diesetting errors.

Hydraulic presses may incorporate tooling safety features. The full force can be set to occur only at die closure. Should a foreign object be encountered high in the stroke, the ram can be programmed to retract quickly to avoid tooling damage.

Lubrication

Hydraulic presses have very few moving parts. Those parts that do move operate in a flood of pressurized oil, which serves as a built-in lubrication system. Should leakage occur, it is usually caused by the failure of an easily replaceable part such as the ram packing or a loose fitting.

Hydraulic presses having guide rods or gibbing may require additional lubrication. The same types of lubrication systems used on mechanical presses are effective.

Mechanical presses with high force capacities are physically much larger than their hydraulic counterparts. Few mechanical presses have been built with capacities of 6,000 tons (53.376 MN) or more. Higher tonnages or more compact construction is practical in hydraulic presses. Hydraulic presses for cold forging are built with up to 50,000 tons (445 MN) or greater capacity. Some hydraulic fluid cell presses have force capacities over 150,000 tons (1,334 MN).

Force Capacities and Stroke Length

The bed size, stroke length, speed, and tonnage of a hydraulic press are not necessarily interdependent. In Figure 5-3 we see a large bed size press having low force capacity. Note the small cylinder size. Uses for these presses include cutting and pressing soft materials such as fabric, and in wood or plastic laminating applications.

Hydraulic presses are built with very high force capacities in machines that have relatively small bed sizes. Applications include net shape applications such as cold forging and extruding. High force capacity machines have large cylinder diameters and massive tie rods.

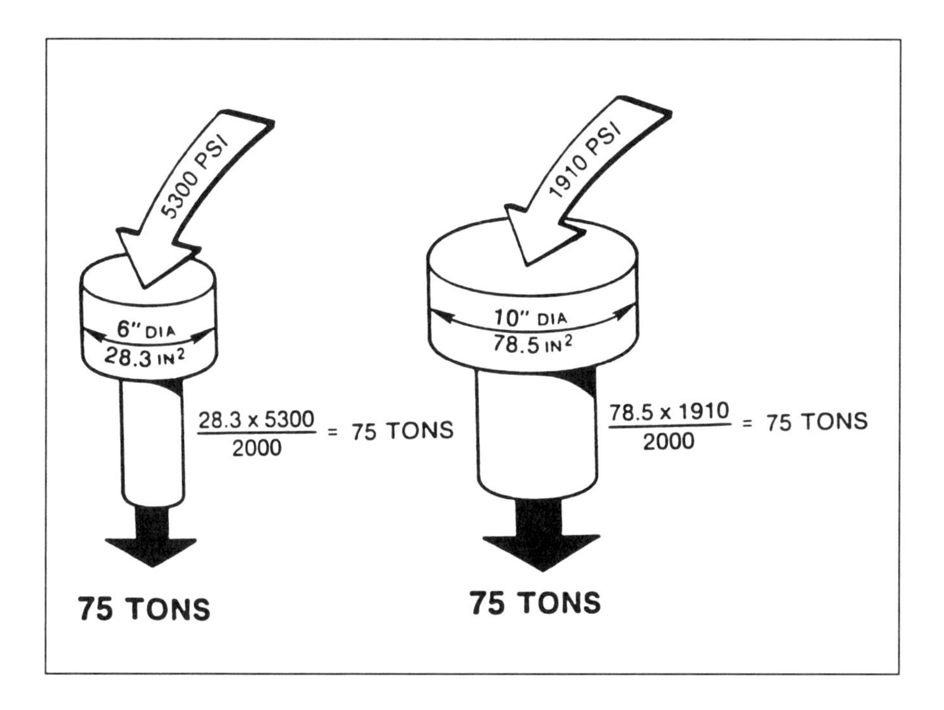

Figure 5-3. *Two pistons having different diameters both deliver 75 tons (667 kN) of force by applying different pressures to each piston.*

Since the stroke length is fully adjustable, long stroke lengths provide for ease of setup and flexibility of application. The full stroke may be used to open the press for installing dies. In production, the stroke length can be set as short as possible to provide for stock feeding and part ejection while maximizing stroking rates.

Figure 5-3 illustrates how two pistons having different diameters both deliver 75 tons (667 kN) of force. The force developed by a hydraulic piston is the product of the area of the piston times the applied pressure.

Thus, 75 tons (667 kN) of force can be achieved by applying 5,300 psi (36,538 kPa) to a 6-inch (152.4 mm) diameter piston. The same 75 tons (667 kN) of force is achieved by applying 1,910 psi (13,168 kPa) to a 10-inch (254 mm) diameter piston.

There is no set rule on the best peak operating pressure for a press design. Obviously, higher pressures permit the use of more compact cylinders and smaller volumes of fluid. However the pumps, valves, seals, and piping are more costly because they must be designed to operate at higher pressure.

The force of a hydraulic press can be programmed in the same way that the movements of the press are preset. In simple presses the relief valve system that functions to provide overload protection may also serve to set the pressure adjustment. This allows the press to be set to exert a maximum force of less than press capacity. Usually there is a practical lower limit, typically about 20% of press capacity.

Programmable controllers are a feature of many modern hydraulic presses. The correct pressure, together with ram travel and other parameters, is stored in memory by job number and automatically preset by the diesetter. For deep drawing operations, the blankholder or hydraulic die cushion force can be varied through the press cycle for best results.

Hydraulic Press Speeds

Most press users are accustomed to describing press speeds in terms of strokes per minute. Speed is easily determined with a mechanical press. It is always part of the machine specifications.

The number of strokes per minute made by a hydraulic press is determined by calculating a separate time for each phase of the ram stroke. First, the rapid advance time is calculated. Next the pressing time or work stroke is determined. If a dwell is used, that time is also added. Finally the return stroke time is added to determine the total cycle time. The hydraulic valve reaction delay time is also a factor that should be included for accurate calculations.

These factors are calculated to determine theoretical production rates when evaluating a new process. In the case of jobs in operation, measuring the cycle rate with a stopwatch is sufficient.

Most hydraulic presses are not considered high-speed machines. In the automatic mode, however, hydraulic presses operate in the 20 to 100 stroke per minute range or higher.

These speeds normally are sufficient for hand-fed work. The speeds are comparable to that of mechanical OBI and OBS presses.

Guidelines for Press Selection

Figure 5-4 (A) illustrates a low force capacity press with a large bed area. Such machines are used for cutting soft materials and laminating work.

Figure 5-4 (B) illustrates a press designed for heavy work such as cold forging. Required here is a compact machine capable of delivering high forces over a small area.

When choosing between a mechanical or hydraulic press for an application, a number of items should be considered. The force required to do the same job is equal for each type of press. The same engineering formulas are used.

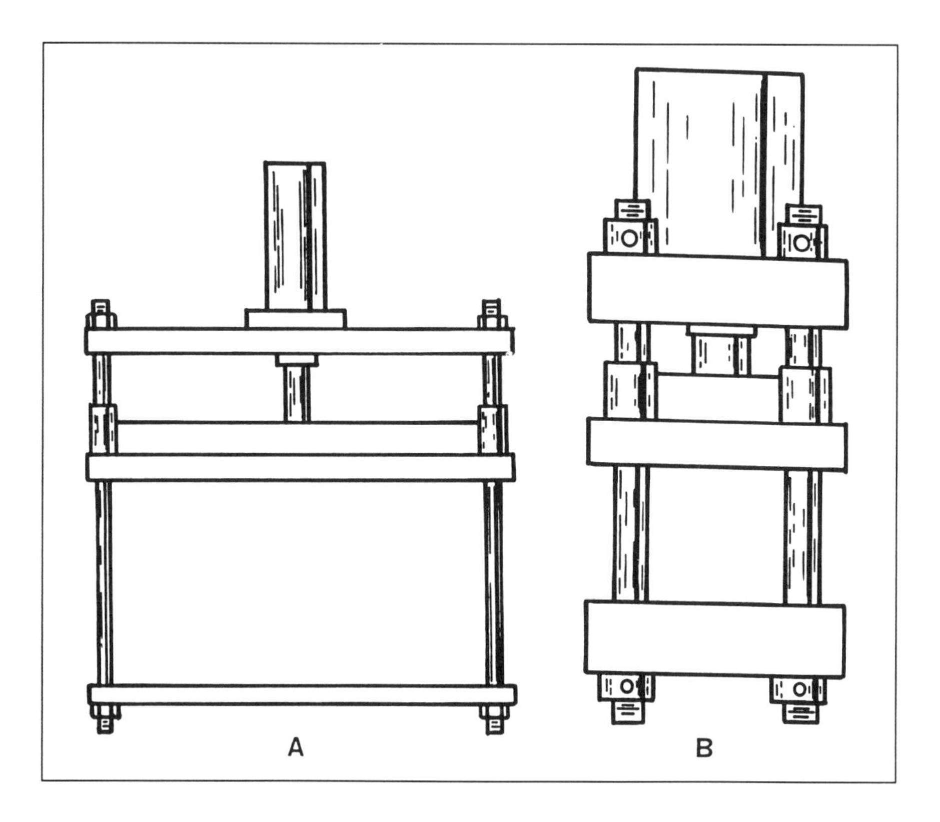

Figure 5-4. *Cutting soft materials and laminating work may require a low force capacity machine with a large bed area (A); heavy work such as cold forging requires a compact machine capable of delivering high forces over a small area (B).*

There is always a possibility that an existing job operated in a mechanical press requires 20 to 30% more force than the rated machine capacity. The problem may go unnoticed, although excessive machine wear will result. If the job is placed in a hydraulic press of the same rated capacity, there will not be enough force to do the job. Always make an accurate determination of true operating forces to avoid this problem.

The forming speed and impact at bottom of stroke may produce different results in mechanical presses than in their hydraulic counterparts. There are some specific limitations. Drop hammers and some mechanical presses seem to do a better job on soft jewelry pieces and impact jobs. The coining action may produce a sharper impression if the impact is present.

In deep drawing, the uniform velocity and full force throughout the stroke may produce different results. Usually better results are obtained in a hydraulic press.

Like the mechanical press, open gap-frame machines provide easy access from three sides. Four-column presses, such as the machine shown in Figure 5-5, ensure even pressure distribution provided that there is no off-center loading.

Straightside presses, such as that illustrated in Figure 5-6, offer the rigidity required to withstand off-center loading.

Accessories

Most hydraulic press builders offer many control options and accessories. These include:

- A distance reversal limit switch preset for the depth of ram stroke for automatic return to the top of stroke position;
- A pressure reversal switch set for the highest force delivered before the ram returns automatically to the top of stroke;
- Automatic or continuous cycling controls used in conjunction with automatic feeding equipment;
- Dwell timers that are adjustable and set to open the press after a preset dwell period;
- Ejection cylinders or knockouts that can be actuated at a preset position, time, or pressure;
- Rotary index tables and other work positioning devices often powered by the press hydraulic system; and

- Hydraulic die cushions that have the advantage of taking up less space than air cushions while offering smoother, more positive resistance.

Press Quality

Since the applications for hydraulic presses range all the way from simple hand pumped maintenance presses to machines having very high force capacities, types of construction and desirable features vary accordingly.

Here are just a few design and construction questions that will provide a basis for comparison of one machine with another.

Frame. Compare the weight if possible. Try to determine the character of the frame construction. If a weldment, look at the plate thicknesses, extent of ribbing, and stress relieving.

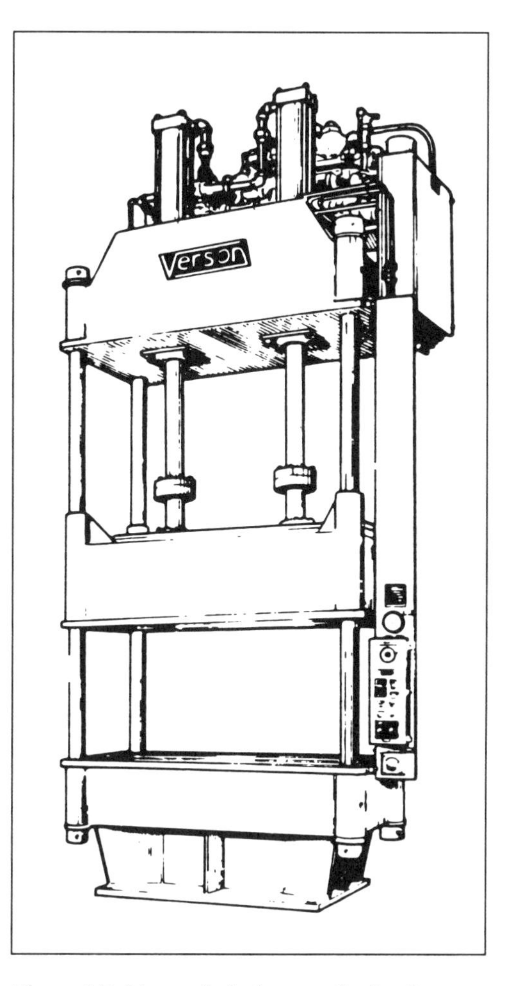

Figure 5-5. *A two-cylinder four-post hydraulic press: note that the hydraulic reservoir, pump, and controls are located on top of the machine. (Verson Corporation)*

Cylinder and slide construction. The cylinder size, type of construction used, and availability of service parts are important. Also determine how the ram travel is guided.

Maximum system pressure. The pressure at which the press delivers full tonnage is important. The most common range for industrial presses is from 1,000 psi (6,894 kPa) to 3,000 psi (20,682 kPa). Some machines operate at substantially higher pressures, which may accelerate wear. Make sure that replacement parts are readily available.

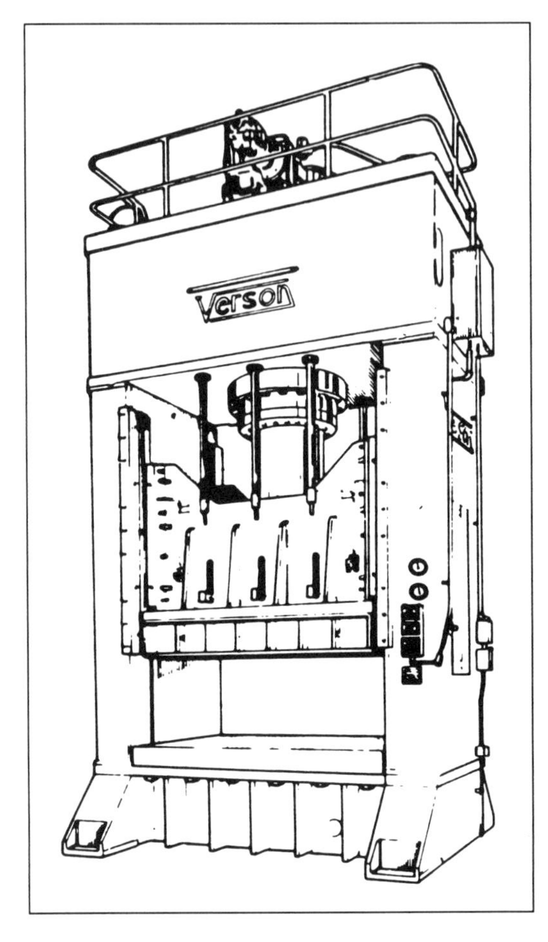

Figure 5-6. A straightside hydraulic press is designed for applications requiring close alignment. (Verson Corporation)

Horsepower. The duration, length, and speed of the pressing stroke are the major factors that determine the required horsepower.

Speed. Take the time to calculate the speed based on the operations you intend to perform. There are wide variations in hydraulic press speeds.

HYDRAULIC PRESS LIMITATIONS

The fastest hydraulic press is slower than a mechanical press designed for high-speed operation. For example, the high speeds, together with short stroke and feed progressions, used for electrical terminal production favor mechanical presses.

While hydraulic presses are available with a reasonably accurate built-in method of stopping the down stroke, generally stop or bottoming blocks must be provided in the tooling. Under production conditions, stroke depth typically can be controlled to within 0.020 inch (0.51 mm), even though readout devices with higher resolution may be provided on the machine.

Hydraulic presses are often provided with controls to reverse the machine at a preset pressure. This feature used in conjunction with stop or bottoming blocks in the die, can result in excellent part uniformity.

Problems with snap-through energy release are common to both mechanical and hydraulic presses. Damage to the hydraulic press structure may result. Severe snap-through shock can damage lines, fittings, valves, and the press electrical controls. Some hydraulic presses designed for heavy blanking have built-in hydraulic damping cylinders on each corner of the machine to arrest snap-through energy.

MODERN DEVELOPMENTS IN HYDRAULIC PRESSES

As hydraulic presses continue to improve, there is no doubt that they will play an increasingly significant role in industrial production. Hydraulic presses are already in more widespread use in Europe than in North America.

Hydraulic press speeds have increased over the last few years. Hydraulic component manufacturers have developed new valves with higher flow capacities, faster response time, and precise flow control capability. But unless a radically different hydraulic circuit design is developed, it is unrealistic to predict that hydraulic press speeds will overtake the mechanical press.

Mechanical press crankshaft-driven feeders are essentially obsolete for new installations. Today, hydraulic presses use the same roll feeders and other auxiliary equipment designed for mechanical presses. Actuation is by one or more microprocessor-based programmable controllers. Important features of such systems are easy programming and multiple job memory capability.

Some hydraulic press manufacturers build snap-through arresting devices into hydraulic presses used for heavy blanking applications. Hydraulic snap-through arresters are also available as add-on devices for retrofitting to existing mechanical and hydraulic presses.

Hydraulic dampers are effective snap-through arresting devices on both mechanical and hydraulic presses. While the dampers are an effective solution, they add to equipment cost, require time-consuming adjustment for different jobs, and increase energy consumption. Snap-through energy control should be achieved through good die timing wherever possible.

Modern control systems permit the press sequence to be programmed for each job. Based on job memory parameters the correct pressure, stroke length, speed and dwell time, retraction force can be set up quickly.

Important improvements on all types of presses are continuing to increase the comfort and safety of the operator. Better illumination, quieter machines, comfortable work positions, semi-unattended operation, and provisions for simplified machine adjustments all add to operator comfort and increased productivity.

Hydraulic presses are increasingly specified for production applications where mechanical presses were once used almost exclusively. Proper selection and use of the machine can be enhanced by a greater understanding of the characteristics of a hydraulic press. The manufacturing engineer should view the press as only one part of a total system that includes tooling, part feeding, personal protection, and part unloading equipment.

Programmable Hydraulic Blankholder Force Control

Hydraulic die cushions are used on both mechanical and hydraulic presses. They have several advantages when compared to an air cushion. These include:

- Much larger forces can be obtained in the same press bed space.
- Timed cushion lock-down or return delay; this feature is used to avoid deforming the part as the press opens.
- The ability to control the instantaneous cushion pressure with a servo valve; this feature can be used to optimize the blankholder force as a deep drawing operation is in progress.

By controlling the hydraulic die cushion pressure with a servo valve, optimization of blankholder force can be achieved. Typically the pressure of air-actuated die cushions increases 10% or more between initial contact and the end of travel. A pressure increase of up to 40% is typical for self-contained nitrogen cylinders and some manifold systems. Metal movement on the blankholder may be severely retarded at the end of the forming cycle by this pressure increase. The result may be failure due to fractures. A programmable hydraulic die cushion can optimize blankholder forces through the forming sequence. For example, the following sequence may be best for producing automotive quarter panels:

- A high blankholder pressure is maintained upon initial punch contact with the blank. This will allow the metal to be impressed with the main character features of the draw punch, lessening the chance of lateral slippage.

- The blankholder force is then reduced to allow metal to be drawn into the die cavity.
- The exact pressure to allow metal flow without objectionable wrinkling is maintained until the punch nears the end of travel.
- Near the end of the draw punch travel, the blankholder force is increased in order to prevent metal movement; this may be required to obtain plane strain stretching of the side-walls in order to reduce springback and stiffen the part.

Hydraulic Control of Mechanical Press Blankholders

Figure 5-7 is a simplified illustration of a method for combining the function of hydraulic overload cylinders and blankholder force control. The hydraulic overload cylinders at the press blankholder connection typically provide 1,000 inches (25.4 mm) of travel in an overload. The pressure source for each overload cylinder is a pilot cylinder or pressure intensifier. Air or oil under

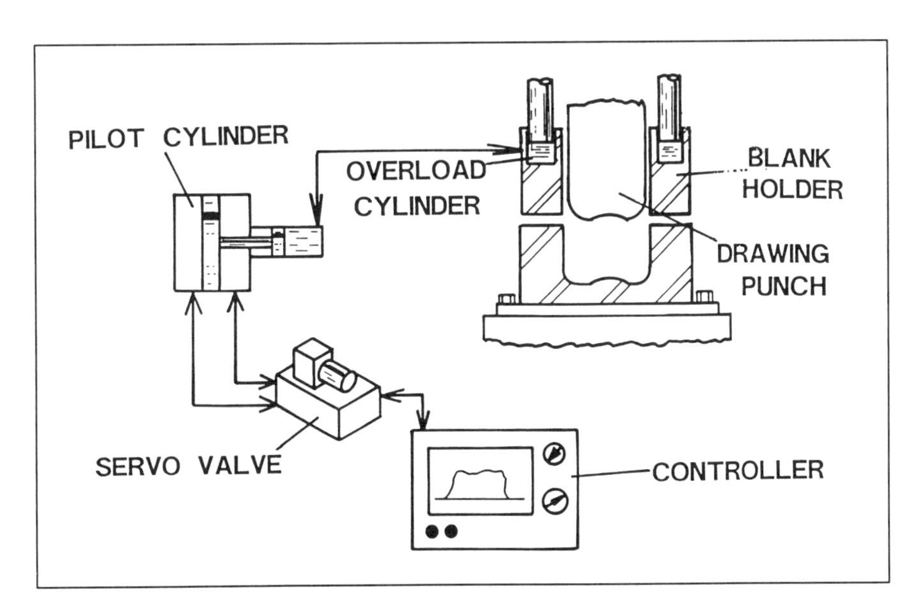

Figure 5-7. A blankholder force control system employing hydraulic overload cylinders in conjunction with a programmable servo control system. (The Ohio State University Engineering Research Center for Net Shape Manufacturing)

pressure is metered to the pilot cylinder or intensifier by a servo valve.

The servo valve is actuated by a programmable controller that varies the blankholder force throughout the forming cycle, using preprogrammed instructions for the part being produced. This system is less flexible than the programmable hydraulic die cushion due to the short cylinder travel available. However, in four-point presses, it can relieve or increase the pressure on a critical corner during the stroke to eliminate a fracture or draw an area more tightly to stiffen the part.

Both the hydraulic die cushion and blankholder force control systems have long histories of use in pressworking. The advent of the modern programmable electronic controller has made these blankholder force control systems much more flexible and simplified their use.

HYDRAULIC FORMING MACHINES AND DIES

An advantage of forming processes, in which hydraulic pressure acts on one side of the workpiece, is that only one half the die is needed. Simple dies may use rubber pads or cast shapes alone as a forming medium to transmit pressure.

The Guerin process uses a thick rubber pad contained within the ram of the press. The die is placed on the lower press platen or bed. This process has long been used to form short runs of parts from thin soft materials such as aluminum.

Specialized dies use oil or water under pressure to act directly on the workpiece. Uses include forming metal into female cavities and bulging special shapes in tubing and deep drawn parts. The tooling costs are generally low and complicated shapes can be produced. Housekeeping problems may be a concern, especially where oil is used.

There are several processes that apply hydraulic pressure to the workpiece through a flexible rubber bladder or membrane. These systems combine many of the advantages of direct fluid application without the mess.

The Verson Corporation Wheelon forming process uses a method of applying direct hydraulic pressure to the rubber forming pad. The blanks are placed over simple male dies, similar to those used in the Guerin process. Figure 5-8 illustrates a Wheelon hydraulic forming machine. The blanks and dies are moved into the press frame on a carrier, and forming pressure is applied by hydrauli-

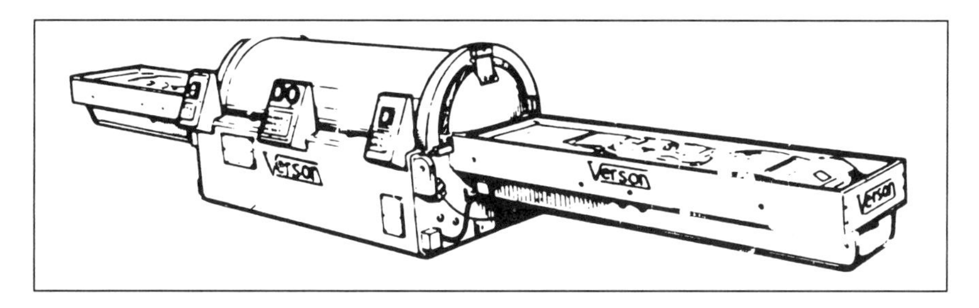

Figure 5-8. The Verson Corporation Wheelon hydraulic forming machine. (The Verson Corporation)

cally inflating a rubber bladder mounted in the immobile roof of the press.

Figure 5-9 shows a cross section of the press frame with a die and blank in place. The rubber bladder is shown in the released and the forming positions. This method is limited in depths of draw to about the same as the Guerin process but, with pressures of 6,000 to 10,000 psi (41.4 to 68.9 MPa) available, practically all wrinkling is eliminated.

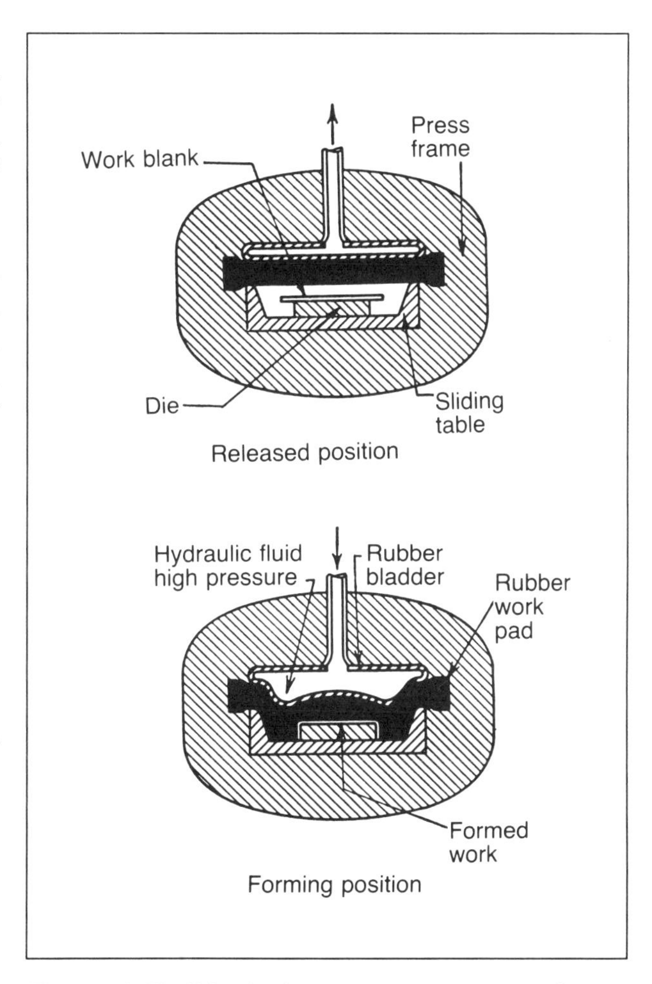

Figure 5-9. The Wheelon forming process sequence of operation. (The Verson Corporation)

The SAAB fluid-form

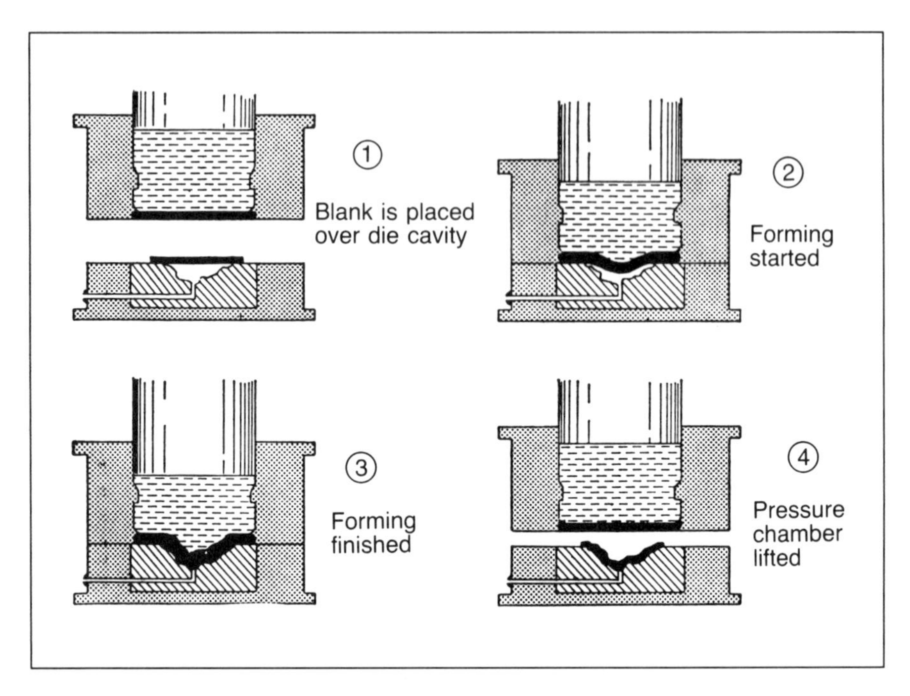

Figure 5-10. Stages of the SAAB fluid-form method of pressworking. (SAAB Fluid Form Division, McMahon and Co., Inc.)

method, illustrated in Figure 5-10, uses a flexible diaphragm punch which assumes the shape of the die. The punch serves as a blankholder. The diaphragm is soft, and behaves much like a fluid at the high pressures used.

In this process, the punch and blankholder of the conventional draw-die press are replaced by a steel cylinder which contains hydraulic fluid. The pressure is developed by the telescoping piston upon press closure. The lower die used for the SAAB fluid-form method is of conventional construction. Air escape holes are provided for the air enclosed in the die cavity.

The process can be adapted to a conventional single or double-action press by the installation of a fluid-form unit. The hydraulic unit is removable, permitting the press to be used for conventional presswork.

The fluid-form unit consists of two main parts, the punch or press chamber and the die holder. The die holder consists of a steel cylinder containing a telescoping piston which is sealed against the cylinder wall by a rubber or bronze packing and is retained in its upper position by a spring. The lower end of the cylinder is closed by a rubber diaphragm which is retained by an annular slot in the cylinder wall. The size of the unit (diameter of the diaphragm) depends primarily on the available press forces. Presses between 100 and 3,000 tons (890 kN to 26.7 MN) or more are suitable for the process.

BIBLIOGRAPHY

1. R. Lown, *Hydraulic Presses in the 80's,* Based on SME Technical Paper, MF82-918, Society of Manufacturing Engineers, Dearborn, Michigan, 1982, as updated by Mr. Lown for public presentations.
2. M. Ahmetoglu and T. Altan, *Improving Quality in Stamping by Controlling Blankholder Force and Pressure,* The Ohio State University Engineering Research Center for Net Shape Manufacturing, presented at FMA/SME Presstech Conference, Rosemont (Chicago), Illinois, October 1994.
3. D. Smith, *Die Design Handbook,* Section 14, "Rubber-Pad and Hydraulic-Action Dies," Society of Manufacturing Engineers, Dearborn, Michigan, 1990.

CHAPTER 6

BASIC DIE OPERATIONS

Basic die operations are divided into several categories, such as cutting, bending, forming, drawing, and squeezing. All require portions of the workpiece to undergo plastic deformation. To accomplish this, the yield point of the material must be exceeded through the use of such techniques as cutting or punching.

Cutting operations are essentially a controlled process of plastic deformation or yielding of the material, leading to fracture. Both tensile and compressive strains are involved. Often, bending or stretching of the work also occurs.

Simple *air bending* involves both tensile and compressive straining of the material in the bend zone. Upon release from the die, residual stresses result in a *springback* of the material. To avoid the effect of residual stresses, the metal in the bend-affected zone may be squeezed or *coined*. Here, the entire thickness of the material in the bend is subjected to compressive forces above the material's yield point.

Stretching is another process used to form metal. Forming processes involving cold working, such as stretching, often improve mechanical properties of the material. Familiar products made by stretching include automotive and aircraft skin panels.

In popular shop terminology, the term *draw die* may be applied to any die where metal is stretched or *drawn* into the die cavity. Many such operations form the part by stretching or bending, rather than drawing. Round cup drawing, where the metal is restrained by a blankholder as it is drawn into the die cavity by a draw punch, is an example of a simple drawing operation.

In cup drawing, the metal thickens or is compressed in the circumferential direction as it is drawn into the die cavity. As the metal is drawn over the die radius it is thinned. Fracture of the metal due to thinning limits the severity of this process. A large body of empirical data describes the limits of this process, which may involve multiple *redrawing* operations.

Operations where the metal's thickness dimension is subjected to pressures that change its thickness or shape are called *coining*.

This term is based on coin minting, the most common type of squeezing operation. Squeezing operations may also be called *cold forging*, depending upon the purpose and severity of the operation.

Understanding the behavior of metal during these basic operations is a key to using modern analytical techniques, which have taken troubleshooting pressworking operations from a trial-and-error process to an engineering science. The analytical tools include *circle grid analysis* (CGA) and the *forming limit diagram* (FLD). Computer aided formability analysis is also used to determine if stampings can be successfully manufactured.

SHEARING AND CUTTING OPERATIONS

Cutting, which includes shearing, is the most common pressworking operation. A single formed stamping, such as a sieve or automobile inner door panel, has many holes, all of which are produced by cutting operations. In the case of the sieve, which has many holes in a regular pattern, the process may be referred to as *perforating*. However, each example uses the same cutting process to produce a single hole.

Figure 6-1 (A) illustrates a punch, making initial contact with the stock. In the case of some high-speed operations, a shock wave (Figure 6-1 [B]) is generated in the punch. In most operations, a compressive strain (Figure 6-1 [C]) stores energy within the punch body as the die closes, much like a coil spring being compressed.

As the press closes, the punch remains in contact with the stock (Figure 6-2 [A]) until the pressure transmitted through the punch exceeds the yield strength of the material in shear. The material continues to yield until complete fracture (Figure 6-2 [B]) occurs. Finally, the compressive strain in the punch is released and the slug is pushed completely into the die opening (Figure 6-2 [C]) at the bottom of the press stroke.

Clearance Required for Various Operations

In a typical cutting operation, the punch penetrates the material to a depth equal to about one-third of stock thickness before fracture occurs. An equal portion of the material is forced into the die opening. The portion of the stock penetrated by the punch and extruded into the die opening will be highly burnished. The fractured material will have a rough surface on both the stock and the slug.

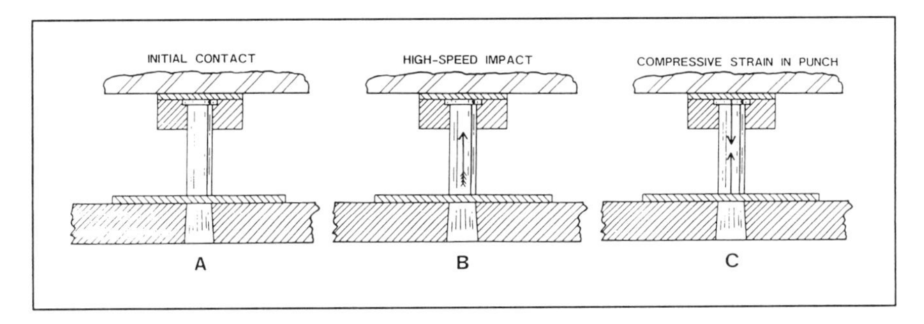

Figure 6-1. A punch making initial contact with the stock (A). In the case of high-speed operations a shock wave (B) is generated within the punch. As the die continues to close (C), energy is stored as compressive strain in the punch body. (Smith & Associates)

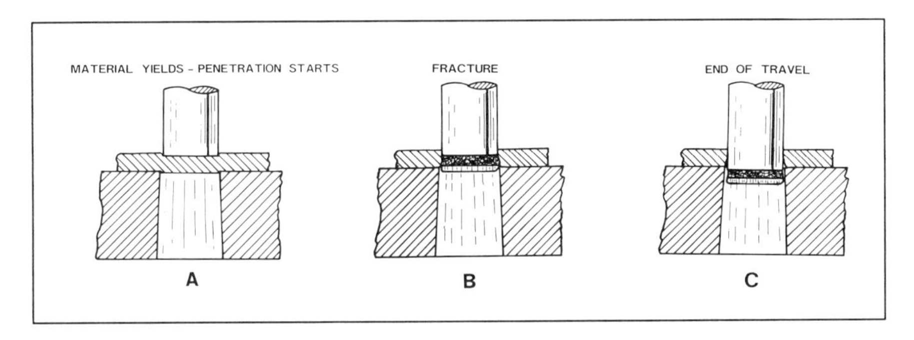

Figure 6-2. The punch remains in contact with the stock (A) until the pressure transmitted through the punch is great enough to exceed the yield strength of the material. The material continues to yield until complete fracture occurs (B). The compressive strain in the punch is released and the slug is pushed completely into the die opening (C) at the bottom of the press stroke. (Smith & Associates)

There are no absolute rules governing the amount of clearance between the punch and die. This clearance is normally expressed as a percentage of stock thickness per side. For mild steel, the clearance is usually between 5 and 12% of stock thickness. In general, tight clearances will result in holes having a high ratio of shear or burnish to fracture, at the expense of accelerated tooling wear.

Plastic deformation of the material occurs throughout the cutting process. Cutting also involves a controlled tensile failure of the material. The fractured portion of the edge will be somewhat rough due to the tearing action. In thick materials, this roughness may be quite pronounced. The fracture, which starts from each side, may not meet evenly. One or more sharp projections could result. Optimizing the die clearance to best suit the material being cut may be required to minimize an uneven fractured edge condition.

Double breakage sometimes is observed when tight clearances are used. Low cutting speeds tend to make the problem worse. The punch may momentarily stall before complete fracture occurs. Shearing again occurs, as evidenced by either a second burnished band or several shiny areas. Finally, the remaining stock thickness completely fractures by continued plastic deformation.

If scattered secondary shiny areas are observed on the inside of the hole or slug, check for a rough torn fracture. The fracture, which travels from both sides of the stock, may not be meeting evenly, leaving a ragged edge. The torn peaks of the fracture may be burnished by the die. Here, the solution often is to increase the punch to die clearance.

Figures 6-3 (A) through 6-3 (C) illustrate the results of three different clearances. Normal clearance typically leads to one-third burnish and two-thirds clean breakage. Excessive clearance may produce die roll at the point of punch entry and a large burr on the underside of the part. Insufficient clearance can result in double breakage, evidenced by two brightly burnished bands.

Pressure required to cut through the stock increases with the ultimate tensile strength of the material. Die cutting is a more efficient process than parting metal by tensile failure. There is no absolute relationship between tensile strength and shear strength. Generally, shear strength is between 60 and 80% of ultimate tensile strength.

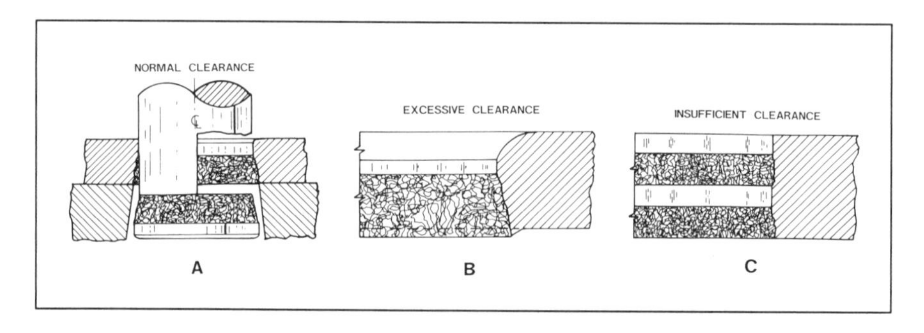

Figure 6-3. Normal clearance (A) results in one-third burnish. Excessive clearance (B) results in excessive die-roll and a burr. Insufficient clearance (C) can result in double breakage. (Smith & Associates)

Excessively tight clearances result in high cutting forces. Another effect is extreme pressure on the cutting edges, which shortens tool life.

Increasing clearance within reasonable limits lowers cutting pressure, extending tool life. A limiting factor is the amount of taper permitted in the hole and the allowable burr height. If the clearance is excessive, higher than normal forces will be required and the hole will have excessive die roll, taper, and burr height.

Calculation of Cutting Force Requirements

Generally, if the ultimate tensile strength of the material is used to calculate the cutting forces based on the area of material being cut, a substantial safety factor exists. For example, AISI-SAE 1010 cold-rolled steel has an approximate ultimate tensile strength of 56,000 psi (386 MPa) and a shear strength of 42,000 psi (290 MPa). The shear strength of the material increases due to the fast strain rates encountered in high speed pressworking. The ultimate strength should be used to provide a safety factor in such cases.

Calculating the force required to cut or shear materials can be accomplished by physical measurement of the total cut length. It may be calculated from the dimensions on the part print. If the part is produced in a progressive die, all pilot holes and work done to cut off the carrier strip must be included.

The die designer or engineer must calculate force requirements to determine the size and type of press required. An assumption that the pressroom can somehow fit a new job into an existing press can be a foolish blunder.

Length of cut, material thickness, and shear strength must be known. The length of cut can be easily determined by many computer aided design (CAD) programs.

Once the total length of cut is known, this distance is multiplied by the stock thickness in order to obtain the square area of material being cut. Finally the total area of cut is multiplied by the shear strength of the material. The same system of units must be used. The formula is:

$$Fs = L \times t \times Ss$$

Where:

L = Length of cut.
t = Thickness of material.
Ss = Shear strength of the material as defined by ASTM tests.
Fs = Force required to shear in the same system of units as L, t, and S.

In North America, many shops still carry out engineering calculations for stampings, using measurements based on the inch for length and thickness. Shear or yield strengths are based on pounds per square inch (psi). The press force required is often stated in short tons, based on 2,000 pounds.

The metric system is in standard use throughout most of the world. Metric pressworking linear and area measurements are given in terms of the meter, centimeter and millimeter. Pressworking forces are either stated in metric tons based on 1,000 kilograms, the kilo-Newton (kN) or mega-Newton (MN). The preferred metric unit for material shear and yield strengths use kilo-pascal (kPa) or mega-pascal (MPa).

Calculating Cutting Energy Requirements

The energy required to cut through metal is often surprisingly small, especially when compared to processes such as sawing, oxy-fuel gas, and laser cutting. This is because the actual shearing of the stock is completed when the punch or cutting steel penetrates the material approximately one-third stock thickness.

Using the foot pound second (fps) system of units, the formula is:

$$E = F \times D$$

Where:

E = Energy in foot pounds.
F = Cutting force in pounds.
D = Distance sheared in feet.

Working in inch-tons may be more convenient. An inch ton equals 166.67 foot pounds.

For example, adding a safety factor, the shear strength of mild steel is approximately 50,000 psi (345 MPa) or 25 short tons per square inch. Cutting a 12-inch (304.9-mm) diameter round blank 0.1875 inch (4.763 mm) thick requires the punch to penetrate only one-third material thickness or 0.0625 inch (1.588 mm), before fracture occurs.

The total area being sheared is 7.0686 square inches, requiring a force of 176.72 tons. However, this force only acts through a distance of approximately 0.0625 inch. The required energy is 11.04 inch-tons or 1840.8 foot-pounds; which equals 2496 joules or watt-seconds. If the press operates at 60 strokes per minute (SPM), one blank per second is cut. Ignoring frictional losses, a motor output of only 3.347 horsepower is required to restore energy to the flywheel.

In high-speed pressworking, the force required may have measured values higher than expected based on the shear strength of the material. This is because as the punches shear the stock more rapidly, the strain rate of the material increases. In high-speed perforating operations, the measured cutting forces may approach the ultimate material strength.

Figure 6-4 illustrates the shear, tensile, and compressive forces that occur during the cutting process. The amount of lateral force varies with the cutting clearance and material. For round and symmetrical holes, the lateral forces balance out. However, the die must be sufficiently strong to withstand high spreading forces. For notching, shearing, and other unbalanced operations, the alignment system of the die must not allow excessive deflections to occur.

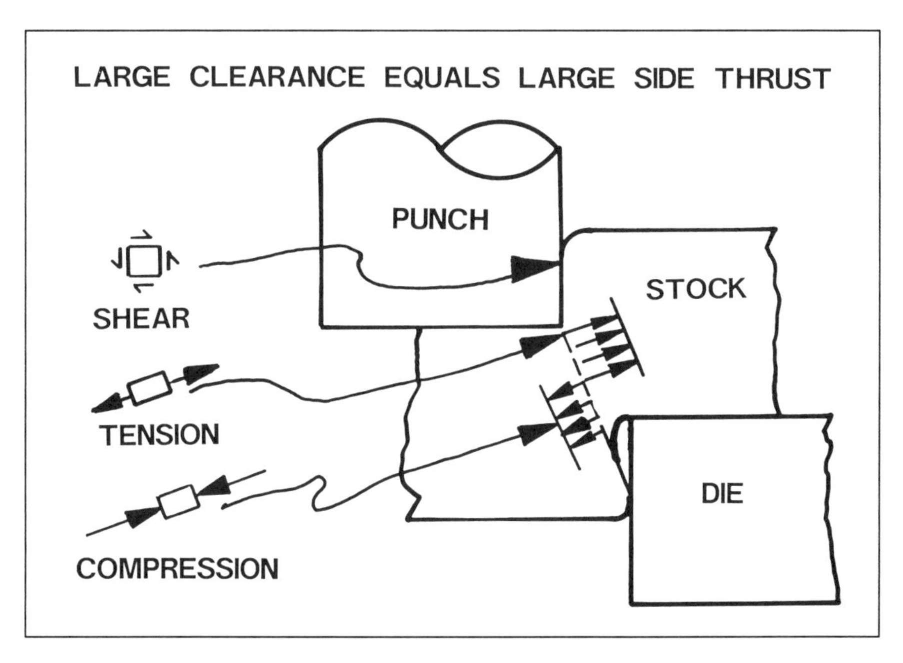

Figure 6-4. *The shear, tensile, and compressive forces that occur during the cutting process. (Danly Machine)*

This side thrust, or lateral forces, can result in excessive deflections of die components, such as punches, heel blocks, and guide pins. As lateral deflection occurs, clearances increase. The lateral pressure can exceed the press force by a factor of three or more due to a wedge-like mechanical advantage. The misalignment can damage the tooling and produce scrap parts.

Shaving Operations

The edges of a conventionally punched hole are not straight. This was illustrated in Figure 6-3 (A). For most applications, punched holes do not have critical requirements for straightness. If a hole with straight sides is required, an operation known as *shaving* may be used.

Figure 6-5 illustrates the sequence of a shaving operation. The punch-to-die clearance is normally much less than for conventional hole punching operations.

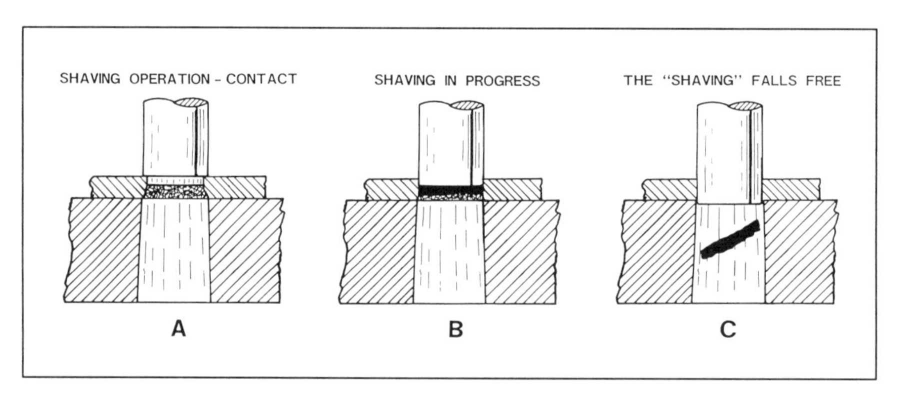

Figure 6-5. *Initial contact (A) at the initiation of a hole-shaving operation. As shaving progresses (B), a chip or shaving is produced that falls free (C) at the end of travel. (Smith & Associates)*

A shaving operation is very much like a machining operation. The punch functions as a cutting tool. Cutters for machining mild steel generally have a back-rake angle of 15 to 20 degrees. Optimizing this angle permits cutting with less force, resulting in longer tool life and better surface finishes.

Each material has an optimum rake angle. Some aluminum alloys may require as much as 45 degrees back rake, while many brasses require no back rake.

Figure 6-6 (A) illustrates a shaving punch provided with a back-rake angle. This provides cutting tool like action, which produces a superior surface finish and may extend punch life. Figure 6-6 (B) illustrates a close-up view of the concept.

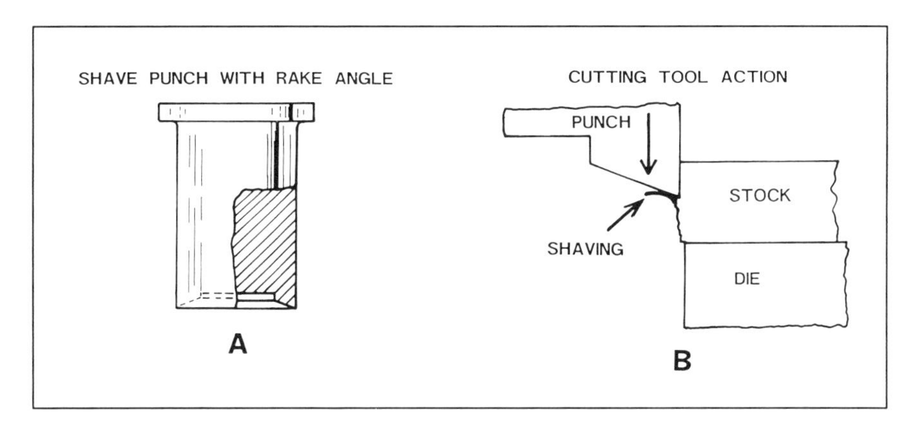

Figure 6-6. A shaving punch provided with a back-rake angle (A) delivers cutting-tool like action (B). (Smith & Associates)

Depending upon the condition of the initial punched hole, two or more shaving operations may be required. Work-hardening of the surface and mill scale can shorten punch life. A pressworking lubricant may aid the process.

Punch-Head Breakage Problems

Punches used in high-speed operations can fail due to head breakage. In this case, the cause is shock loading due to high-speed impact on the stock. Rapid strain rates that occur in high-speed stamping typical of the production of perforated metal may increase the punch cutting load from the normal shear strength to a value approaching the ultimate tensile strength.

High stripping loads often contribute to head breakage. Another punch head failure mode is breakage due to extreme compressive forces and load reversal when cutting heavy stock. This is illustrated in Figure 6-7.

Figure 6-7 (A) illustrates a view of a punch head in a retainer having a hardened tool-steel backing plate. Typically, the backing plate is heat treated to 38 to 45 RC for toughness.

As the punch cuts through heavy metal, compressive strain is developed in the punch body, backing plate, and die shoe. This results in deflection or bending of the punch head.

As the punch breaks through the material, there is a recoil action resulting in tensile strain on the punch head. Repeated flexure of the punch head can cause crack formation (Figure 6-8) at the juncture of the punch head and body. Continued operation under

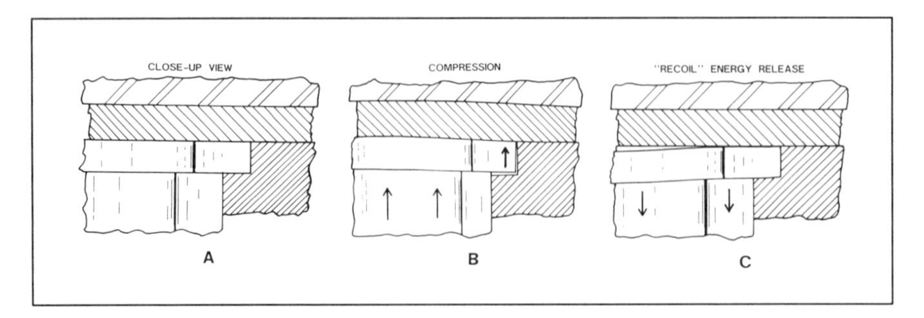

Figure 6-7. A punch head in a retainer (A): compression of the punch and head occurs during cutting (B) resulting in recoil upon compressive strain release (C) as the punch breaks through. (Smith & Associates)

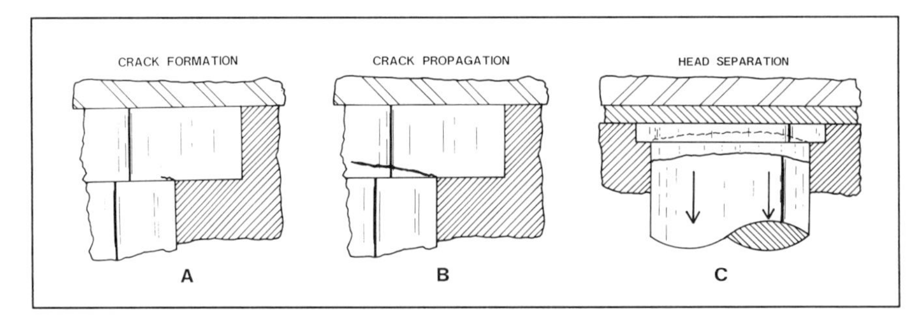

Figure 6-8. Stress concentration at the juncture of the punch head and body can produce crack formation (A). Continued operation results in crack propagation (B), ending in complete separation of the punch head (C). (Smith & Associates)

high compressive loading can cause the crack to grow larger, often separating the punch head from the body. The punch body may shatter, causing sharp fragments to become airborne at high velocity. Serious injury to pressroom personnel can occur. All pressroom injuries are avoidable by proper process planning, process control, employee training, and good pressworking practices.

Figure 6-9 illustrates solutions that eliminate all but the most serious head breakage problems. The first modification is to grind a slight back angle on the head, permitting the punch to be compressed into the backing plate without head flexure. The second improvement is to provide a generous radius at the head-to-body juncture to reduce the stress concentration in this area.

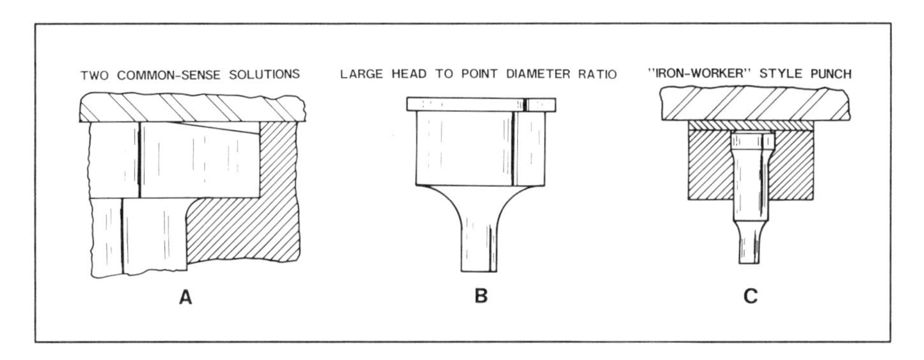

Figure 6-9. *A back angle on the punch head and a large radius at the juncture of the punch head and body (A) reduce stresses; a large body to point diameter ratio (B) is an effective but expensive solution; an "Ironworker" style punch (C) retains the punch while avoiding stress concentrations in any area of the head. (Smith & Associates)*

Another solution calls for the body and head diameter to be much larger than the point. There is a major drawback. A large-diameter, expensive tool-steel blank is required, which is not only expensive, but also limits how closely together punches can be spaced.

The ironworker style punch has been used for many years in portable equipment that punches rivet and bolt holes in heavy plate and structural members. The design is based on reducing stress concentration as much as possible in the body style. The design is both economical to produce and seldom fails under heavy punching loads. A disadvantage, when applying the design to pressworking tooling, is the need for a nonstandard retainer.

Cutting Force Reduction Techniques

The peak pressure required to cut through material can be reduced by grinding shear angle(s) on the punch or die. If the cutout portion is discarded, the shear is placed on the punch.

Figure 6-10 illustrates three methods of reducing cutting forces by grinding angular shear on a punch point. Shear angles result in the slug being curved or bent. If the punched-out portion is the part, the shear must be ground on the die to limit part distortion.

A simple method is to grind an angle on a punch point. While this method reduces peak punching forces, the shear is unbalanced and results in a lateral force.

Balanced Vee shear is an improvement. This method has two advantages. By balancing the load with symmetrical shear, lateral

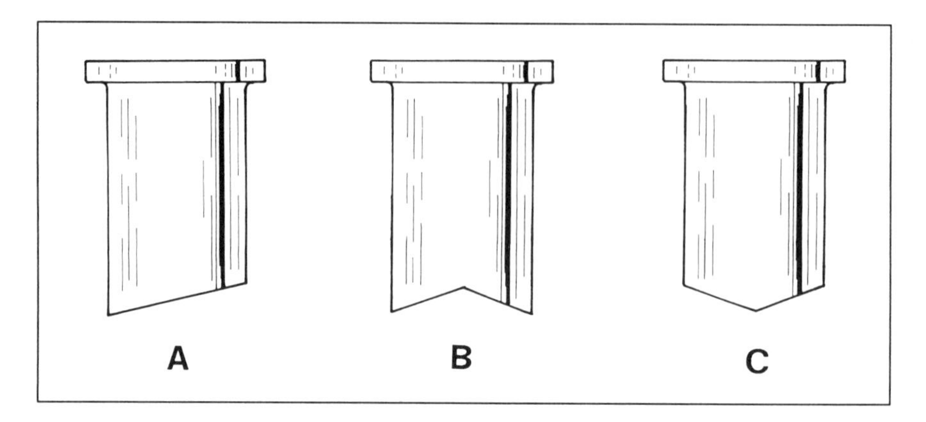

Figure 6-10. Cutting forces can be reduced by adding shear to the punch; unbalanced shear (A) should be avoided; Vee shear (B) is balanced, and reduces the load on the die button, although not suited for heavy loads; pointed balanced angular shear (C) reduces the cutting load, although the side-force on the die may be higher.

force is avoided. The hollow Vee design also starts cutting at the edges of the slug and finally completes the fracture at the center. This may help reduce the spreading force on the die.

A disadvantage is that pressure is concentrated in the center of the punch and, when cutting thick high-strength materials, this punch may split at the center of the Vee.

For especially heavy punching operations, a pointed punch (Figure 6-13[C]) can be used to reduce peak punching loads. If this design is used, the die may need to be made larger to withstand any increase in lateral loads created by the slug's bending-action. In both punch styles (B) and (C), the total energy required is greater due to the addition of bending forces. However, the peak force required is less.

All types of shear ground on the punches will result in curled or bend slugs. If the slug is the part, this may be unacceptable. In such cases, angular shear may be provided on the die. Distortion of the scrap skeleton is usually not a concern. The goal is to produce flat parts (Figure 6-11).

In addition to providing angular shear on the punch or die, the entry of individual punches may be timed to reduce cutting forces. In the majority of cases, where the punches penetrate the stock one-third of stock thickness when fracture occurs; the entry of the punches may be stepped in increments of approximately one third stock thickness.

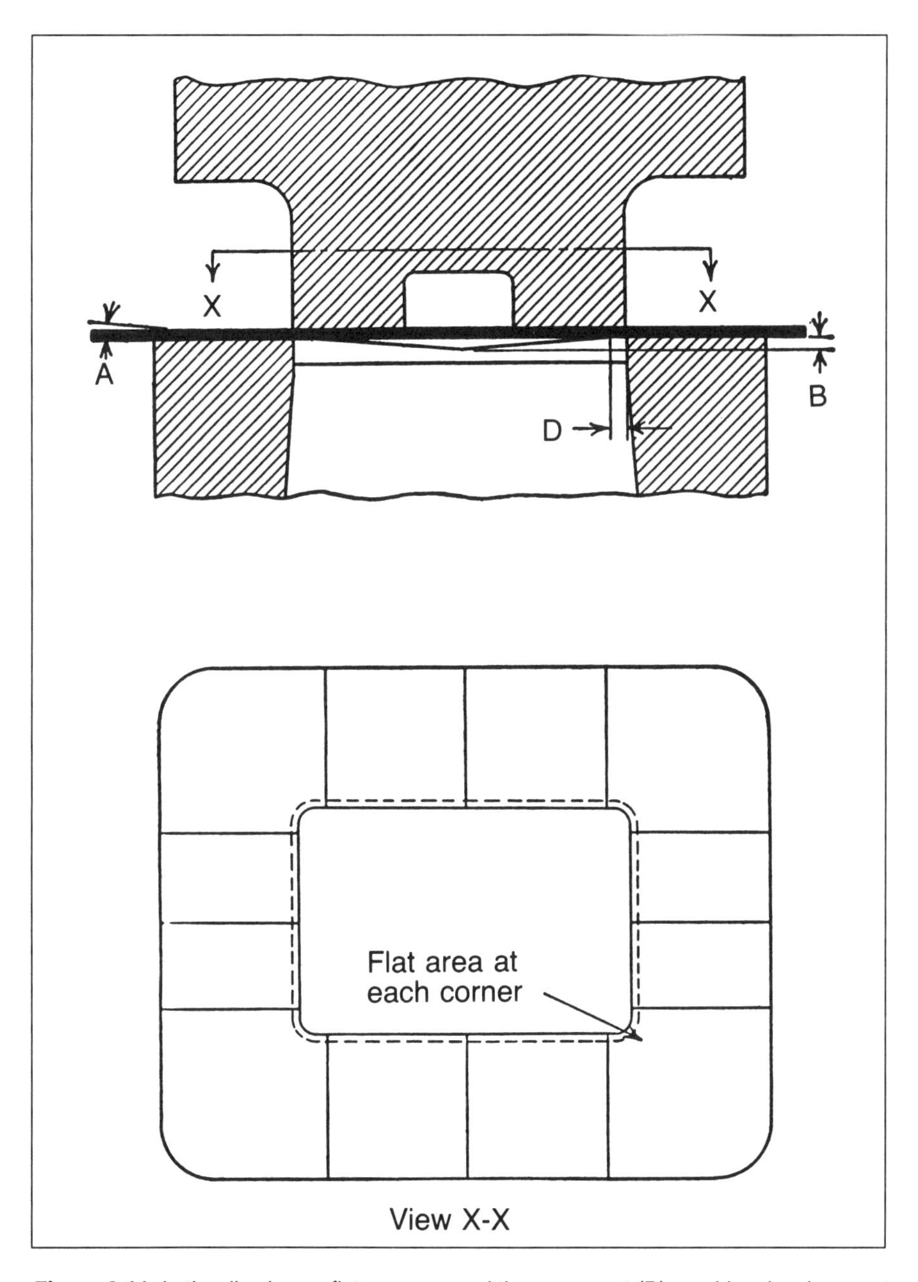

Figure 6-11. *In the die shown, flat areas around the corners at (D) provide a level support for the stock, and prevent slippage at the start of punch engagement. With the amount of shear (B) and angle of shear (A), cutting progresses from the outside corners to the center, producing a flatblank. As a general rule, the amount of shear (B) is approximately equal one to three times stock thickness.*

In many cases, optimizing punch and die shear, together with stepping punch entry, can reduce peak cutting forces one-half to two-thirds the untimed value. It is important to note that the total flywheel energy per stroke is not reduced.

The process of optimizing cutting forces can be aided by force monitoring and waveform signature analysis. These methods are valuable process control tools.

In the case of heavy cutting operations, the press members are strained or deflected. Releasing this stored energy is a pressworking problem known as snap-through. Dies with multiple punches have many opportunities to reduce this damaging energy release.

It is very important to maintain records of the optimum die timing for each job. The information will save trial-and-error work when the die is resharpened, and aid in timing new jobs.

Hole and Part Size Changes

The size of a punched hole is often smaller than that of the punch. The final step in cutting with a punch and die is withdrawing the punch from the stock. When this occurs, some residual stresses caused by metal displacement are relieved. Generally, this results in shrinking the hole as the punch is withdrawn.

The punched out portion is approximately the same size as the die opening, with the exact size depending on the material and the severity of the punching operation.

PRESSURE PADS AND STRIPPERS

A stripper's function is to strip or remove the material from the punches, in the case of cutting dies. Pressure pads, often used for this purpose, take many forms depending on the amount of stock control required.

For high precision dies, especially those involving long slender punches, the stripper may be guided with pins and bushings. The stripper can be used to guide the punches by means of hardened bushings.

The simplest type of stripper is the tunnel or fixed variety. Its two main advantages are economy and reduced force requirements. When springs or nitrogen cylinders are used, the pad force requirements must be added to the cutting force requirements.

Positive or Tunnel Strippers

Figure 6-12 illustrates a simple hole punching die employing a positive, or tunnel, stripper. Close-up views are illustrated in Figure 6-13. Before stripping action can occur, the punch must lift the stock against the top of the stripper. The stock then falls to the die surface as the punch is withdrawn.

The tunnel or fixed stripper is popular for simple heavy punching operations as well as low-cost short-run dies. One problem is that distorted stock tends to deflect the punch upon withdrawal, and may break or chip the punch face.

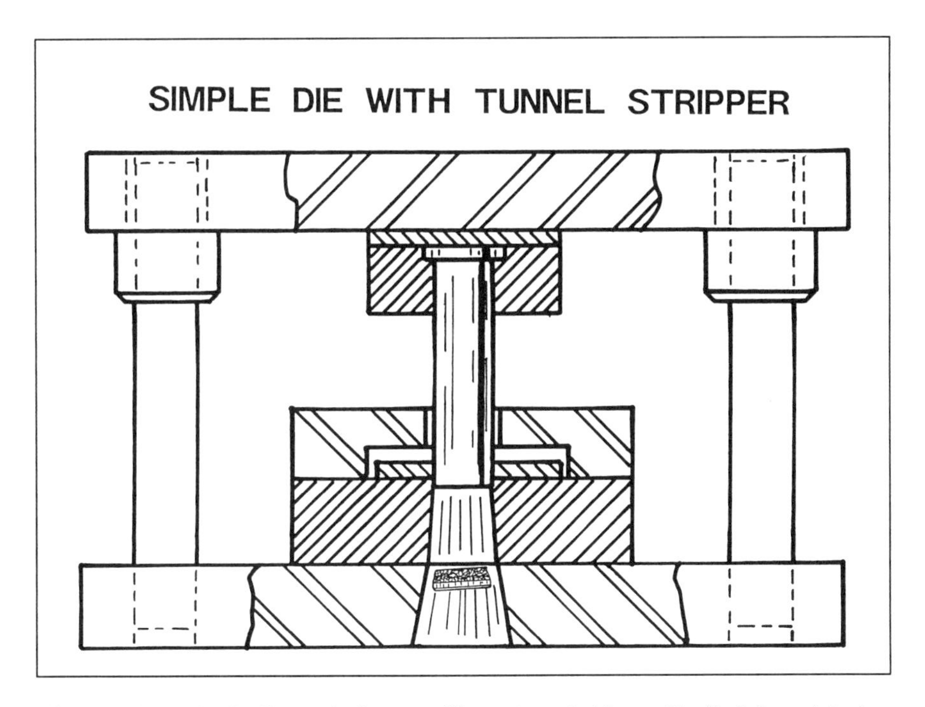

Figure 6-12. A simple die employing a positive or tunnel stripper. (Smith & Associates)

Stripping Forces

The amount of force required to strip the stock from punches varies greatly. The stripping requirements can vary from 1 or 2% of the cutting force to 20% or more. Stripping loads tend to be increased by tight cutting clearances because there is a higher ratio of burnish to fracture, therefore more area gripping the punches.

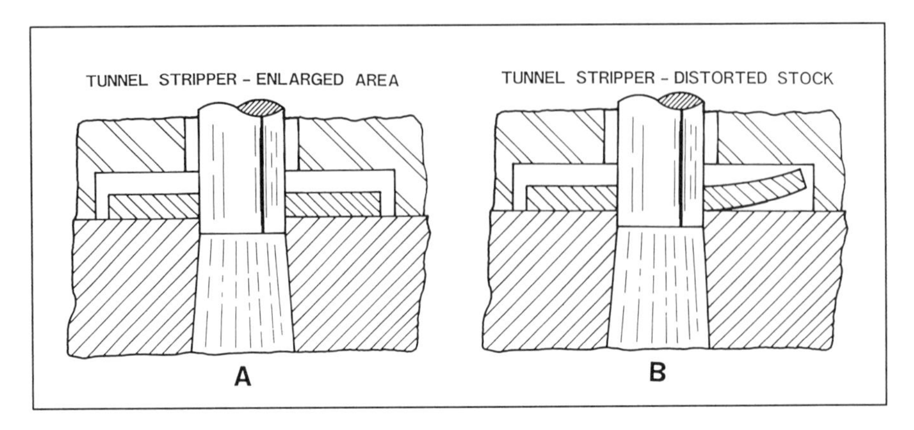

Figure 6-13. *Close-up view of the tunnel stripper illustrated in Figure 6-12. The stock is lifted and stripped (A) upon withdrawal of the punch. Distortion of the stock (B) can break or chip the punch face.*

Tight cutting clearances produce more residual stress in the material around the punches than does greater clearance. One result is that the stock will grip the punch tightly. Also, there may be more hole size closure upon punch withdrawal.

Proper die clearances serve to both reduce cutting and stripping forces. Sufficient lubrication is also helpful. If cold welding and galling occurs on the punch surface, stripping forces will be greatly increased, and die resharpening required much sooner.

Elastomer Strippers

Special die rubber or polyurethane strippers are frequently used on low-cost tooling. For simple punch shapes, having round shanks, a variety of styles are commercially available.

Special shapes of elastomer strippers and pressure pads can be cast from two-part elastomer compounds. Often a prototype part serves as a part of the mold. These materials are also useful for fabricating nonmarking automation grippers and other part transfer components.

COMPOUND DIES

The term *compound die* usually refers to a one-station die, designed around a common vertical center line in which two or more operations are completed during a single press stroke.

Usually, only cutting operations are done, such as combined blanking and piercing.

A common characteristic of compound-die design is the inverted construction, with the blanking die on the upper die shoe and the blanking punch on the lower die shoe. The pierced slugs fall out through the lower die shoe. The part or finished blank is retained in the female die, which is mounted on the upper shoe.

Compound dies are widely used to produce pierced blanks to close dimensional and flatness tolerances. Generally, the sheet material is lifted off the blanking punch by a spring-actuated stripper, which may be provided with guides to feed the material. If the material is hand-fed, a stop positions the strip for the next stroke.

The blank normally remains in the upper die, and is usually removed by a positive knockout at the top of the press stroke. Ejection of the blank from the die by spring-loaded or positive knockout does not require angular die clearance. This simplifies die construction, and assures constant blank size through the life of the die.

Example of Compound Washer Die

A compound die for making a washer is shown in Figure 6-14. One press stroke punches the center hole and blanks the piece from 0.015-inch (0.38-mm) cold-rolled steel strip. The piercing punch is attached to the upper die shoe, while the blanking punch is attached to the lower die shoe.

In this design, the piercing punch contacts the material slightly ahead of the blanking die. The part is stripped from both the blanking die and piercing punch by a positive knockout. The blanked strip is lifted off the blanking punch by a spring-loaded pressure pad.

The part must be removed from the upper die at the top of each stroke, a potential disadvantage of compound dies. The part is usually knocked out with a press-actuated knockout bar.

Small parts knocked out of the upper die may be ejected by a timed blast of air. Larger parts can be removed by a shuttle unloader that enters the die opening as the ram ascends. The unloader is normally driven by the press ram, although air, hydraulic, or servo motor driven units may be used. Accomplishing positive part removal during each press stroke may limit the

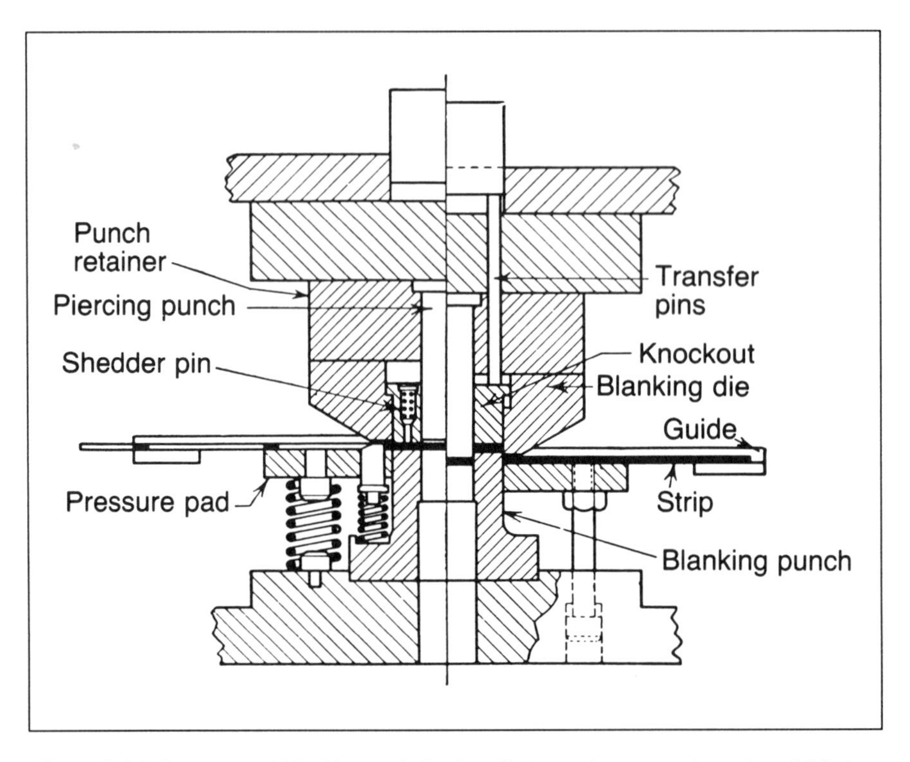

Figure 6-14. A compound blanking and piercing die to produce a washer; dies of this type are widely used to produce accurate flat blanks. (Cousino Metal Products Company)

speed of the operation. For low production jobs, manual removal with appropriate safeguarding precautions may suffice.

CUT-AND-CARRY OR PUSH-BACK OPERATIONS

Cut-and-carry operations are quite useful for producing parts with high flatness requirements. The cut-and-carry, or push-back, operation is an upside-down compound die.

The feature of pushing the part back into the carrier strip or scrap skeleton provides a positive means to get the part out of the die without an auxiliary unloader. Cut-and-carry operations are often used as stations in progressive dies.

The cutting station makes use of a counterforce pad. The pressure is usually supplied by springs or nitrogen cylinders. Figure 6-15 illustrates a typical cut and carry operation.

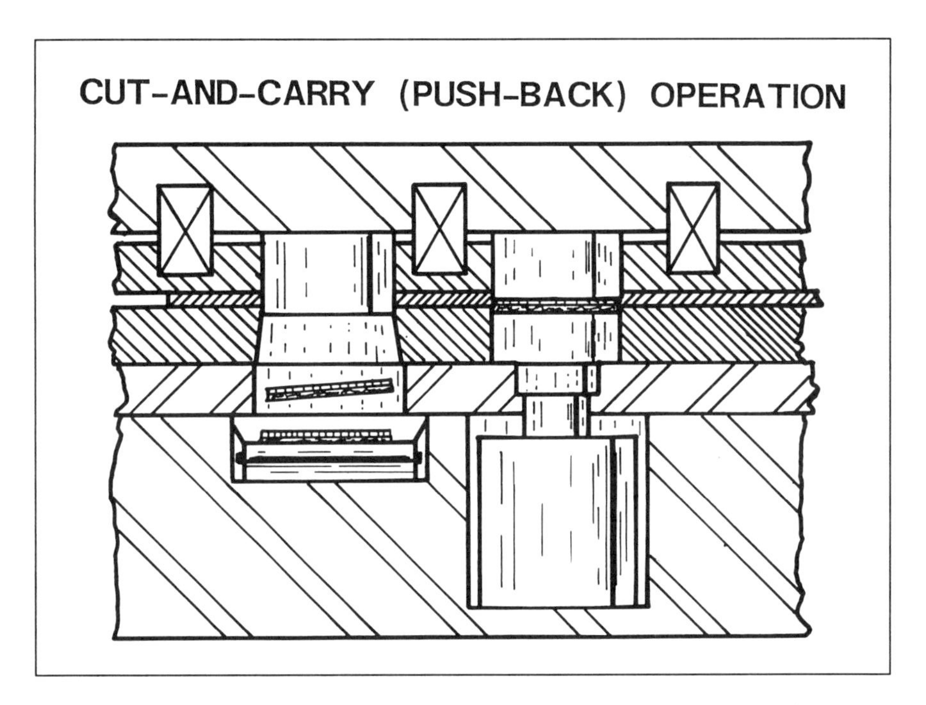

Figure 6-15. Section through a typical cut and carry die. (Smith and Associates)

The part may progress through several other stations before being pushed out of the carrier strip in a knockout station. The knockout station must have approximately double the clearance of the cutting station to accommodate normal part expansion and pitch growth errors. Otherwise, slivers would be a problem.

Successful cut-and-carry stations must completely fracture the part from the material and always contain it within the carrier opening. Figure 6-16 shows three close-up views of the fractured edge in the cut-and-carry station. Should the part not be completely fractured, a double breakage, evidenced by a second shear (shiny) band, will occur in the pushout station.

If the part is shoved completely out of the carrier, it will expand and the carrier opening will shrink due to normal material elasticity. Forcing the part back into the opening normally results in distortion and cold-welding problems.

The pushout station should be robust enough to cut the stock in case of a misfeed or improper coil start-up procedure. In other

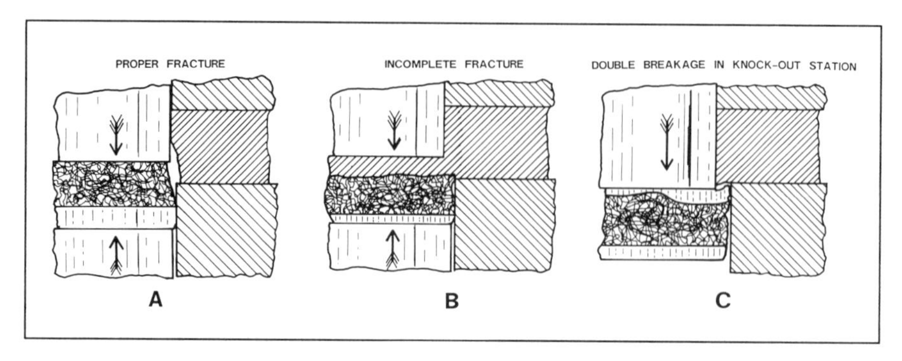

Figure 6-16. *A close-up view of the fractured edge (A) in the cut-and-carry station illustrating proper fracture. Incomplete fracture (B) will result in a secondary or double breakage (C) in the knockout station. (Smith and Associates)*

words, the die should be designed to withstand some degree of abuse.

Three main variables are available to control the process. The main cut-and-carry success factors are:

- Completely fracturing the part from the material; and
- Always containing the part within the carrier opening.

The variables are, in order of effectiveness:

- Controlling the depth of punch entry;
- Amount of die clearance; and
- Controlling material properties.

Variation of the spring, or nitrogen, pressure used to supply the counterforce is not an effective variable. Should too low a pressure be used, the part flatness will be affected.

The depth of punch entry is a means to accommodate range of material properties. Changing die clearances can be accomplished only by modifying or changing the die details. The correct clearance is often determined by experimentation in critical operations. Minimizing cutting forces with optimal clearances normally results in correct operation.

Material properties will have some effect on the ratio of shear to fracture. In general, harder materials will fracture more easily than softer ones.

Cut-and-carry or pushback operations is a proven method to enhance part flatness while providing a convenient means to carry the part out of the die. Complete fracture must occur, and the punch and counterforce must be configured to prevent the part from leaving the carrier in the cut-and-carry station.

BENDING METALS

Bends are made in sheet metal to gain rigidity and produce a part of the desired shape. The simplest type of bending is air bending, so called because the die does not contact the outside of the bend radius. The part to be bent is simply supported on either side of the bend and a force is applied to a punch in the center to accomplish the work.

Figure 6-17 illustrates a beam supported at two points, with a load applied at the midpoint. The load produces compressive stresses in the inner material being bent and tensile stresses in the outer material. If the force applied does not exceed the yield strength of the material, the beam returns to its original shape upon removal of the load. However, if the stress exceeds the material yield strength, the beam retains a permanent set or bend when the load is removed. Springback or elastic recovery occurs until the residual stresses in the bend are balanced by the stiffness of the material.

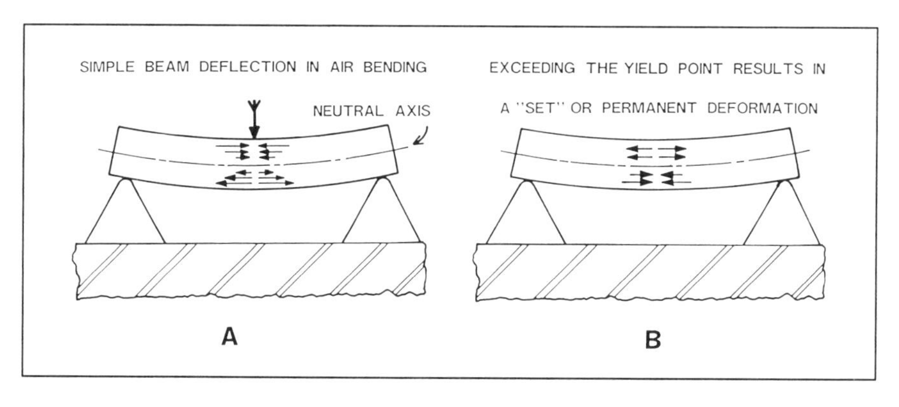

***Figure 6-17.** Simple beam deflection in air bending. If the applied force (A) exceeds the material yield strength, the beam retains a permanent set or bend (B) when the load is removed. (Smith & Associates)*

Material in the bend zone is not stressed equally. Material in the inner and outer surfaces is stressed the most, and the stress gradually diminishes toward a neutral axis between the two surfaces. At that point, the stress is zero and there is no length change.

Springback and Residual Stresses

Whenever metals are formed, some springback occurs. The root cause of springback is residual stresses, an inevitable result of cold working metals. For example, in a simple bend, residual compressive stress remains on the inside of the bend, while residual tensile stress is present on the outside radius of the bend.

When the bending force is released, the metal springs back. Springback occurs until the residual stress forces are balanced by the material's stiffness, which is a function of the material's modulus of elasticity. This explains why materials with a high modulus of elasticity (as compared to tensile strength), such as mild steel, spring back less than materials with a lower modulus but having comparable tensile strength, such as hard aluminum alloys.

The most common method to compensate for springback is to bend the material beyond the desired angle a sufficient amount and allow it to *spring back* to the desired angle after elastic recovery occurs. This method of springback compensation is termed *overbending*.

Many complex factors determine the amount of springback that will occur in a given operation. As a result, data for a specific material and forming method is often developed under actual production conditions to aid process control and future product development.

Press brake tooling for air bending, such as that illustrated in Figure 6-18 (A), is quite simple. Air bending is one of the most common press brake operations. This method of bending also requires minimum tonnages for the work performed. Exact repeatability of ram travel is required to maintain close repeatability of the bend angle.

There are several causes of bend-angle variation. These include:

- Changes in stock yield strength;
- Variation in stock thickness;
- Machine variations due to temperate changes; and
- Machine deflection, especially in long press-brake dies.

Compensation for any change of condition affecting the angle of bend may require adjustment of the ram travel. Press-brake bed deflection may require shimming to compensate for the deflection.

Coining has the advantage of producing sharp accurate bends with less sensitivity to material conditions than air bending. The disadvantages include force requirements many times higher than air bending, and accelerated die wear.

Figure 6-18 (B) illustrates a press-brake die designed to coin the bend in order to obtain a precise angle of bend. This coining action eliminates the root causes of springback, which are the tensile and compressive residual stresses on opposite sides of the bend. This is accomplished by bringing the entire thickness of the metal in the bend area up to the yield point of the material.

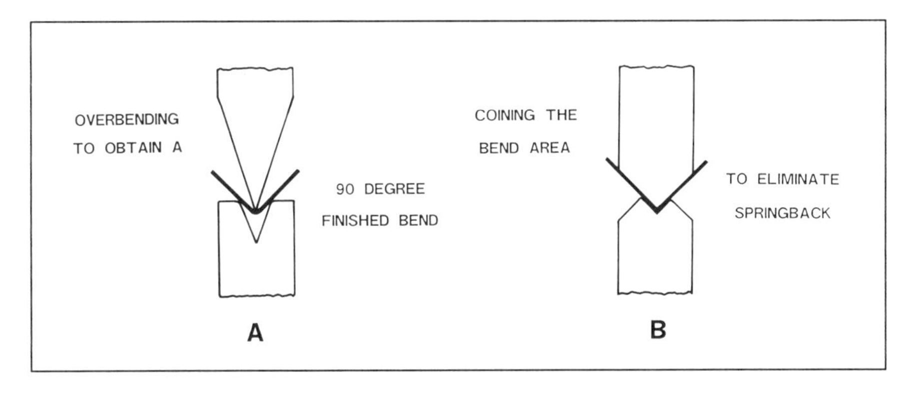

Figure 6-18. *Simple tooling (A) of the type used to air-bend sheet-metal parts in press brakes. Coining the bend to obtain sharp accurate bend (B) is a method which requires high tonnages, but produces sharp accurate bends.*

Examples of Die Bending Operations

Figure 6-19 illustrates sectional views through wipe flanging dies. In this design, the flange steel *wipes* the metal around the lower die. In the conventional design, the wiper steel coins the top portion of the bend to control springback. The top thickness of the bend can be squeezed beyond the material yield point by careful adjustment of the die shut height. Only the top portion of the bend is coined.

If excessive coining pressures are applied, the metal at the top of the bend will be extruded, resulting in a weak and distorted bend condition. An improved flanging method is to relieve the radius in the flange steel so it does not contact the top of the bend radius. One way is to relieve the flange steel at approximately a 20-degree angle tangent to the radius. Another method is to machine the flange steel to a radius somewhat larger than that of the outside of

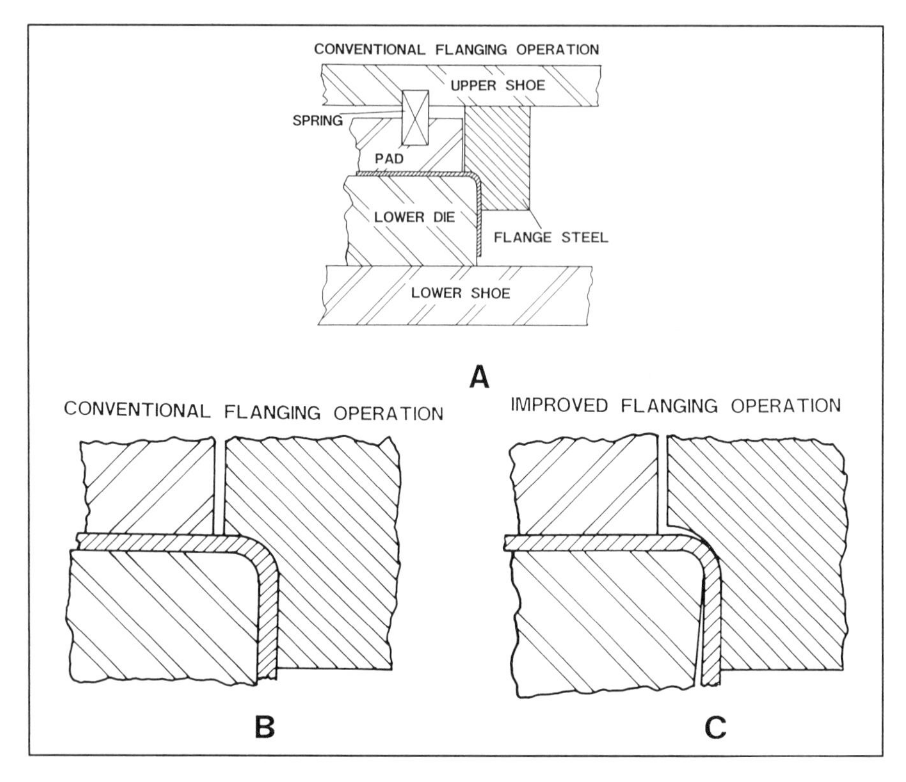

Figure 6-19. *Sectional view (A) of a wipe flanging die. A popular method is to coin the top of the bend (B) to control springback. A better method is to coin the side of the radius (C) and provide a relief angle on the lower die steel.*

the bend. The flange steel is positioned so that the tightest point is 45 to 60 degrees beyond the top of the radius.

The side of the form steel is relieved five or more degrees to permit the material to be overbent. This method is several times more effective than coining the top of the bend. Another advantage is that the bending process is not as sensitive to variation due to press adjustments and material conditions.

Rotary-Action Bending

A patented rotary-action bender known as the Ready™ bender combines the low-tonnage requirements of air bending with the accuracy and multiple bend capability of wipe-flange tooling.

Figure 6-20 illustrates a ready bender making initial contact with the stock. As the upper die travels downward, the stock is

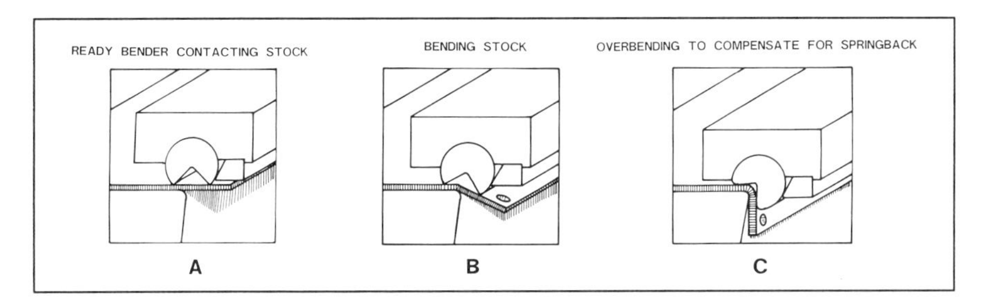

***Figure 6-20.** A Ready™ bender making initial contact with the stock (A); as the die closes, the bender both clamps and bends the stock (B); the stock is over-bent at bottom of stroke (C) to compensate for springback. (Ready Tools, Inc.)*

clamped and bent by the rotating bender. As the die closes, the bending action progresses. A relief angle in the lower die permits the rotary member to overbend the stock at bottom of stroke to compensate for springback control.

Rotary-action benders can bend angles up to 120 degrees. The rotary member is usually made of tool steel heat treated for long wear. The bending pressure is typically 50 to 80% less than that required for conventional wipe bending. The lower pressure permits many types of prepainted materials to be fabricated without damaging the finish. Rotary benders can also be constricted of nonmetallic materials, such as hard thermoplastics, for work with prefinished materials.

The bend angle adjustment with conventional wipe-flange tooling is usually made by adjusting the flange-steel up or down by shimming. In the case of rotary-action benders, the bend angle is adjusted by moving the assembly containing the bender in the horizontal plane relative to the lower die member or anvil. Attempts to obtain a tighter bend by excessive lowering of the press shut height can result in tooling damage.

Types of Flanges and Process Limitations

Flanges on metal parts fall into three basic types. The simplest is the straight flange, which is a straight bend. Concave flanges are called stretch flanges because the metal formed into the flange is stretched. Convex flanges are known as shrink flanges since the metal is compressed or shrunk. In addition, there are combinations of these types of flanges that can be performed in a single operation. The flanging processes are limited by the severity of deformation that can be accomplished before fractures or wrinkles occur.

Figure 6-21 illustrates a number of different flanges. The problems associated with straight flanges — the most common type — are springback and scoring. Close tolerance bends should not be specified unless necessary to enhance the appearance or function of the part. Frequent adjustment of the springback control method being used may be needed because of material variations in cases where close bend angles must be maintained.

Forming stretch flanges involves stretching the metal during the bending operation. The greatest amount of stretch occurs at the edge of the flange, and is essentially zero at the bend radius.

The metal in a shrink flange is compressed causing it to shorten in length. The amount of shrinkage is greatest at the edge of the flange and diminishes to zero at the bend radius.

Irregular or curved flanges tend to have less springback problems than straight flanges. A reverse flange is a combination of a stretch and shrink flange. Another combination flange is the joggled flange. While the majority of the flange is straight, the corners are stretch and shrink flanges respectively.

A flanged hole is a type of stretch flange. Flanged hole applications include locating bosses, holes for tapped threads, openings for heat transfer tubes, and nonchafing passages for wires. Flanging an opening in a stamping can greatly increase the part's rigidity.

Edge splitting can be a problem accompanying stretch flanging, as illustrated in Figure 6-22. The likelihood of splitting depends on the material properties and the edge condition resulting from shearing or trimming. Tensile stress can be lessened by reducing the flange length or by providing notches or scallops to reduce the flange strength.

A simple general equation expresses the strain at the edge of the stretch flange, where most failures begin.

$$e_x = \frac{R_2}{R_1} - 1$$

Circle grid analysis (CGA) is an excellent method for determining the actual amount of strain at a flanged edge. A simple comparative test for materials to be flanged may be performed by expanding a drilled, deburred hole with a lubricated conical punch to determine the forming limit.

Edge conditions such as burrs and rough fractured edges reduce stretch flange formability. Such edge conditions result in excessive cold working of the metal.

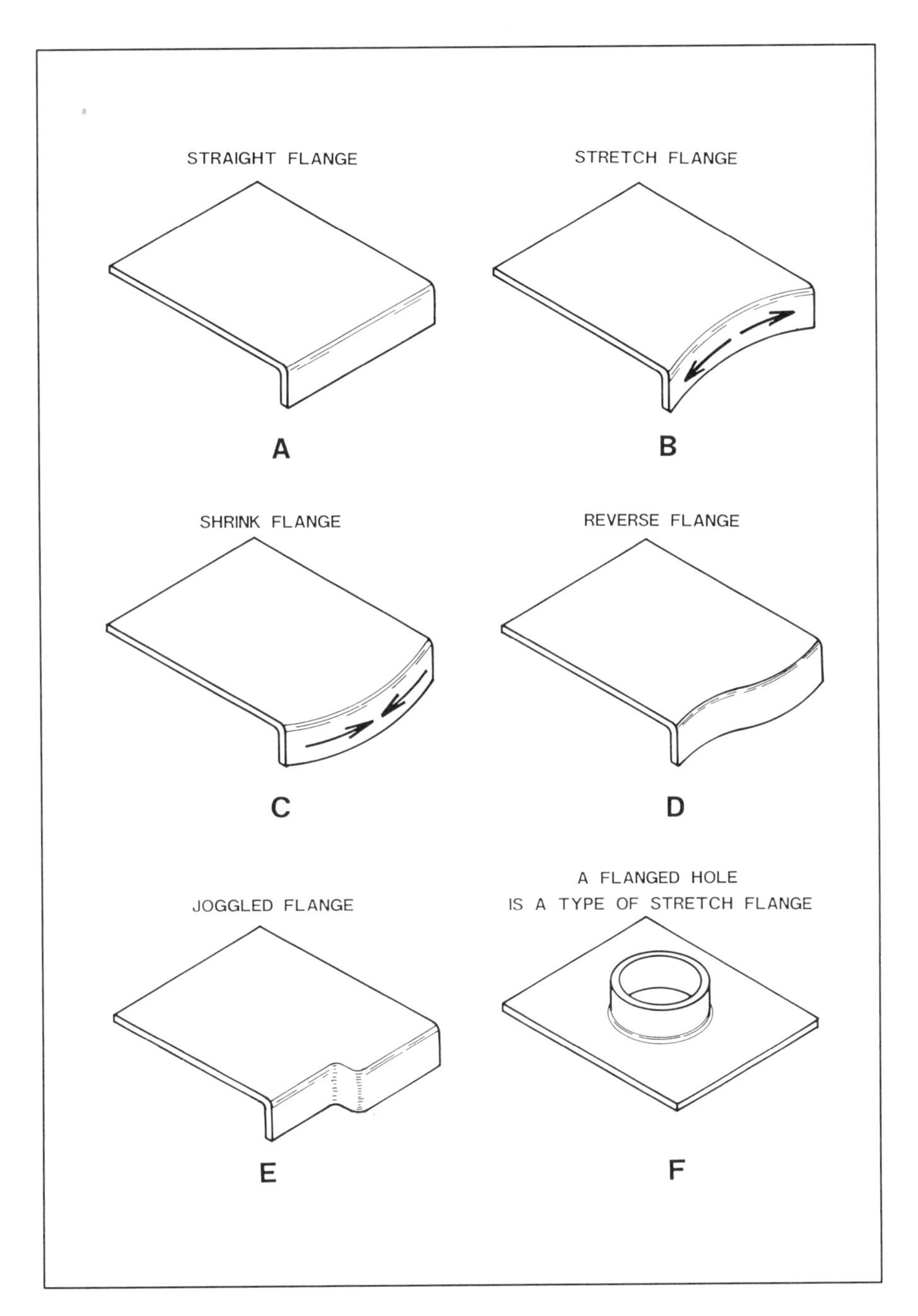

Figure 6-21. *Examples of types of flanges: a straight bend or flange (A); a stretch flange (B); in shrink flanges the metal is compressed (C); a reverse flange is a combination of a stretch and shrink flange (D); a joggled flange (E); a flanged hole is a type of stretch flange (F).*

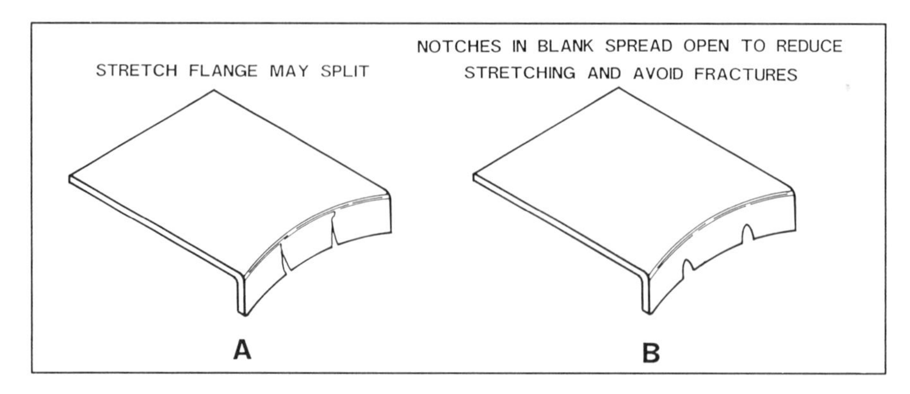

Figure 6-22. *Edge splitting can be a problem (A) when stretch flanging. The stress during flanging can be reduced by providing notches or scallops (B).*

Minimum Bend Radii

Minimum bend radii vary depending on metal type. Most annealed metals can be bent to a radius equal to the thickness, although some softer metals can be bent to an inside radius one-half metal thickness. The minimum bend radius is reduced by short-bend radius lengths. This is not a practical consideration provided that the minimum bend radius length is eight or more times the metal thickness.

The rolling direction in sheet or strip metal limits the minimum bend radius. An angle of 90 degrees between the bend axis and the direction of rolling allows most, but not all, metals to be bent to the smallest possible radii.

Bend Allowances

For close work the exact length of metal required to make a bend is often determined by trial and error. The assumed neutral axis (Figure 6-17) varies depending on the bending method, location in the bend, and type of stock being bent.

Direction of grain in a steel strip relative to the bend has a slight effect on the length of metal required to make a bend. Bending with the grain allows the metal to stretch more easily than bending against the grain, but produces a weaker stamping. Bend allowance depends more on the physical properties of the material, such as tensile strength, yield strength, and ductility than on the metal from which it is made.

An important factor in determining the neutral axis is how the bend is accomplished. Less metal is required for a bend made by a tightly wiped flange than for an air bend on a press brake. Wiping the flange tends to stretch the metal.

Exact bend allowance is the arc length of the true neutral axis of the bend (above, the neutral axis metal is stretched; below, metal is compressed). The problem is that neutral axis can only be approximated. Many manufacturers assume the neutral axis is 0.33 of stock thickness from the inside radius of the bend for inside radii of less than twice stock thickness. For inside radii of two times stock thickness or greater, the neutral axis may be assumed to lie approximately 0.50 of metal thickness from the inside radii. One reason for relatively less metal being required to make a tight bend is the sharp radius tends to be stretched slightly.

For 90-degree bends, the radius of the assumed neutral axis is multiplied by 1.57 (the number of radians in 90 degrees) to determine the amount of metal required to make the bend. For bends not exactly 90 degrees, multiply the number of degrees of bend times 0.0175 (the number of radians in a degree) and substitute the result for the coefficient 1.57.

Tabular data for the length of material in radii, including metric equivalents, are available. Formulas for developing these tables are based on extensive experimental data. Such tables were a necessity before electronic calculators became available.

To produce close-tolerance stampings, stock thickness variations must be held to a minimum. Figure 6-23 illustrates sections through a typical bending or forming die. If stock thickness is correct, the part will be properly formed. Stock that is too thin will produce underbent angles.

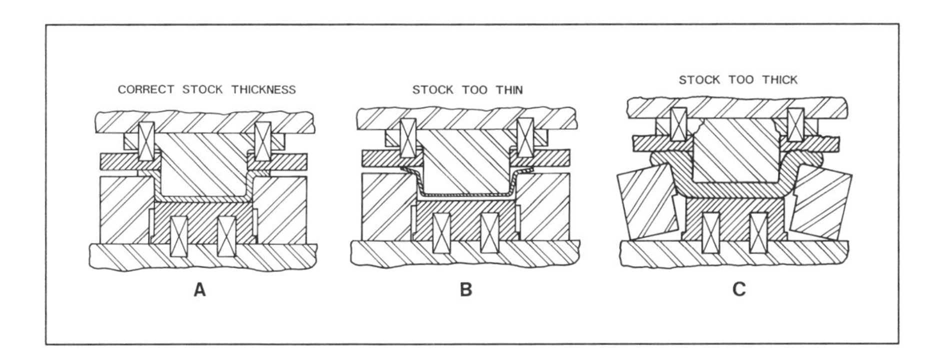

Figure 6-23. *Sections through a typical bending or forming die. Provided that the stock thickness is correct, the part will be properly formed (A). If the stock is too thin (B), the angles will be underbent. Damage can result (C) from forming stock that is too thick.*

Severe die damage can come from attempting to form stock that is too thick for the die clearances. Large lateral forces are developed that can greatly exceed the applied press tonnage. The force is multiplied by wedge-like action. This type of damage often occurs should operator inattention or an automation malfunction result in two blanks in the die at one hit.

Hemming Operations

Hems are used primarily to provide a smooth rounded edge and for attaching two sheet metal parts. They are an effective way to eliminate a dangerous sheared edge. Hems are used extensively in automobiles to join inner and outer closure panels.

A sharply bent flattened hem can be used with materials having high ductility. A tear drop, or rounded edge on the hem can be used for materials without the ductility to form a flattened hem. The same minimum bend radii considerations based on material formability apply to hems used for attaching other sheet metal parts.

Flanges and hems can be made in one operation. However, to simplify tooling, hemming is often performed on a part flanged in a previous operation.

The sequence of a typical hemming operation after bending a 90-degree flange in the outer panel is to place the inner panel in the flanged outer panel and bend the outer panel an additional 45 degrees. The partially assembled panels are next transferred to the hem die where the final assembly takes place by flattening the prebent hem.

Typical hemming pressures, including seaming pressures, are generally seven times the forming pressures required for 90-degree bends. They may be as high as a ratio of 40:1. Variables are: stock thickness, tensile strength, size of area to be flattened or hemmed, and tightness of the hem.

Bending Pressures

The amount of pressure required depends on stock thickness, length of the bend, die width, whether a lubricant is used, and the amount of wiping, ironing, or coining present.

A simple formula to determine V-bending forces for air bending in a die, such as that illustrated in Figure 6-18 (A), is:

$$F = KLSt^2/W$$

where F = bending force required, in lbf (N)

K = die-opening factor: varies from 1.20 for a die opening of 16 times metal thickness, to 1.33 for a die opening of 8 times metal thickness

L = length of bent part, inches (mm)

S = ultimate tensile strength, psi (Pa)

t = metal thickness, inches (mm)

W = width of the V-channel, or U-forming lower die, inches (mm)

The equation for deriving bending pressures is valid for V-shaped dies only. For channel forming and U-forming, multiply the result by 2. In forming a channel with a flat bottom, a blankholder is necessary. Multiply blankholder area in square inches by 0.15 to get the required tonnage, and add to the bending force derived from the equation.

The force required to bend or flange a sheet as shown in Figure 6-24 is:

$$F = 0.167 \frac{SLt^2}{A} \text{ theoretical}$$

$$F = 0.333 \frac{SLt^2}{A} \text{ for wiping dies}$$

where F = bending force required, lbf (N)

K = a constant that varies from 0.167 for large die radii and clearances to 0.333 for sharp die radii and high plastic working stresses

t = sheet metal thickness, inches (mm)

L = length of bend, inches (mm)

r1 = punch radius, inches (mm)

r2 = die radius, inches (mm)

C = die clearance, inches (mm)

S = Ultimate tensile strength, psi (Pa)

The flanging pad must grip the part firmly to ensure the part remains in tight contact with the male die half during flanging. A rule of thumb for determining sufficient pad force is:

Pad force = SLt/3

where F = force, lbf (N)
S = ultimate strength of the material, psi (Pa)
L = flange length, in. (mm)
t = material thickness, in. (mm)

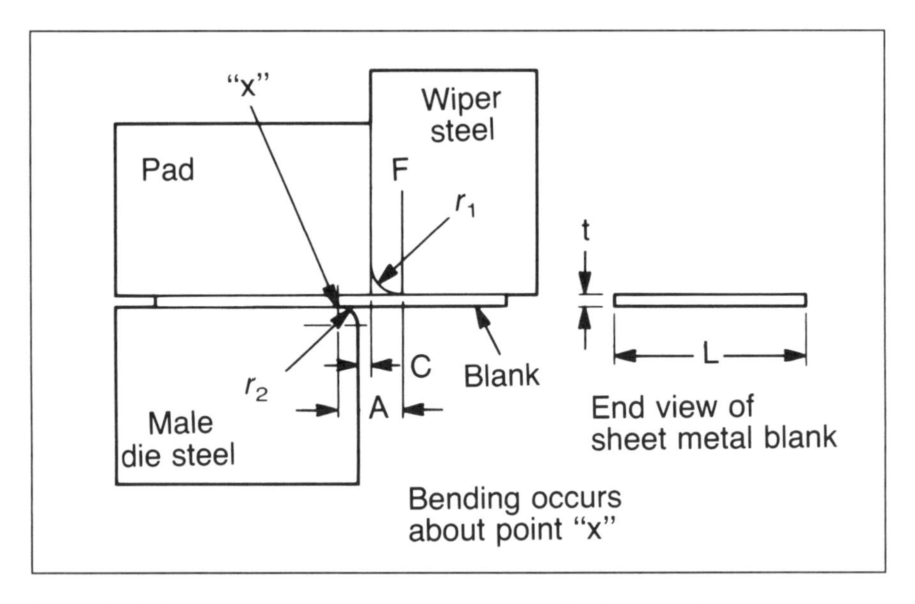

Figure 6-24. *Factors that are entered to determine flanging force requirements.*

DRAWING AND STRETCHING METALS

The term *drawing* is loosely applied to many press-forming operations. Strictly defined, drawing is sheet metal deformation process in which the following three actions occur simultaneously:

- The metal is drawn into a die cavity by a draw punch.
- The metal is restrained by a blankholder to control wrinkling while allowing smooth metal flow.

- Compression of the metal occurs at approximately a 90-degree angle to the direction in which the metal is stretched.

Deep Drawing a Cylindrical Cup

Figure 6-25 illustrates the forces involved in deep drawing a metal cup. It is important to note that all the drawing force is transmitted by the draw punch to the cup's bottom.

The cup drawing process starts with a flat round blank, which is subjected to both radial tension and circumferential compression. The metal thickens as it flows toward the draw radius. Deep drawing is unique because of the deformation state of the metal restrained by the blankholder.

In general, the metal flow in deep cup drawing may be summarized as follows:

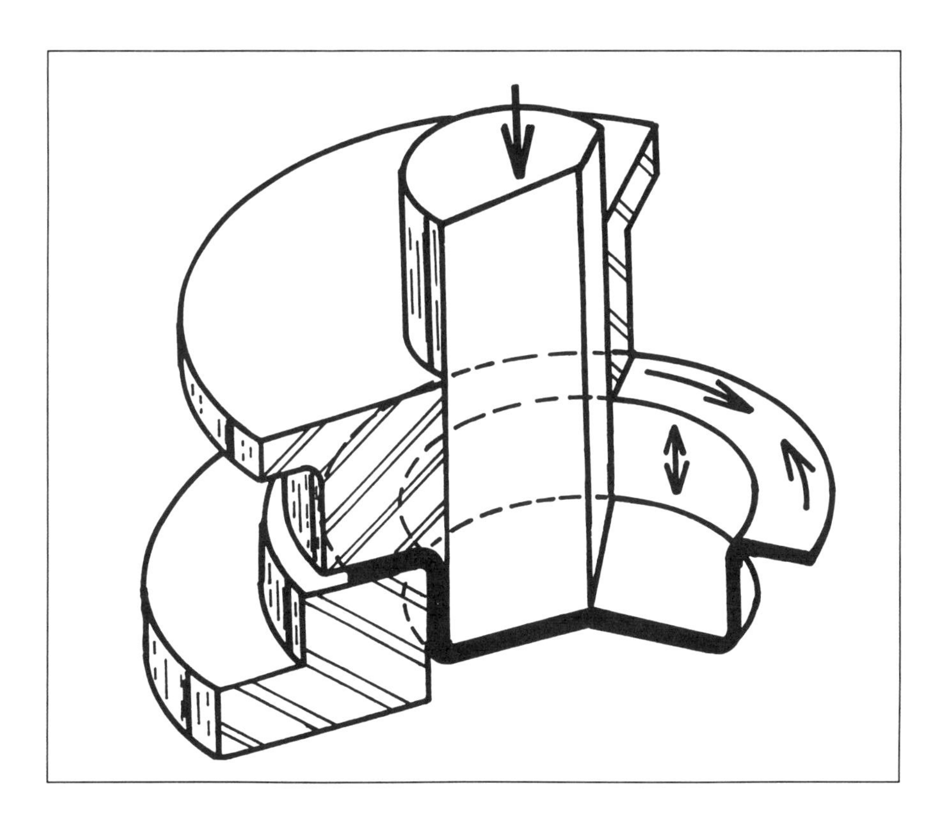

Figure 6-25. Sectional view of a simple draw-die illustrating the tensile forces, and circumferential compressive forces involved in deep cup drawing. (Smith & Associates)

- Little or no metal deformation takes place in the blank area that forms the bottom of the cup.
- The metal flow taking place during the forming of the cup wall uniformly increases with cup depth.
- The metal flow at the periphery of the blank involves an increase in metal thickness caused by circumferential compression (Figure 6-25).

The success of a drawing operation depends upon several factors, including:

- Formability of the material being drawn;
- Limiting the drawing punch force to less than that which will fracture the shell wall; and
- Adjusting the blankholder force to prevent wrinkles, without excessively retarding metal flow.

The tensile force transmitted through the sidewalls draws the metal restrained by the blankholder over the draw radius and into the die cavity. All of the force needed to overcome friction and accomplish the deformation is transmitted through the cup sidewalls.

Figure 6-26 illustrates a properly drawn cup-shaped shell and two stages of failure of the drawing process. It shows a deep-cup drawing in which very little deformation occurs over the bottom of the punch. Nearly all deformation occurs in the metal restrained by the blankholder. The maximum force requirement for the

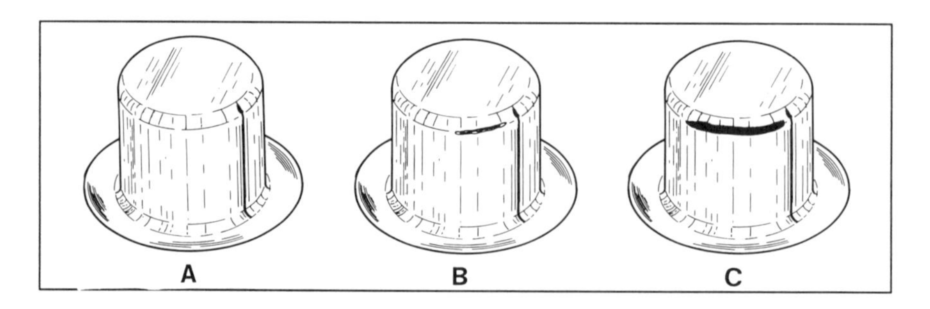

Figure 6-26. *A successfully drawn cup-shaped shell (A); failure of the drawing process due to localized thinning (necking) at the punch radius (B); severe failure (C) resulting in a fracture. (Smith & Associates)*

drawing process is limited by tensile failure of the material in the sidewall. As this limit is approached, the metal will neck, or thin, excessively in a localized area usually near the punch radius.

Many complex interactions occur during the cup-drawing process. Actual force required depends on the cross-sectional area of the cup wall and the yield strength of the material as it is worked. Should the process fail, some or all of the following factors may be the root causes:

- The ductility or drawability of the stock may be too low.
- The blankholder force may be too high.
- Scoring or galling may be present on the die surfaces.
- The blankholder geometry and draw radius may not provide for metal thickening and smooth flow into the die cavity.
- Incorrect or insufficient drawing lubricant.
- The depth of draw or percentage of blank reduction may be too great.
- One or more redrawing operations may be necessary to obtain the desired depth of draw.
- Annealing may be required between redrawing operations when using materials that work harden rapidly.

When drawing operations fail, often the material is immediately blamed. The logic is that if the vendor can supply some material that will run, it should be possible to do so consistently. Material formability properties do vary from lot to lot, even within the same coil. It is poor economy to specify expensive deep-drawing quality material, when good die design and maintenance will permit the use of less costly commercial-grade stock.

Provided that there is proper sidewall clearance in the die, the punch force will not exceed the ultimate tensile strength of the material cross section in the wall. Since some thinning occurs as the ultimate tensile strength is approached, using this figure for force requirement calculations usually provides a substantial safety factor. It should be noted that excessive blankholder forces can cause any cup drawing operation to fail.

The yield strength may be used to produce results that correspond more closely to measured values.

Example of Drawing Force and Energy Requirements

Earlier in this chapter, we calculated the force and energy requirements for cutting a low-carbon blank 12-inch (304.9 mm) in diameter and 0.1875-inch (4.763 mm) thick. The required energy is surprisingly small, only 1840.8 foot-pounds (2496 J). At 60 strokes per minute (SPM), ignoring frictional losses, 3.347 horsepower will sustain the process.

Several arbitrary assumptions are made. This heavy 12-inch diameter 0.1875-inch thick blank is to be drawn into a flanged cup having a diameter of 6.000 inches (152.4 mm) and a depth of draw of 2.000 inches (50.8 mm). The yield strength of the material is 40,000 psi (275 MPa). Here we will assume that the force and energy required is based on working the cup wall at or near its yield strength. It should be noted that the yield strength is normally specified as a minimum value.

The total cross-sectional area of the wall is 3.534 square inches. Based on the yield strength of 40 ksi, 70.69 tons of punch force is needed. While the force is much less than required to cut the blank, it must act through a distance of 2.000 inches (50.8 mm). The work or energy required per part is 141.37 inch-tons or 23,562 foot-pounds (31,950 J).

The energy input per part is 12.8 times greater to draw the part than that required to cut the blank. While the 60 strokes per minute rate may be too high for optimum formability, neglecting frictional losses, 42.84 horsepower would be required to restore energy to the flywheel.

In pressworking processes, materials are worked in a plastic state. High internal friction is present. Nearly all energy required for the process is converted to heat.

Converting Mechanical Energy to Heat

Studies conducted on drawing automotive suspension components at Ford Motor Company by the author confirmed that 80 to 90% of this energy exits the die as latent heat in the stamping. Nearly all mechanical energy-to-heat conversion occurs within the stamping itself during deformation. The remainder comes from surface friction at the part-to-die interface.

The severely strained portions of the part exit the die at a higher temperature than the surface temperature of the corresponding die surfaces. The energy input was determined by calculations of force versus distance using waveform signature analysis to determine the energy input.

The temperature profile of the part and die was measured by a small thermistor probe. The temperature data was used to determine the heat energy in the part and the temperature gradient of the heat energy conducted through the die.

There is close agreement of theoretical versus measured values of mechanical to heat energy. In general, large dies function as effective heat sinks, when operated at press speeds from 8 to 20 SPM.

In the dies studied, ion nitriding of the nodular-iron die surfaces essentially eliminated the metal pickup or galling problems that occurred as the die warmed during operation. However, cooling by chilled water piped through the die is useful for some severe drawing operations.

Water-based pressworking lubricants assist in cooling through evaporation, and should be used wherever possible.

Stretch Forming

Figure 6-27 illustrates a stretch forming operation to produce a dome-shaped part. The edges of the blank are securely clamped with a lock bead. Only the metal in the punch area is deformed.

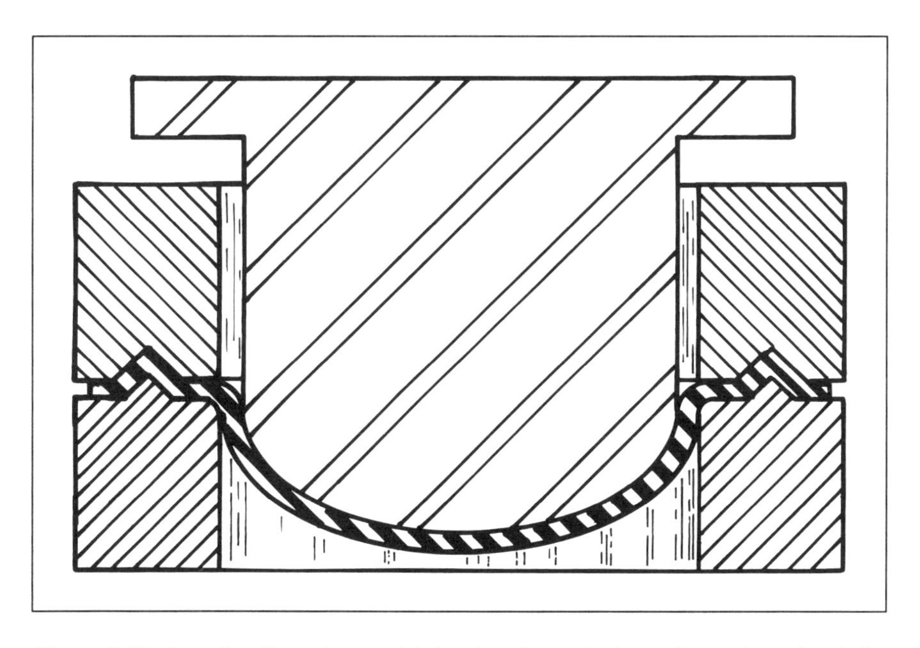

Figure 6-27. *A section through a stretch forming die producing a dome-shaped part: the edges of the blank are securely clamped with a lock bead, and the metal in the punch area is deformed by biaxial stretch. (Smith & Associates)*

Both the width and length dimension of the metal are stretched. This type of forming is known as biaxial stretch.

Stretch forming is a very common operation. The forming of automotive, appliance, and aircraft panels is a widespread application. Typically 3 to 10% strain is required to obtain the mechanical properties needed for proper stiffness.

Like deep-cup drawing, stretch forming involving severe deformation depends on good material properties, proper lubrication of the punch, and correct die maintenance. To obtain enough stretch to realize good part stiffness, it is important to maintain enough blankholder force to prevent metal slippage through the lock bead. The blankholder force required throughout the press stroke is greater than the pressure required to form the lock bead upon initial die closure.

Stretch-forming process failure is evidenced by excessive localized thinning or necking, often leading to fracture. Surface roughness of both the die and material should be minimized. While smooth surfaces still have high friction, roughness may result in scoring part fracture. Good lubrication is an important success factor in operations involving severe deformation.

Plane Strain

If the metal is formed by stretching in one direction only, the operation is called plane strain or simple stretching of the metal. This forming method, illustrated in Figure 6-28 (A), allows limited strain or elongation before a fracture occurs.

Bending and Straightening Operations

Figure 6-28 (B) illustrates a forming method in which the metal is bent and straightened as it passes over the blankholder radius. Bending and straightening operations permit large deformations with very little thickness change.

A simple type of die employing bending and straightening is used to make U-shaped cross sections with right angle flanges. Parts of this type are used as stiffeners attached to flat panels in many applications. Examples are automotive body frame rails and cross members, which are assembled by welding.

Bending and straightening operations are employed in conjunction with processes such as drawing and plane strain in the production of complex stampings. A simple example is the stamping of rectangular shells.

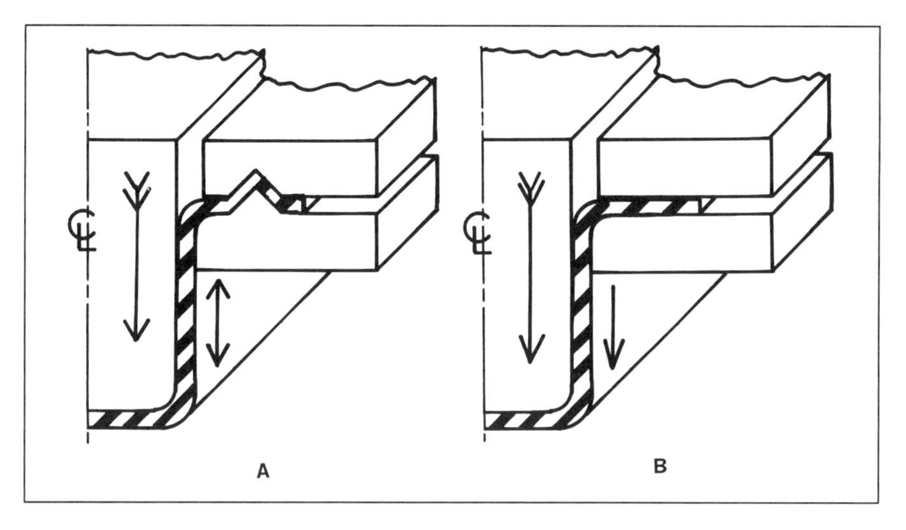

Figure 6-28. Method (A) illustrates a plane strain condition in which the metal, restrained by a blankholder lock bead, is stretched and thinned; in method (B), the metal is bent and straightened, permitting larger deformations and little thickness change. (Smith & Associates)

Combined Drawing and Bending

Stamping a rectangular shell involves both cup drawing and simple bending. True drawing occurs at the corners only. The metal movement at the sides and ends metal movement involves bending and unbending. Figure 6-28 (B) illustrates a section through a portion of a die where bending and straightening occurs as the metal passes over the radius.

The stresses at the corner of the shell are compressive, and thicken the metal moving toward the die radius. After the metal has been drawn over the radius, the forces are tensile. The metal between the corners is in tension on both the side wall and where restrained by the blankholder. This portion of the operation involves deep cup drawing illustrated back in Figure 6-25.

Unlike circular shells, in which pressure is uniform on all diameters, some areas of rectangular and irregular shells may require more pressure than others. The metal at the corners of the blank is compressed and will thicken. Both thickening and some wrinkling of the metal at the corners is normal. Figure 6-29 illustrates the appearance of a box-shaped drawn shell made from a rectangular blank. The metal at the corners often increases in thickness up to 25% or more.

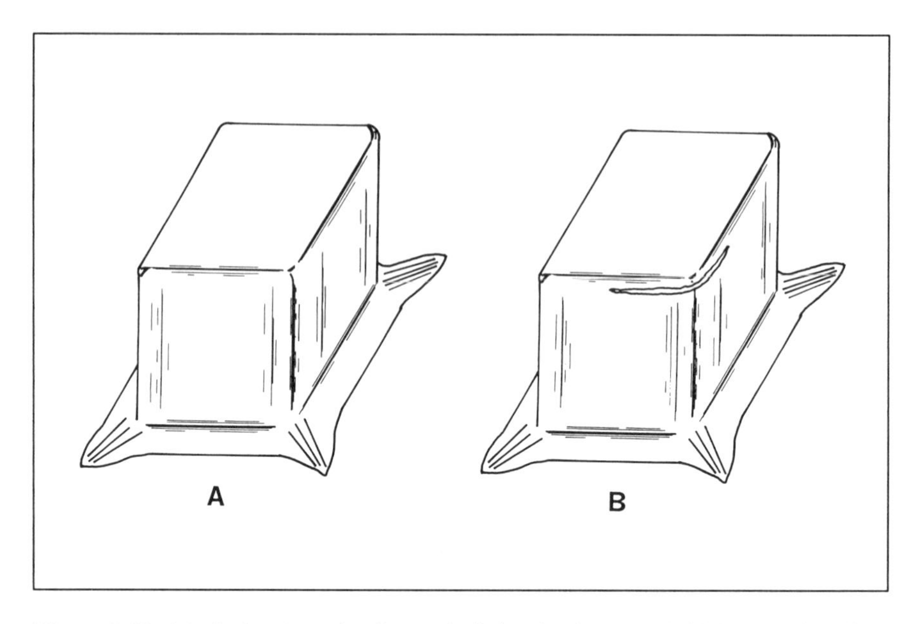

Figure 6-29. A typical rectangular drawn shell showing how metal thickens and tends to wrinkle at the corners (A); the normal location of a necking or fracture failure (B). (Smith & Associates)

A common error in constructing and maintaining rectangular shell drawing dies is to machine the blankholder surfaces perfectly flat. Clearance should be provided in the blankholder to allow the metal to thicken. Usually, this is done with a pneumatic hand grinder when the die is tried out.

A skilled die tryout technician will optimize the metal flow by making a series of trial parts and reworking the blankholder as needed. In some cases, it is necessary to increase draw ring and punch radii with the product designer's approval.

Minor product changes are often highly beneficial to reduce or eliminate the occurrence of fractures. The corner is the usual location of a fracture in a rectangular drawn shell. The localized thinning, or necking, leading to a fracture is the same failure mode that limits the severity of deep-cup drawing.

After the die tryout work is complete, the die is oil stoned and polished. If long service life is required in severe applications, ion nitriding, hard chromium plating, or other processes to reduce surface wear and friction, may be beneficial.

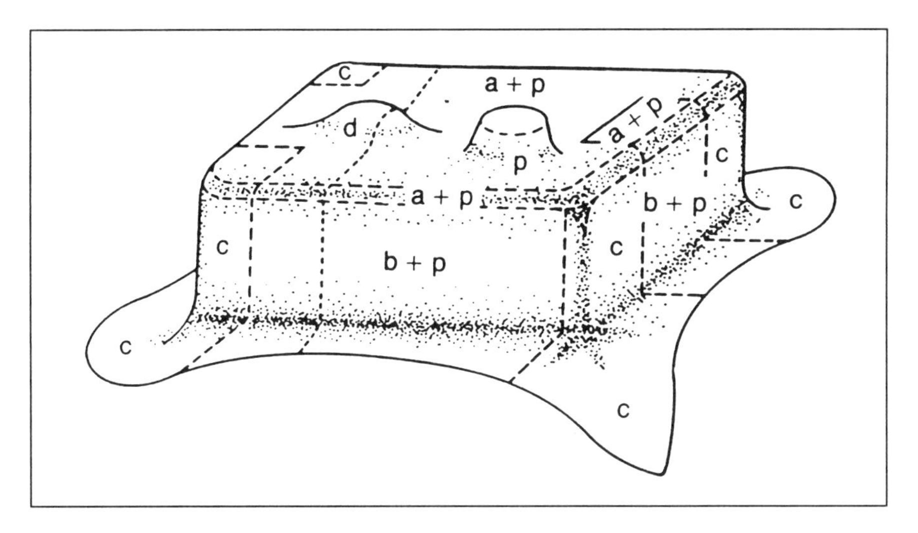

Figure 6-30. A complex stamping is produced by a variety of forming operations, each of which may be analyzed separately.

Complex Stampings that Combine Basic Forming Operations

Figure 6-30 illustrates a complex stamping produced by a variety of forming operations. A total of five basic operations are shown. These include simple bending (A), bending and straightening (B), cup drawing (C), dome or biaxial stretch (D), as well as plane strain stretch (P).

Each of the distinct types of forming operations in Figure 6-30 may be analyzed separately. Computer-aided formability analysis can predict manufacturability without trial and error.

STAMPING ANALYSIS TECHNIQUES

Forming flat sheet metal into complex radically deformed stampings can appear to involve skills and processes that are more an art than a science. The fact is, modern stamping design and development techniques permit the product designer to work with manufacturing and tooling engineers to design parts that can be manufactured with certainty.

Stamping designs should be based on data of successful prior designs and formability analytical methods. Uncertainty concerning the manufacturability of complex stampings often results in added expense and delay factors such as:

- Trial production on temporary tooling to prove process feasibility;
- Delays in marketing the product containing the stamping while the process or product design is changed;
- Specifying more operations than should be needed as a safety factor; and
- Choosing alternative processes and materials such as molded plastics.

Assuring Easily Manufactured Stamping Designs

Easy-to-use computer software programs are available to assure that proposed stamping designs can be manufactured with certainty. Other powerful tools are CGA, which ties in with the forming limit diagram (FLD). Using these tools avoids costly trial-and-error guesswork.

Computerized Analysis Techniques

Software is available to analyze the amount and type of deformation in a stamping design. Such computer-aided analysis ties in nicely with CAD design of stampings, and should be applied early in the product design process. The CAD math data, which describes the part, is used for computerized formability analysis. Computerized analysis falls into several categories, such as:

- Simple sectional analysis programs;
- General analysis programs, which fully model the part, typically based on finite element analysis; and
- Programs that analyze the stamping based on the type of deformation occurring in individual areas.

The sectional analysis programs are useful for determining the amount of strain present in a specific area of a stamping. Here, anticipated problems with a design can be checked easily.

A moderately-priced personal computer has sufficient computational capacity to run sectional analysis programs to determine strain conditions. Estimating the effect of surface friction on metal movement is a useful feature of nearly all computerized formability analysis programs.

To fully model the part, using the finite element or finite difference method, general analysis programs are required. Stamping the whole part is simulated in three dimensions with a single computer program. There are many complex interactions that occur during the stamping simulation, so a mainframe or super computer may be necessary.

Sectional analysis is good for identifying and troubleshooting a number simple forming conditions. General analysis programs require a lot of computing power and time to calculate the interaction of many complex variables occurring throughout the forming process.

A simplified approach is to break down complete stampings into local regions that can be analyzed individually. In this way, a stamping such as that shown in Figure 6-30, is analyzed as individual zones that interact in a predictable manner as they are formed. Some good programs include an expert systems approach.

CIRCLE GRID ANALYSIS

The circle grid analysis (CGA) technique permits measurement of the deformation that occurs when forming stampings. First, a grid is stenciled on the surface of the blank by dye transfer or electrochemical etching. This grid deforms with the blank and allows point-to-point calculations of the deformation that occurred during the stamping operation.

Press Shop Applications

Figure 6-31 illustrates a bumper jack hook. The grid of circles, placed on the blank, shows differing types and severity of deformation required to form the finished stamping.

Note that most of the circles are deformed very little, while a few, especially one near the lip, show pronounced elliptical patterns. A stamping of this type will often fracture at the location where the edge is most severely stretched.

Distribution of stretch is useful information by itself. Location of high-stretch concentrations and the direction of maximum stretch often are sufficient to suggest solutions to forming problems. However, CGA uses a numerical rating system for the circle deformation.

The system of rating forming severity is based on measuring deformation of the circles and plotting the measurements on a

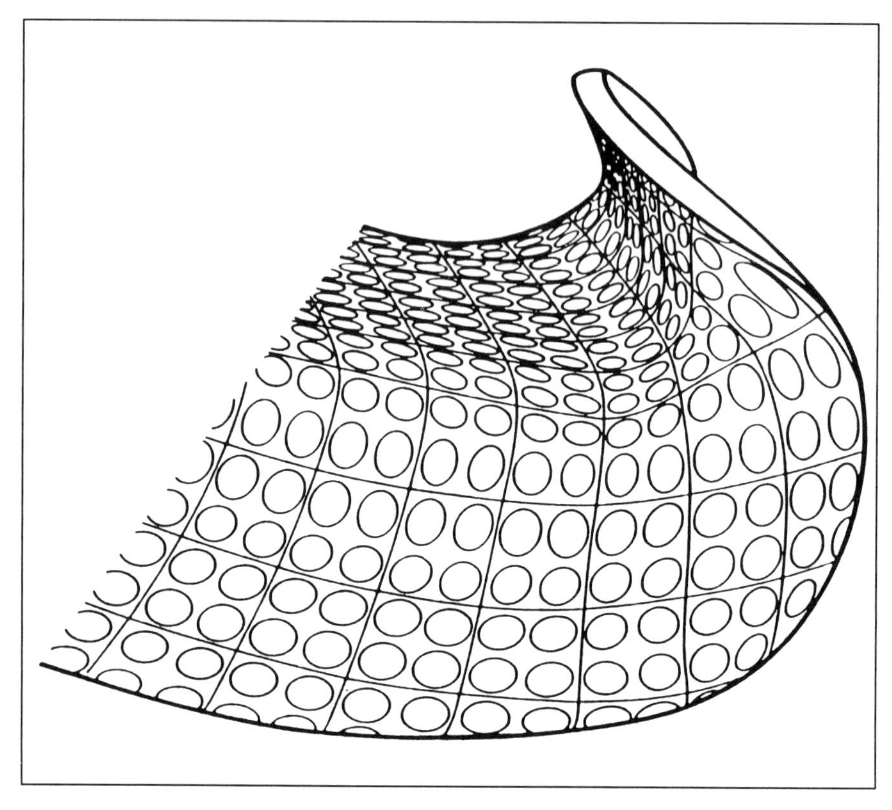

Figure 6-31 A bumper jack hook formed from a circle gridded blank. (National Steel Corporation)

graph. Grid measurements are easily made with a transparent mylar tape imprinted with a calibrated scale, (Figure 6-32). The tape is flexible and can be laid around a radius or tucked into a tight corner. Calibration of the tape eliminates any need to calculate stretch. The tape is used to measure the major (length) axis of the ellipse first and then rotated to measure the minor (width) axis.

Stretch Combinations

Many combinations of major and minor circle deformations can be found on different stampings. Figure 6-33 illustrates five types of deformation. One special case is shown on the right, where both the major and minor axes of the ellipse are equal — the circle

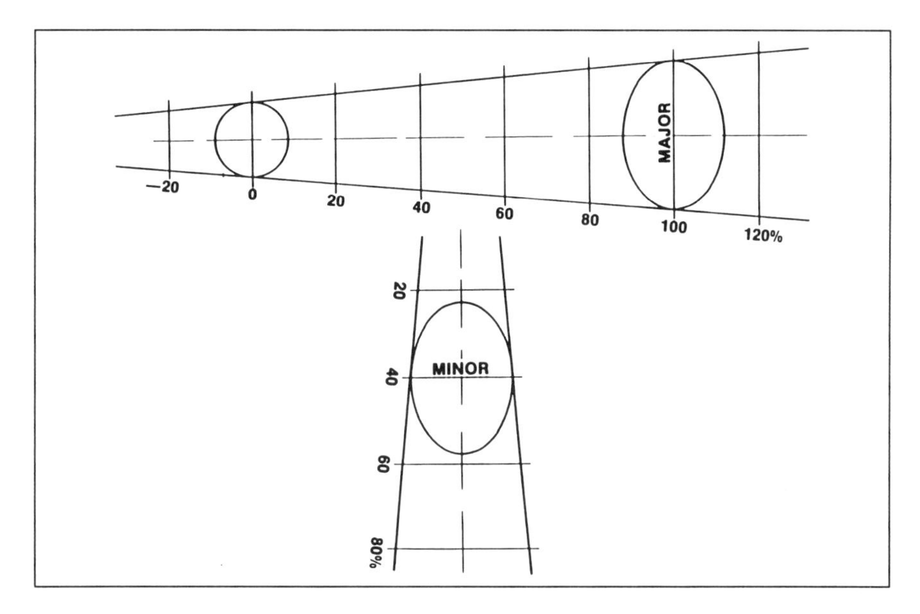

Figure 6-32. Measuring circle deformation with a mylar tape overlay. (National Steel Corporation)

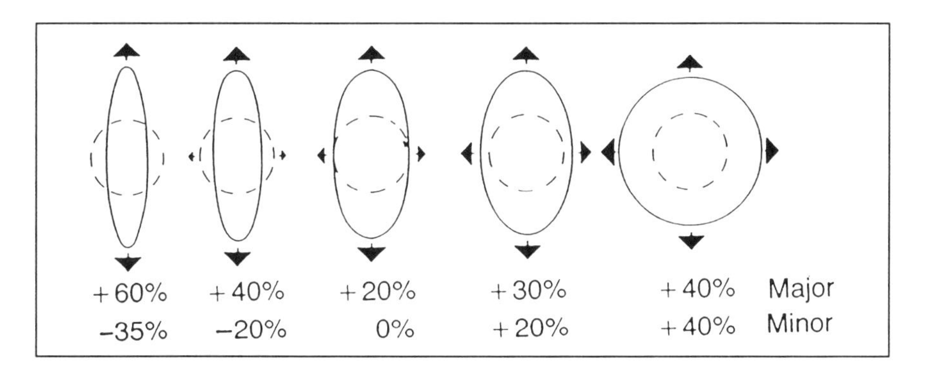

Figure 6-33. Examples of deformed circles. (National Steel Corporation)

becomes a larger circle. This is called balanced biaxial stretch and would probably be observed on a gridded part formed in the die that was shown in Figure 6-27.

The middle case also is special. Here the minor stretch component is zero. Called plane strain, this stretch condition is found over edge radii or across character lines. The die illustrated in

Figure 6-28 (A) forms the part by plane strain, which involves stretching the metal in one direction only.

The two examples on the left side of Figure 6-33 illustrate large major elongations while the minor stretch is negative. Circle deformations of this type are observed in the sidewalls of drawn cups, (Figure 6-26), and the corner sidewalls of rectangular drawn shells (Figure 6-29). This combined compression and elongation, indicates that the metal is subjected to both circumferential compression and tensile stretching as it is pulled toward and over the draw radius.

Due to the variety of combinations, a method for plotting them on a single graph is necessary. The plotting technique used in Figure 6-34 allows both the major and minor stretch for each circle to be plotted as a single point.

The major stretch is plotted on the vertical axis, while the minor stretch is plotted on the horizontal axis. Circles that plot on the left side of the diagram have negative minor stretch, while circles that plot on the right of the diagram have positive minor stretch.

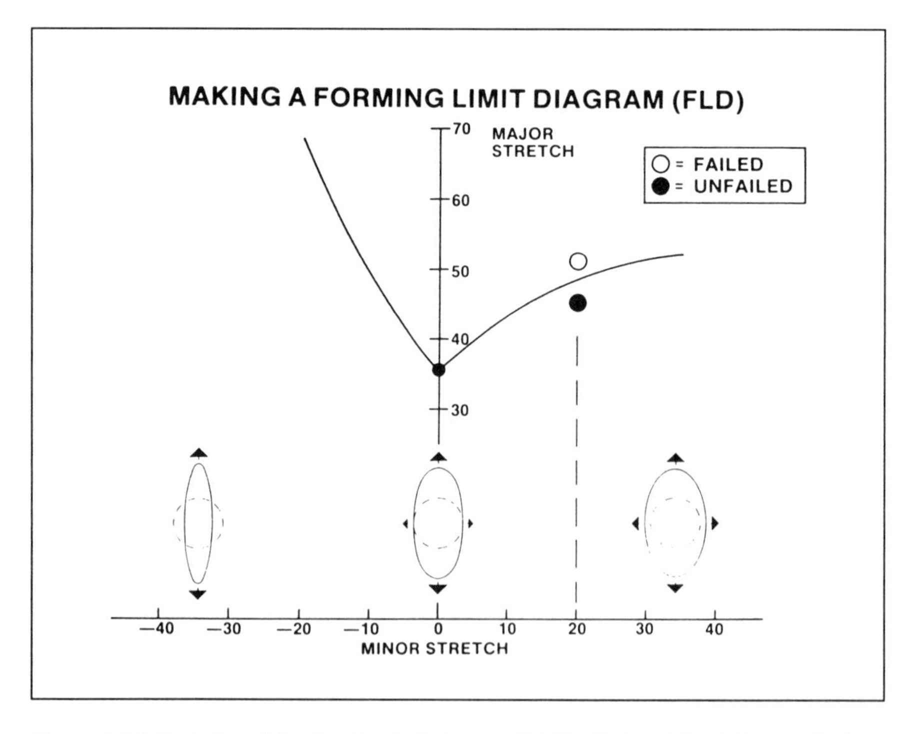

Figure 6-34. Evolution of the forming limit diagram (FLD). (National Steel Corporation)

Three of the ellipses from Figure 6-33 are plotted on this diagram. Note that the case of plane stretch (minor stretch equal to zero) is plotted on the vertical axis.

Figure 6-34 illustrates an unsymmetrical V-shaped curve, which is the forming limit. Circles plotted below this curve show no evidence of necking or fracture, while those above it fail. A graph developed in this way is called a forming limit diagram (FLD). The point where the FLD intersects the major stretch axis is called FLDo. Here only plane strain deformation is occurring. To initially develop this diagram, many samples of failed versus unfailed circles from the same material must be plotted.

The shape of the FLD (Figure 6-34) is constant for most low-alloy sheet steel used in the automotive, appliance, agricultural, container, and similar industries. The curve moves up and down the axis for different coils of steel. Thus, the location of the curve can be described by specifying the intersection of the curve's FLDo with the minor-stretch axis.

Figure 6-35 illustrates how the FLD can raise or lower for different steel sheets. The level of the FLD — as specified by FLDo — is a characteristic of the sheet steel. For example, a thinner sheet of steel would have a lower FLD than a thicker sheet of steel. Also, a higher strength steel would have a lower FLD than a lower strength steel.

Using CGA as a Process Control Tool

CGA is a powerful process control tool. Data on key areas of the stamping that are near the forming limit should be checked periodically to determine the effect of die wear on formability. Should a production stamping process start to experience problems, a blank of the material can be quickly gridded and analyzed. The CGA results can be compared with the historical data for the part and steel formability specifications.

Should the part always be found to run well within the safety zone, often a less costly steel or lubricant can be used. If only a few areas on the stamping are close to failure, a blankholder improvement or minor product change often will ensure manufacturability of the product.

The circle grid analysis system is excellent for training apprentices. By making tooling, lubricant, and material changes and then observing the metal deformation changes, cause-and-effect patterns can be readily discerned.

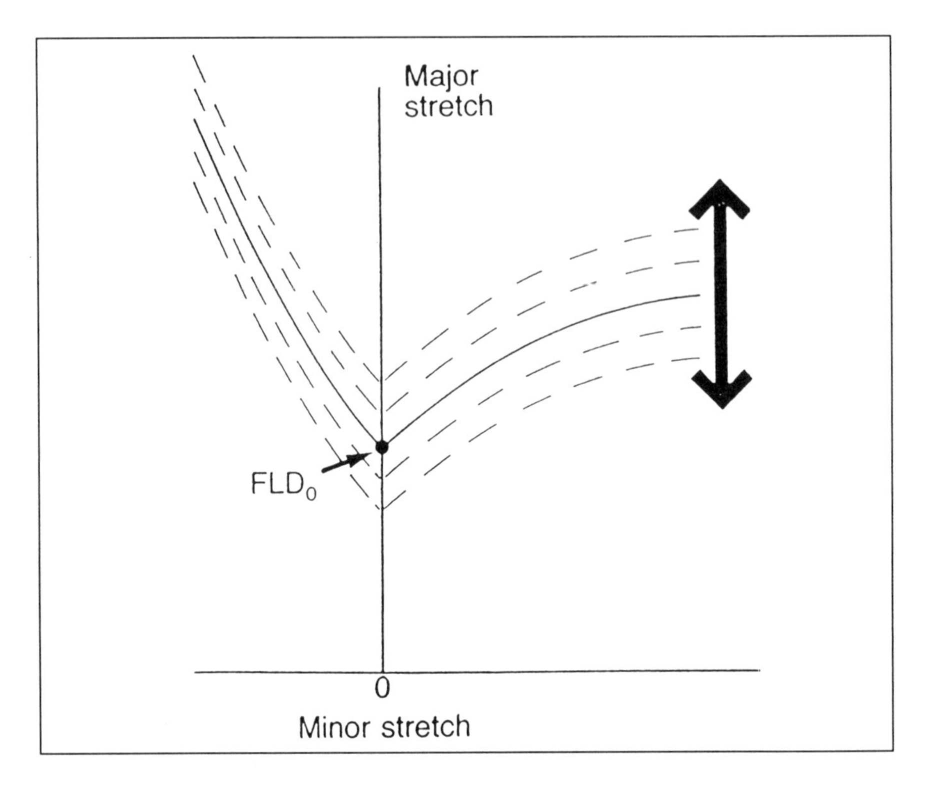

***Figure 6-35.** Different steels move the FLDo point. (National Steel Corporation)*

BIBLIOGRAPHY

1. D. Smith, *Die Design Handbook*, Section 10, "Displacement of Metal in Drawing," Society of Manufacturing Engineers, Dearborn, Michigan, 1990.
2. *Sheet Steel Formability*, The Committee of Sheet Steel Producers, American Iron and Steel Institute, Washington, D.C., August 1, 1984.
3. S. Keeler, *Circle Grid Analysis (CGA)*, National Steel Corporation Product Application Center, Livonia, Michigan, 1986.
4. M. Tharrett, *Computer Aided Formability Analysis*, presented at an SME Die and Pressworking Tooling Clinic, Dearborn, Michigan, August 25-27, 1987.
5. D. Smith, *Die Design Handbook*, Section 10, "Displacement of Metal in Drawing," and Section 11, "Product Development for Deep Drawing." Society of Manufacturing Engineers, Dearborn, Michigan, 1990.

6. D. Smith, *Die Design Handbook,* Section 4, "Shear action in Metal Cutting," Society of Manufacturing Engineers, Dearborn, Michigan, 1990.
7. D. Smith, *Using Waveform Signature Analysis to Reduce Snap-Through Energy,* SME Technical Report MF90-11, Society of Manufacturing Engineers, Dearborn, Michigan, 1990.
8. D. Reid, *Fundamentals of Tool Design,* Third Edition, Chapter 7, "Design of Pressworking Tools," Society of Manufacturing Engineers, Dearborn, Michigan, 1991.
9. D. Smith, *Quick Die Change,* Chapter 29, "Control the Process with Waveform Signature Analysis," Society of Manufacturing Engineers, Dearborn, Michigan, 1991.
10. D. Dallas, *Pressworking Aids for Designers and Diemakers,* Society of Manufacturing Engineers, Dearborn, Michigan, 1978.
11. C. Wick, J. Benedict, and R. Veilleux, Editors, *Tool and Manufacturing Engineers Handbook,* Volume 2, Fourth Edition — Forming, Society of Manufacturing Engineers, Dearborn, Michigan, 1984.
12. A. Mohrnheim, "Strain Relationships in Cupping and Redrawing of Tubular Parts," *The Tool and Manufacturing Engineer,* March 1963.
13. Arnold Miedema, Reference Guide for the videotape training course, *Progressive Dies,* Society of Manufacturing Engineers, Dearborn, Michigan, 1988.
14. Michael R. Herderich, *Experimental Determination of the Blankholder Forces Needed for Stretch Draw Design,* SAE Paper 900281, Warrendale, Pennsylvania, 1990.
15. S. Keeler, *From Stretch to Draw,* SME Technical Paper MF69-513, Society of Manufacturing Engineers, Dearborn, Michigan, 1969.

CHAPTER 7

HIGH-SPEED PRESSES AND DIE OPERATIONS

The press speed in strokes per minute (SPM) defining high-speed operations is not universally agreed upon. As a general rule, high-speed operation involves press speeds of 300 SPM or greater. Machine size and part configuration are factors. Speeds at which the inertia and vibration of moving machine and die parts become an important consideration is one definition of high-speed operation. The upper end of the speed range for small high-volume parts can exceed 2,000 SPM.

Gap-frame presses are used for some high-speed work. However, the more costly precision high-speed straightside presses provide better alignment and are more widely used. Multiple-slide forming machines comprise a separate class of machine widely used for high-speed presswork.

Common factors in presses designed for high-speed operation include compact robust construction. Special attention is given to the close fit and lubrication of all bearing surfaces.

Coil stock fed progressive dies are widely used in high-speed presses. Here, a number of basic die operations are performed in sequence. Usually, one or more parts are completed per stroke. The peak force required in each station occurs at different points during the stroke. Some unbalanced loading is probable. Excellent alignment of both the press and die is critical to quality work and low maintenance costs.

A variety of products, ranging from razor blades and tiny electrical connectors to motor laminations and beverage container lids, are produced by these processes. The multiple-slide forming machine is especially versatile. It is used to produce products ranging from surgical needles to assembled worm-gear hose clamps.

When analyzing high-speed success factors, the press and die should be considered as a single system. Maintaining low deflection of press components under load is always a goal of good press design. Press frames and ram guiding systems designed for low deflection provide good alignment for the die. The benefits to the stamper are reduced die and press maintenance costs as well as consistent part quality.

DYNAMIC BALANCE AND INERTIAL FACTORS

In high-speed progressive die operation, layout and timing of die stations are governed by sequencing the strip layout for the best die functioning. Location of the various stations often results in instantaneous ram tipping as the die closes.

The resulting instantaneous misalignment should be considered in the process design. Often, offsetting die placement in the press will correct or minimize the problem. A perfect load balance, at any instant in time, is desirable in progressive die operations, but difficult to realize in practice. Operating the tool in a stiff machine having low deflection is highly beneficial to tooling life and quality.

To avoid vibration, high-speed press crankshafts often have counterweights to provide dynamic balance. The counterweights, attached to the crankshaft, function much like those in automotive engines.

When the stroke is short compared to the crankshaft diameter, the shaft has eccentrics to actuate the pitmans. Comparatively large diameter eccentric shafts are used to maximize rigidity. High-speed presses are virtually all single-end drive machines. A very rigid eccentric shaft is especially important in two-point presses to minimize any angular misalignment due to twist.

Some high-speed presses are dynamically balanced for the reciprocating mass of the slide and upper die. This is done by providing a mass having the same moment of inertia as the slide and upper die, but driven in the opposite direction.

The upper die weight should be minimized. One way of doing this is by using aluminum alloy die shoes rather than steel or cast iron. While aluminum is more costly, it is easier to machine. Compared to steel, aluminum has the advantage of higher heat conductivity and greater vibration-damping capacity. Greater thermal expansion and a much lower modulus of elasticity are disadvantages.

Electronic die protection is nearly always used as both a process protection and control tool in high-speed work. Human reaction time is slow, and the operator's attention will wander.

In-die electronic misfeed detection is widely used to avoid or limit mishit damage. A tonnage meter wired to stop the press in an overload is also a desirable process protection feature. To assure minimum tooling damage, stopping time of the press must be less than one revolution. If air-actuated clutches and brakes are used, they should be designed for rapid actuation and stopping.

Hydraulically actuated clutches and brakes are available with very short actuation and release times. If needed for tooling protection, the added cost is a worthwhile investment.

Snap-through

Some high-speed work, especially lamination blanking operations, results in a sudden release of energy, called snap-through, as the punches break through the stock. This energy is stored in the press and die as strains or deflection.

Presses for this class of work should be designed with massive members relative to the size of the machine to provide good stiffness, or conversely, as little deflection as possible for the press force rating. One way that the press user can limit deflection, is to specify the press bed to be no longer than necessary.

Snap-through excites a vibration or oscillation in the press. While steel has the greatest stiffness per unit volume of the materials used to build presses, cast iron and iron alloys provide superior vibration damping. A heavy cast iron frame is a desirable feature that contributes to damping press vibration.

High-speed Press Bearings

The ram guiding and rotating shaft bearings are critical to maintaining machine alignment in high-speed presswork. Working loads, shock, and inertial forces must not result in significant machine misalignment. Two bearing designs are in common use in high-speed presses. Some press builders favor ball and roller bearings. The more common plain sleeve and flat bearings also find widespread use in high-speed applications.

Hardened steel ball and roller bearings can operate with very little clearance. Their application in pressworking is most familiar in the ball-bearing guide pins used to align dies. Rollers are favored for flat bearings such as slide gibbing, although some slide alignment systems make use of round guide posts. One design makes use of the press tie rods to assist in guiding the slide with recirculating ball bearings.

Ball and roller bearings provide good alignment and long service life provided that proper lubrication is maintained and offset loading is avoided. The machine is not infinitely rigid. Ram tipping under normal operation may result in high localized loads and accelerated wear. Ball and roller bearings are not adjustable to take up wear. Maintenance costs involving bearing replacement can be very high.

The plain bearings used in larger presses depend on a film of oil to avoid metal-to-metal contact. Moving bearing surfaces supplied with filtered oil under pressure develop both hydrostatic and hydrodynamic oil film lubrication. The metal surfaces do not come into contact with each other. This lubrication method is used for many critical high-speed bearings. The automobile engine is a familiar example of plain bearings used in a very demanding application.

Plain bearings have advantages in high-speed presses. Like the automobile engine, clean filtered oil under pressure is always supplied to each bearing. So long as the film of oil is maintained, essentially zero wear occurs. Plain bearings will withstand substantial shock loads without metal-to-metal contact occurring. Another advantage over ball or roller bearings is that eventual press rebuilding costs are much lower.

EXAMPLE OF A MODERN HIGH-SPEED PRESS

Figure 7-1 illustrates many key features of a modern high-speed straightside press. The massive one-piece frame (1) is of cast iron alloy construction. The mass and damping capacity of the frame is important to reduce the overall press vibration level.

A combination hydraulic clutch and spring-applied brake (2) provides rapid engagement and stopping. Achieving a reduced stopping angle at high speeds is a necessity if electronic tooling protection is to be effective. An automatic flywheel brake speeds up power lockout to permit rapid access for work in the die area.

Hydraulic preload is applied to the slide adjusting screws to remove all clearance. This feature (3) eliminates what otherwise could be a significant source of process variability. A hydraulic cylinder at each connection supplies pressure for hydraulic slide lockup. The hydraulic system (4) lifts the slide to a fixed open position to provide access to the die for inspection, troubleshooting, and threading of stock. The hydraulic system returns the slide to the original shut height position against fixed mechanical stops.

Slide alignment (5) is provided by two large-diameter hydrostatically guided pistons. Pressurized oil is supplied to these bearings as well as the large wrist pin bearings. In addition four hydrodynamic guide posts (6) are provided between the bed and slide at the material pass line level.

Motorized shut height adjustment (7) is a standard feature. A mechanically driven shut height indicator is provided to assist in making accurate shut height settings.

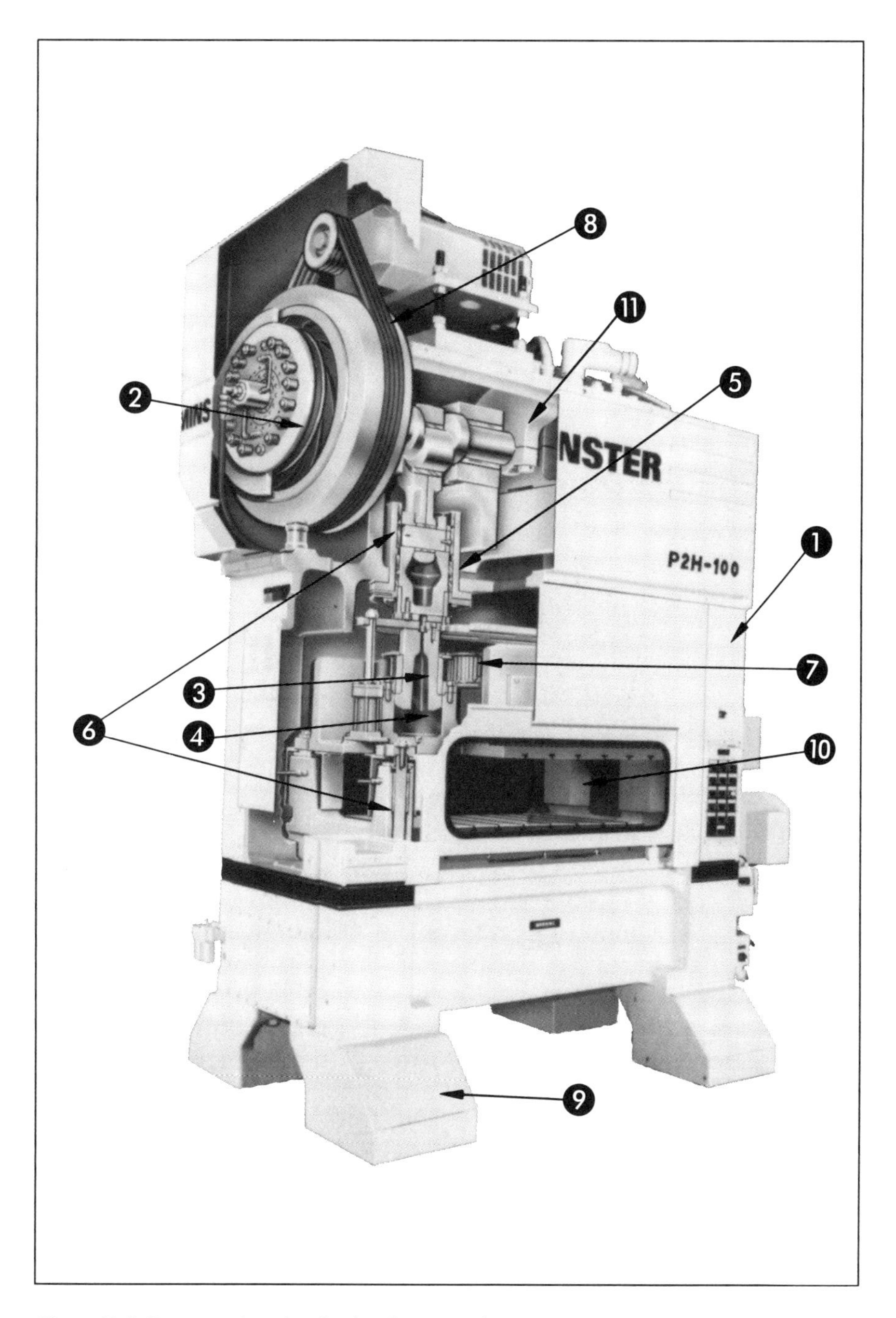

Figure 7-1. *Cutaway view showing key features of a modern high-speed press. (The Minster Machine Co.)*

The press drive motor (8) is equipped with an eddy-current variable speed drive. The motor is totally enclosed and fan cooled.

The press feet (9) can be factory equipped with integral press shock mounts. The press is designed to accept an integral lift-type sound enclosure (10) which is both mechanically and electrically interlocked. A counterweight (11) on the eccentric shaft provides dynamic balance.

The combination of an inherently vibration-damping one-piece frame, shock mounts, and acoustical enclosure greatly reduces sound and vibration levels in the area. Other equipment may include quick die change bolster rollers and die clamps (Figure 7-2).

HIGH-SPEED PROGRESSIVE DIE OPERATIONS

A high-volume application for high-speed presswork is the production of electrical motor and transformer laminations. Most electrical steels contain a high percentage of silicon in order to minimize energy losses. In addition, a hard oxide layer is present on the material which acts as an electrical insulator. High-silicon lamination steels are very abrasive. High-volume production requires precision tungsten carbide tooling and accurate press alignment.

Combined Rotor and Stator Lamination Die

Figure 7-2 illustrates a progressive die to stamp a rotor lamination 2.813 inch (71.45 mm) in diameter, and a mating stator lamination 4.875 inch (123.82 mm) in diameter.

In the first station, a number of holes are punched. These include two pilot holes, the center hole with an alignment notch, the electrical winding holes in the rotor, and an alignment notch for the stator. In station two, the rotor is blanked from the strip. The conductor holes for the stator are punched in the next station.

Station four is an idle station where only pilots engage the strip. This station ensures good alignment and provides for stronger die constriction. Finally, the stator lamination is blanked from the strip in the last stage.

The two blanking punches have a carbide ring fastened to a steel body. All punches and cutting edges are solid or inserted carbide. Hardened-steel retainers and hardened-steel backing plates are used to hold and support all dies and punches.

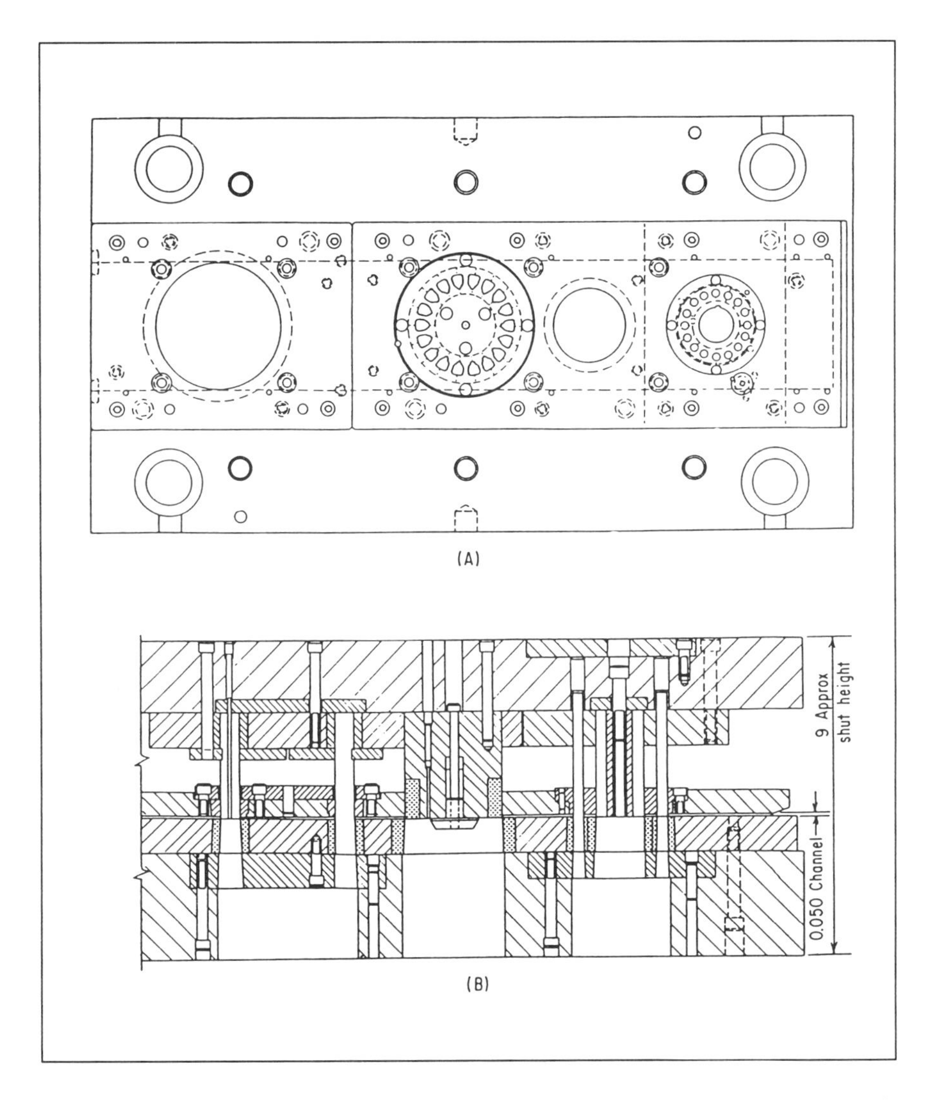

Figure 7-2. *Carbide progressive die to produce one electrical motor rotor and stator lamination per stroke. A plan view (A) is shown. A sectional view of the first three stations (B) illustrates important construction features. (Harig Manufacturing Corp.)*

A tunnel or fixed stripper strips the punches from the stock. This type of construction is less costly than a spring-loaded stripper. The use of fixed strippers is satisfactory, if stock control or punch alignment is not a problem. However, a precision guided movable spring-loaded stripper is often preferred for this type of die.

Multiple Parts Per Hit

Dies that produce more than one part per hit have many advantages over a single out die. These include:

- Higher productivity;
- Production of symmetrically opposite or complementary parts at the same time;
- Balanced die loading to avoid ram tipping or lateral thrust;
- Improved material utilization, especially in the case of round blanks or parts.

Example of a Five Out Progressive Die

The progressive die illustrated in Figure 7-3 produces five spark plug seat gaskets per stroke. The strip pattern illustrated has alternate rows of three and two gaskets across a strip of 0.020 inch (0.51 mm) thick AISI-SAE 1008 sheet steel 5.75 inch (146 mm) wide.

The strip advances 1.25 inch (31.8 mm) per stroke. The sectional view illustrates key features of good die design. An idle station between each operation allows space for robust die section construction. Most stations have a spring-loaded pressure pad attached to the upper die shoe for accurate stock control.

In station one, the blanks are cut from the strip except for a small tab on each side attaching them to the carrier skeleton. Station two draws the blanks into cups and a bead is formed around them in station three. The flat bottoms are punched out of the cups in station four. The radiused portion of the cup bottoms is flanged into straight tubes in station five.

A curling operation is performed in station six, which is closed tightly together in station seven. Finally parts are cut off and dropped through the lower die shoe in station eight. All the punches and the die inserts are backed up by a 0.375 inch (9.52 mm) thick hardened and ground steel plate. Spring-loaded lifters push the part out of the die cavities. The stock guide rails are spring-loaded to hold the strip up when it is advanced to the next station.

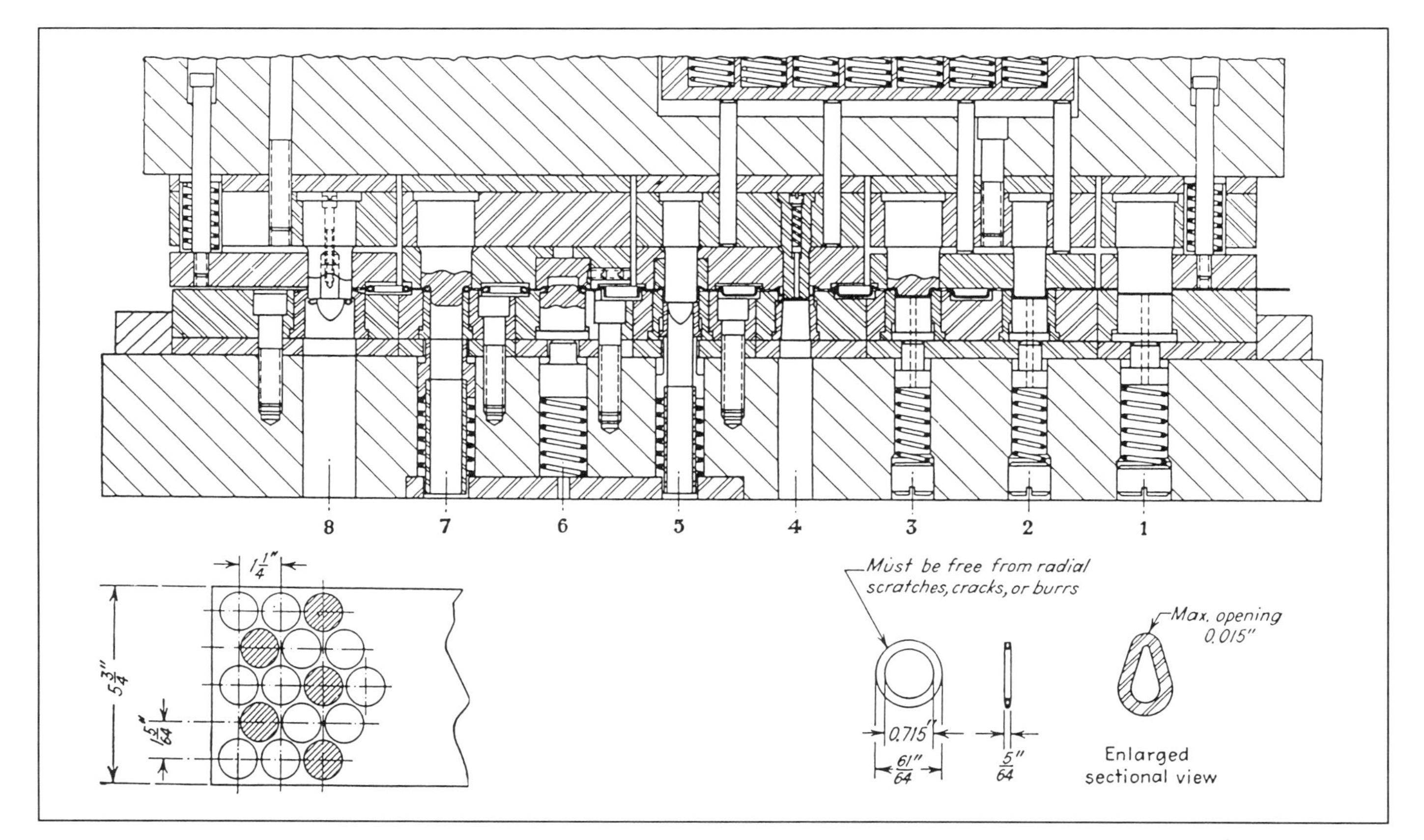

Figure 7-3. A progressive die for producing spark plug seat gaskets. (AC Spark Plug Division, General Motors Corp.)

BIBLIOGRAPHY

1. D. Smith, *Die Design Handbook*, Section 20, "Tools for Multiple Slide Forming Machines," Society of Manufacturing Engineers, Dearborn, Michigan, 1990.
2. D. Hemmelgarn, *Flexibility in the Stamping Room*, The Minster Machine Company, Minster, Ohio.
3. D. Hemmelgarn and D. Schoch, *Tuning Your System to Increase Productivity*, The Minster Machine Company, Minster, Ohio. A patented method for measuring vibration severity as a process control tool is described.

CHAPTER 8

TRANSFER PRESS AND DIE OPERATIONS

Transfer presses have several distinguishing features that suit them for many types of medium- to high-volume work. Most operations use precut blanks, although there are combined operations in which the first station is a coil-fed blanking die.

Many different sizes of transfer presses are used. The first type of transfer press, the eyelet machine, was originally designed to make small metal eyelets for shoes. Today some small transfer presses are still called eyeleting machines, although a great variety of parts are produced on them.

Large transfer presses have force capacities of 3,500 tons (35.136 MN) or more. Transfer press operations have several common factors. These include:

- Individual dies used for each operation;
- Reciprocating transfer feed bars on each side of the press fitted with fingers that move the parts between the dies; and
- Feed bars synchronized with the press ram motion.

Figure 8-1 illustrates a very large specialized transfer press. The automotive and appliance industries are the principal users of this type of machine. All stamping operations required to complete large parts as automotive hoods and roof panels are done in such presses. Flat blanks are destacked and automatically fed into the right end of the press. The transfer feed bar fingers move the parts from die to die. Completed stampings emerge from the left end where they are placed in storage racks or conveyed to the assembly operation.

TRANSFER FEED MOTION

Two types of transfer motion are used. The simplest system uses dual-axis motion. Only in-and-out motion is used to grasp the part. The second axis of motion transfers the part from die to die.

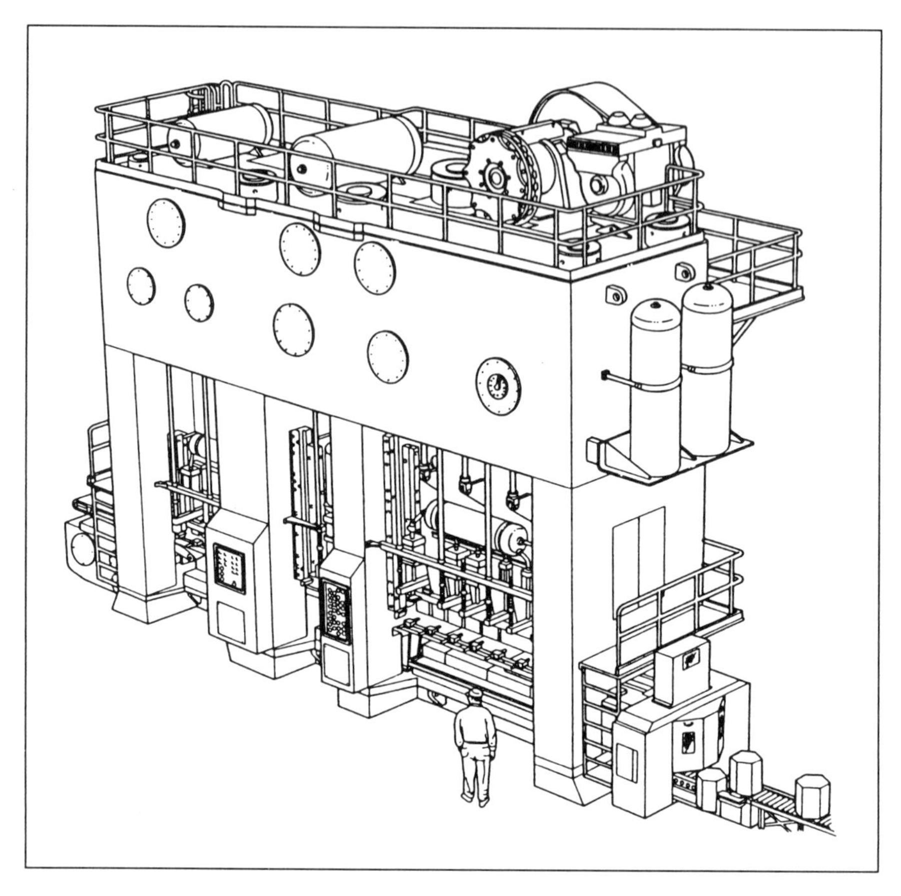

***Figure 8-1.** A four-column, three-slide transfer press with a force capacity of 4,600 tons (40.922 MN). (Verson Corporation)*

Figure 8-2 illustrates the motion of a tri-axis transfer feeder bars for indexing parts between dies in the transfer press. In older designs, the transfer feeder bars are mechanically driven synchronously with the slide motion. The fingers inserted into the transfer feeder bars hold the parts during indexing. In some cases the fingers use pneumatic jaw clamps to grasp the parts. Wherever possible, simple scoops that rely upon gravity are used.

The application of dual-axis transfer feeder bar motion is limited to relatively flat parts having only shallow formed features. The lack of up-and-down motion results in the parts being dragged across the top of the lower die surfaces when transferred. The system works very well within these limitations. The advantages,

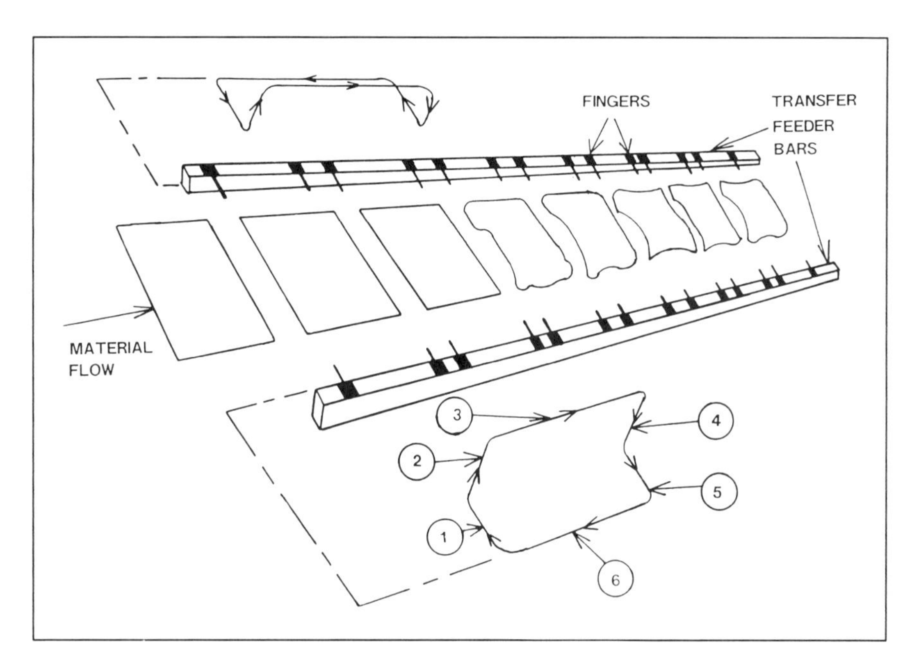

Figure 8-2. Transfer press feeder bars and fingers showing motion diagram sequence: (1) clamp (catch part); (2) lift up; (3) advance; (4) lift down; (5) unclamp; (6) Return. (Auto Alliance International)

compared to a tri-axis system, are lower initial cost, less maintenance expense, and faster cycle times.

Tri-axis transfer is needed for parts having deeply formed features. Here, the part must be lifted out of the die cavity before being transferred to the next die.

Transfer Drive Methods

There are several basic systems for actuating transfer feed motion. The original method, which is still in use, is to drive the transfer directly from the press crankshaft. Gears and cams transform rotary motion to the reciprocating action needed for part transfer.

Another mechanical drive system popular for both new systems and retrofit applications uses plate cams attached to the press ram. The mechanism is driven by cam follower rollers.

Both pneumatic and hydraulic cylinder driven systems are built into multistation dies. Many cleaver designs have been locally fabricated in press shops. The hydraulic systems are more costly, but provide precise control. The motion of pneumatic systems tend to require frequent adjustment.

Fitting hydraulic transfer mechanisms to existing dies is a well-developed technology. Success factors include the use of electronically controlled hydraulic servo-systems that have precise programmable motion. Many new transfer designs are powered by electrical servo motors. In the author's opinion, this is the best technology for most new designs. The technology includes a combination of a high-output servo motor, microprocessor-based electronic control, and solid-state power supplies. These are all mature technologies that are in widespread use in other industrial equipment such as machine tools and robotics.

EXAMPLES OF TRANSFER PRESS OPERATIONS

Swing-Out Transfer Bar Carrier Assembly

When a transfer mechanism is retrofitted to an existing press, exchanging dies can be complicated by the need to remove one or more feed rails or transfer bar carrier assemblies. The feed rail or transfer bar carrier assembly can also make minor die maintenance work in the press very difficult.

Swingout transfer bar carrier assemblies can be employed to permit easy die maintenance and changeover. These patented transfer swingout assemblies can be mounted to the press column, bolster or to a die subplate.

Figure 8-3 illustrates a swingout transfer bar carrier assembly. The dies, used to produce a speaker basket stamping, are mounted on a common subplate assembly having parallels to permit scrap removal by means of in-die conveyors. The scrap is discharged out of the rear of the press. The last transfer station places the finished stamping onto a gravity chute, which conveys the part into a hopper.

Transfer finger part sensing permits the press to be automatically stopped in the event of multiple parts in one station, scrap interference, or loading problems. The transfer is driven by plate cams attached to the press ram. Spring-loaded clutches protect all three motion drives. Electronic proximity switches initiate a press emergency stop in the event that an overload should trip a clutch.

The two bolster outriggers and bolster shown in Figure 8-3 are equipped with hydraulically actuated rollers to permit subplated dies to be changed by a means of a powered die cart or fork truck.

Figure 8-4 illustrates a swingout transfer bar carrier assembly that mounts to a die subplate for transfer dies used to produce

Figure 8-3. A swingout transfer bar carrier assembly provides complete access to the die for maintenance or die removal. The pivoting transfer is mounted to the column of a 220-ton (1957-kn) Niagara 84-inch (2.134 m) wide straightside press. (HMS Products Co.)

stampings of automotive seat parts. The dies are run in a conventional straightside press.

A bolster-mounted swingout transfer bar carrier assembly is illustrated in Figure 8-5. The simulator platform is fabricated of steel plate and carefully machined to duplicate the dimensions of the press bolster.

The HMS transfer systems are popular for retrofitting to existing presses. The dies and transfer fingers must be accurately located. The fingers are doweled in place to assure repeatability. The transfer feed bars may be exchanged to run different products. Electrical servo motor drive is supplied for the die-to-die transfer if the dies are widely spaced.

Multiple Slide Straightside Presses

The straightside press shown in Figure 8-1 is an example of a highly specialized custom-built machine known as a Verson Transmat™. Total machine weight may exceed 2,500 tons (2268

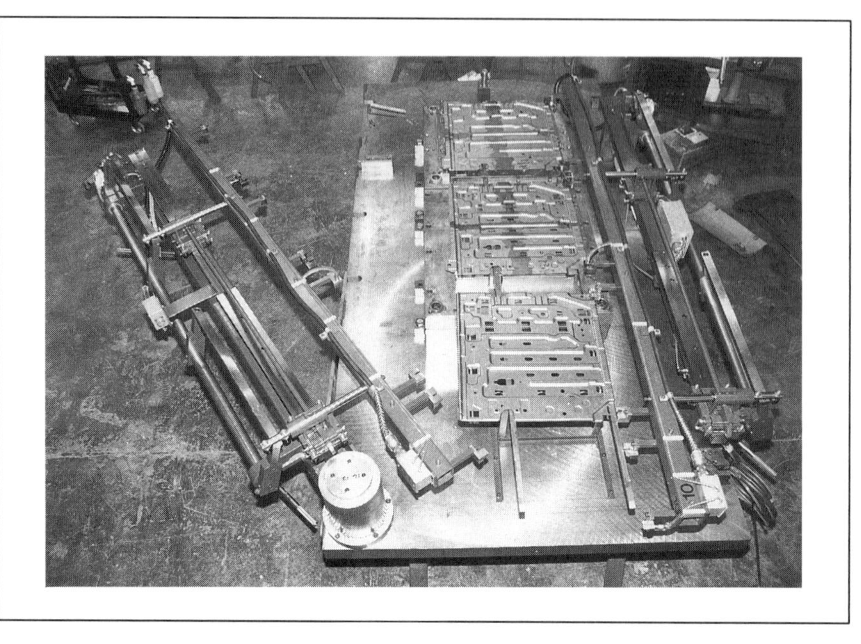

Figure 8-4. A swingout transfer bar carrier assembly that mounts to a die subplate. (HMS Products Co.)

T). The three slides and four columns are customized for the type of work to be performed. Large multiple slide presses have a number of advantages, including:

- By making a machine having a number of columns and slides, shipping the disassembled press over highways with specialized transport trailers is practical.
- Each slide can be designed to provide the required force.
- By using multiple slides, ram tipping can be minimized.

Design Considerations to Avoid Ram Tipping

The Verson Transmat is an old, but dependable, design. New Verson designs are built with electrical servo drives. Verson servo drives can also be retrofitted to older mechanically driven transfer presses. The center slide in the illustration is designed for a heavy single-station forming operation.

Placing a die having high-force requirement, such as a stretch form or reforming operation, on the end of a long press slide can result in severe ram tipping. Die damage, accelerated wear, and

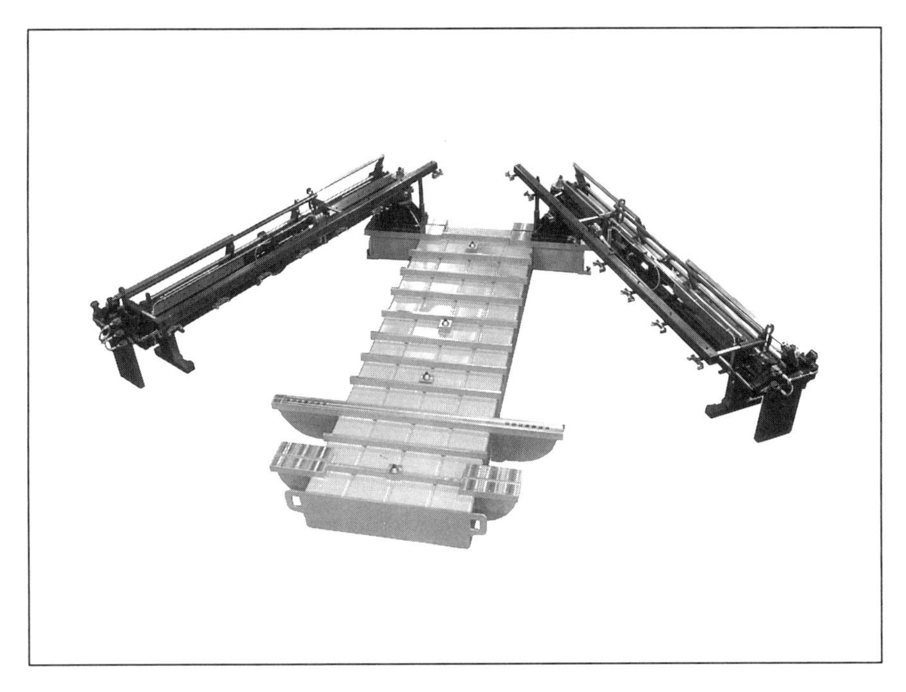

***Figure 8-5.** A swingout transfer bar carrier assembly suitable for bolster mounting. It is attached to a simulator platform used to tryout the completed transfer assembly with the lower dies in place. The cone-type die locating devices used to ensure correct die positioning are seen in the center of the simulator platform. (HMS Products Co.)*

quality problems can result. This is often an unexpected problem in applications where large single- and double-slide transfer presses are specified.

The load on the slide should be as balanced throughout the press stroke. This is an important consideration in the Verson design. Some builders and users may have been unaware of this problem. When troubleshooting transfer press problems, ram tipping is sometimes found to be the cause of poor quality work and excessive press or die wear. A solution may be to add a compensating load on the lightly loaded end of the slide. Nitrogen cylinders or rubber die springs may prove successful to provide the counterforce needed to balance the load.

Automatic Die Change At Auto Alliance

Auto Alliance is an integrated automotive stamping and assembly facility located in Flat Rock, Michigan. Jointly owned by Mazda

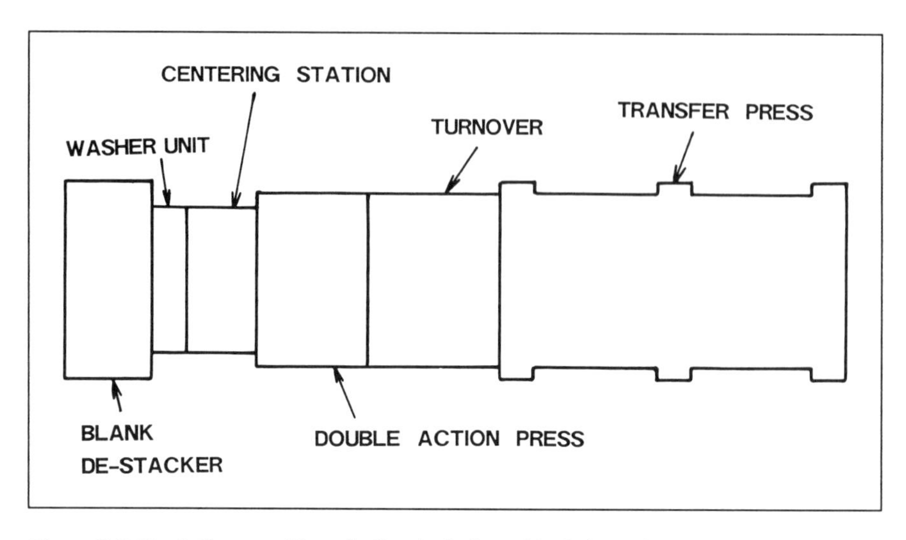

Figure 8-6. *Block diagram of transfer line, including a blank destacker, a double action press with a mechanically linked loader and extractor, a part turnover, and a 2649 ton (23.565 MN) transfer press. (Auto Alliance)*

and Ford Motor Company, the company has large transfer presses to produce stampings that are assembled into Ford and Mazda automobiles.

To reduce inventory storage costs, automatic die change (ADC) is used to enable short part runs. Figure 8-6 illustrates a stamping line designed for ADC.

The events that occur during an ADC follow an exact sequence. Briefly, both the loader and extractor are fully retracted and the die cushion is lowered. All press slides are lowered to bottom dead center and the upper die clamps are released. The slides then rise to top dead center and the computer selected die height adjustments are made for the new job to an accuracy of ±0.004 inch (0.10 mm), while the new loader mechanism cups and jaws are selected automatically.

Next, the moving bolster unclamps and the safety gates rise. Once the safety gates are locked in their up positions, both moving bolsters with dies in place simultaneously move out of and into the press. The bolster is clamped hydraulically to secure its position inside the press, all pneumatic lines are pressurized, and the safety gates close.

The slides automatically lower to the present die heights and stop at bottom dead center to permit the upper dies to be automatically clamped. The slides are then raised to the home position and both the loader and extractor are repositioned. Finally, the die cushion is pressurized, bringing it to the operating height. This completes the automatic die change sequence for the double action press.

Moving Press Bolsters

Two moving bolsters are supplied for each operation, which include the double action press, turnover, and both transfer press slides. This layout allows the external staging of each operation while production is still running. Once a die change is required, the moving bolsters simultaneously move the new job into the press as the old job moves out. Figure 8-7 illustrates how this is done.

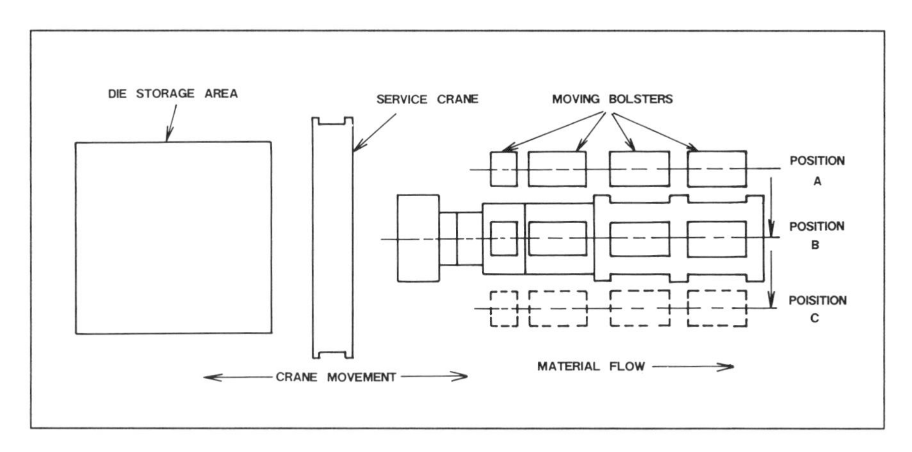

Figure 8-7. *Layout of die storage area in relationship to transfer press line equipped with air moving bolsters. Two moving bolsters are supplied for each operation. As the bolsters at position A move into the press, the bolsters inside the press (position B) move out of the press to position C. The die change is completely automated, and requires less than five minutes. (Auto Alliance)*

Without the ability to externally prestage the moving bolsters, automatic die changes would not be possible. There are four elements involved in external prestaging:

- Installing die locating pins, die cushion pins, and die lifter pins;
- Setting dies and the lower hydraulic clamps;
- Installing the fingers; and
- Installing the turnover fingers.

Transfer Feeder Bar Changeover During ADC

After the transfer press die cushion and die lifters are automatically lowered, the transfer feeder bars are moved to their maximum width setting, disconnected, and lowered to a rest stand for solid support during bolster movement.

As illustrated in Figure 8-8, each transfer feeder bar can be separated into six pieces. The short pieces, known as bar connectors, are in the area between the uprights of the press and remain in the press. The other three pieces move in and out of the press on the moving bolsters for each slide.

Figure 8-8 illustrates the transfer press feeder bar layout. Many automatic functions must work smoothly to ensure automatic bolster movement and feeder bar exchange during ADC. Success factors that ensure correct ADC include:

- Strict procedures to ensure that all employee prestaging activities are performed correctly;
- Precise preprogrammed machine instructions;
- Closely followed equipment maintenance schedules; and
- Training that turns all employees into a team.

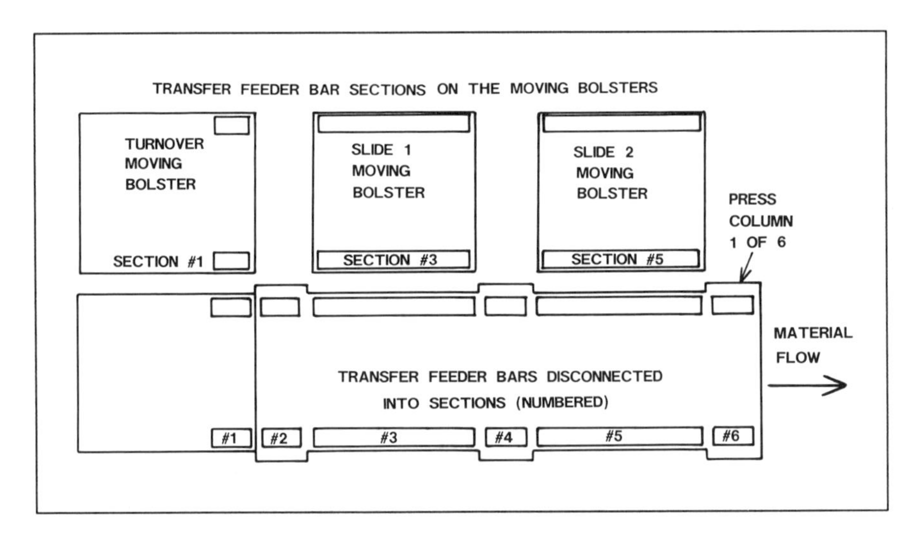

Figure 8-8. Top view of transfer press layout illustrating transfer feeder bars. The bars are uncoupled, the bolsters moved and the new bars recoupled by pneumatic cylinders during ADC. (Auto Alliance)

Employee Training

Great emphasis must be placed upon employee training. Proper training and education is essential to operating and maintaining a large complex transfer press operation. All employees are taught what to recognize, understand, and watch closely.

Program work sheets outline the sequence of work elements that achieve each task. Special instruction sheets are available when an unplanned occurrence or condition arises requiring physical interaction not covered in the program work sheets. This generally relates to some form of mechanical or electrical problem encountered.

Class instruction and examinations are given to prove competency before an employee is certified at a given job. Extensive training programs such as this provide Auto Alliance with a well trained, educated, and versatile work force. Without such training, large transfer press operations cannot be carried out safely, production schedules cannot be maintained, and quality parts cannot be produced.

DIE DESIGN FOR TRANSFER PRESSES

With a few important exceptions, dies used in transfer presses are designed much the same as tooling for individual and tandem press line operations. Important design considerations for successful transfer press operation include:

- All dies maintained at common pass or load height;
- Guide pins, heel block projections and set-up blocks incorporated in the upper die to avoid interference with the transfer fingers;
- All dies under the transfer press slide maintained at a common shut height; and
- Clearance in the dies for the transfer press fingers to pickup and place the parts.

The motion path of the transfer press fingers should be available to the die designer. It is essential that any interference problems be corrected in the die design process.

If die components such as cam slides and part locators are found to interfere with transfer finger motion during new die tryout, the required modifications will be costly. Manual transfer parts may be required that can involve expenditures to meet point-of-operation safeguarding requirements and inefficient operation.

Establishing and maintaining a common shut height for all dies under the slide is an absolute requirement. The total press deflection under load including bed and slide bowing must be compensated for in determining the actual shut height for each die.

BIBLIOGRAPHY

1. C. Wick, J. T. Benedict, R. F. Veilleux, *Tool and Manufacturing Engineers Handbook*, Volume 2, Fourth Edition — Forming, Society of Manufacturing Engineers, Dearborn, Michigan, 1984.
2. D. Smith, *Quick Die Change*, Chapter 21, "Automatic Die Change at Mazda," Society of Manufacturing Engineers, Dearborn, Michigan, 1991.
3. J. Hoening, *Fitting of Transfer Mechanisms to Existing Dies*, SME Technical Paper TE87-790, Society of Manufacturing Engineers, Dearborn, Michigan, 1987.
4. D. Smith, video training series, *Quick Die Change*, Society of Manufacturing Engineers, Dearborn, Michigan, 1992.
5. D. Smith, *Adjusting Dies to a Common Shut Height*, SME/FMA Technical Paper TE89-565, Society of Manufacturing Engineers, Dearborn, Michigan, 1989.
6. D. Smith, *Quick Die Change*, Chapter 11, "Operating Dies at a Common Shut Height," Society of Manufacturing Engineers, Dearborn, Michigan, 1991.

CHAPTER 9

PRESS INSTALLATION AND SENSING SYSTEMS

INSTALLING AND LEVELING THE PRESS

Press sizes range from small bench presses that can be carried with one hand to multiple-slide transfer presses weighing over 1,500 tons (1361 metric tons). No matter the size, however, achieving long trouble-free service starts with proper installation of the machine.

No mounting method is correct for every application. Small gap-frame open back inclinable (OBI) and open back stationary (OBS) presses do not require a critical foundation alignment system. Usually, placing them on felt or rubber isolating pads on a sound reinforced concrete floor is sufficient (Figure 9-1). The felt pads compensate for slight floor irregularities that can cause machine misalignment. Mounting pads also reduce vibration transmission to the floor and prevent lateral press movement.

Straightside presses and large gap-frame presses (Figure 9-2) have more critical alignment requirements in that any irregularities in the mounting surfaces can be transmitted to the machine bed. Also, presses subjected to snap-through loads from cutting thick metal have more critical mounting requirements than similar machines used for drawing or embossing.

If the press is placed on a reinforced concrete floor, the thickness and strength of the concrete are important factors. However, the soil condition under the concrete is also a very important consideration. Well-drained undisturbed clay over sound bedrock can support heavy loads. Foundations placed on bedrock are ideal. Building a pressroom over unstable fill or wet soil can require costly measures to ensure a proper press foundation. Seeking the advice of an engineer who specializes in designing foundations for heavy machinery can avoid costly errors.

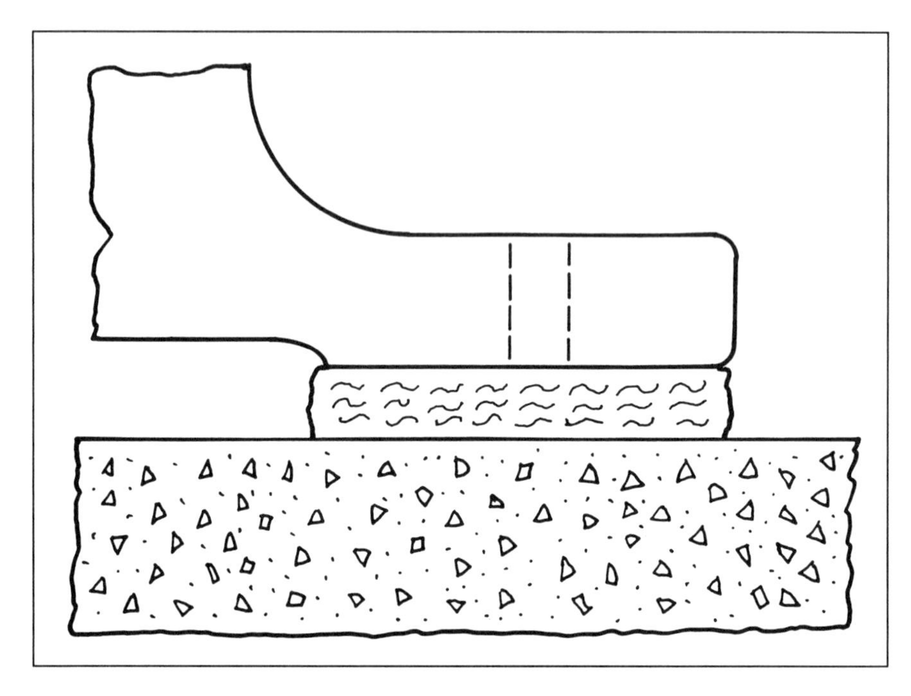

Figure 9-1. *A press foot resting on a felt pad. (Smith & Associates)*

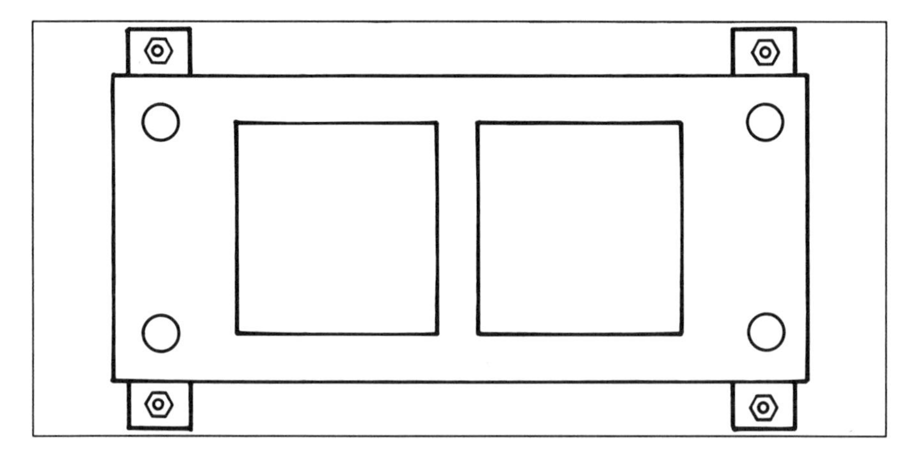

Figure 9-2. *Plan view of a typical straightside press bed. The bed itself is made of gray iron, iron alloy, cast steel or welded plate construction. Four round openings are provided for the tie rods that hold the entire press assembly in compression. Openings are provided in the press bed for die cushions or the discharge of scrap. The press bed is supported on the foundation by four feet. Large multiple slide presses may have two feet per press column, so such a machine can have six or more mounting points. (Smith & Associates)*

Placing the Press on the Foundation

Figure 9-3 illustrates an end-view of a straightside press bed resting on a thick concrete floor. If there is no provision for anchoring the press to the floor, lateral motion or walking (Figure 9-4) of the press may occur. Another problem is that all four mounting feet may not rest solidly on the floor (Figure 9-5), resulting in one corner being unsupported. Press beds should be level and free of skewing or twist if high-quality presswork is to be accomplished.

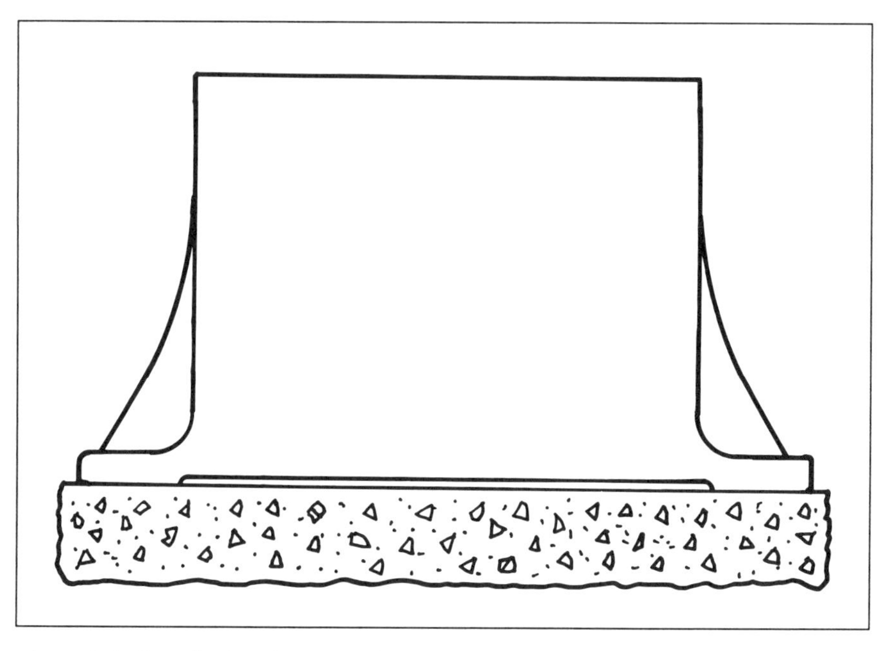

Figure 9-3. Press bed resting on a thick concrete floor. (Smith & Associates)

The terms *level* and *twist* or *skew* are often confused. Level refers to a horizontal condition, and is not an absolute requirement for correct machine operation. However, twist or skew in the press bed results in misalignment of the machine members. If the bed is skewed, the alignment error is transferred to other machine members and the die as well. The clearance in machine parts such as gibbing, gears, and bearings will be changed, possibly resulting

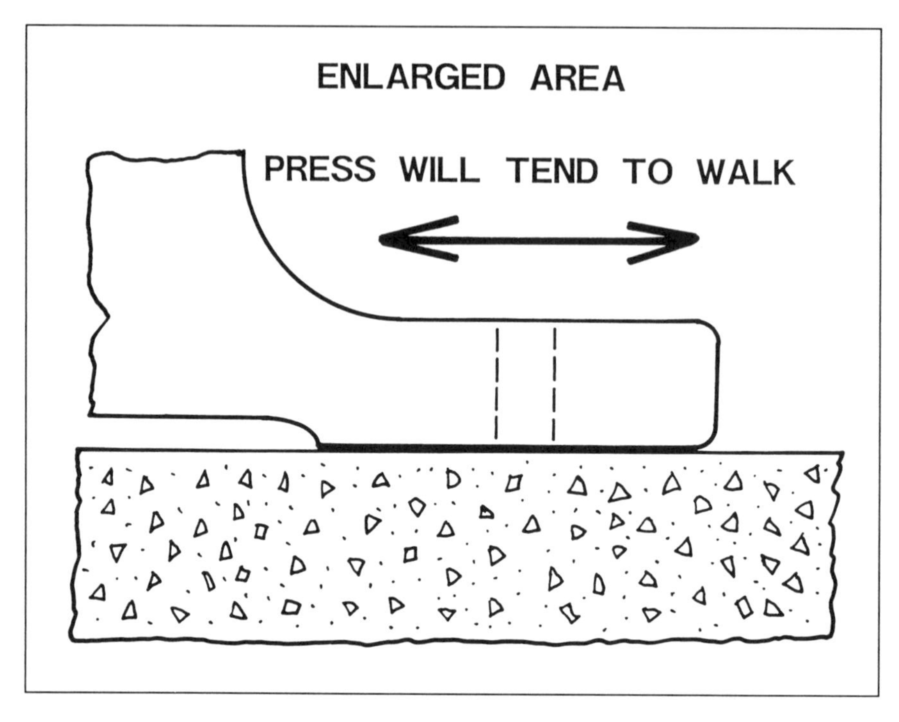

Figure 9-4. If press beds are not anchored to the floor, lateral movement or walking may occur. (Smith & Associates)

in binding machine parts and die misalignment. The end result is accelerated press and die wear, as well as quality problems with the stampings.

Figure 9-6 illustrates a precision machinist's level. It differs from a carpenter's level in that it is much more precise and can be easily calibrated in the field. The basic accuracy of the level is 10 seconds of an arc per division.

Placing the instrument on a level surface plate or coordinate measuring machine table will serve to test both the level and test surface for true level adjustment. First, place the level on the test surface and note the position of the bubble. If it is in the center, turn the level end-for-end. If the bubble remains exactly in the center, the level is correctly adjusted and the test surface is level as well.

If the bubble moves in the same direction an equal amount when turned, the level is correctly adjusted, but the test surface is out-of-level by the amount indicated. Should the bubble not move equal amounts, the cover screw can be removed to gain access to

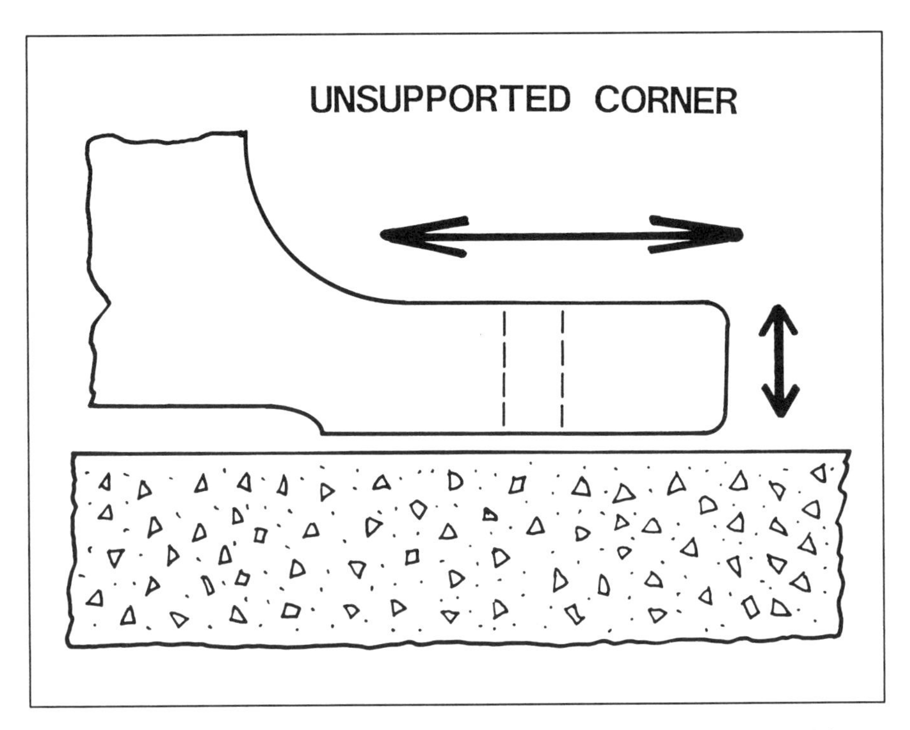

Figure 9-5. *Placement of a straightside press on a concrete foundation may result in one mounting foot being unsupported. (Smith & Associates)*

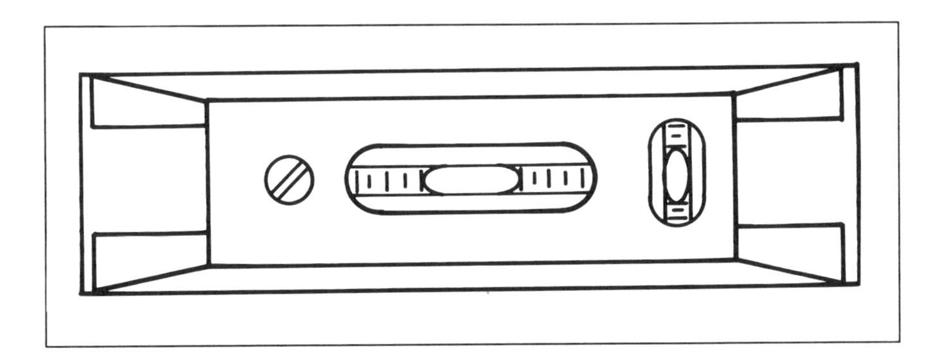

Figure 9-6. *A precision machinist's level. The base is made of iron alloy that has been carefully processed for stability and scraped flat. The top cover of the level is made of a plastic insulating material to minimize body heat transfer to the iron base during handling. The main level vial is precisely ground and graduated in divisions of 0.0005 inch per foot, or 0.042 mm per meter. An auxiliary vial shows lateral position and assists in horizontal setting of the machine bed. (Smith & Associates)*

the adjustment screw. Turn the screw a slight amount and again turn the level end-for-end while observing the result. When the bubble offset is equal when turned, the level is correctly adjusted. The test surface can then be leveled, and the machinist's level double-checked for exact adjustment.

Testing and Adjusting the Press Bed

Figure 9-7 illustrates a machinist's level placed on a straightside press bed. The level at the bottom of the figure indicates a level condition, while the one on the other side of the bed is out-of-level,

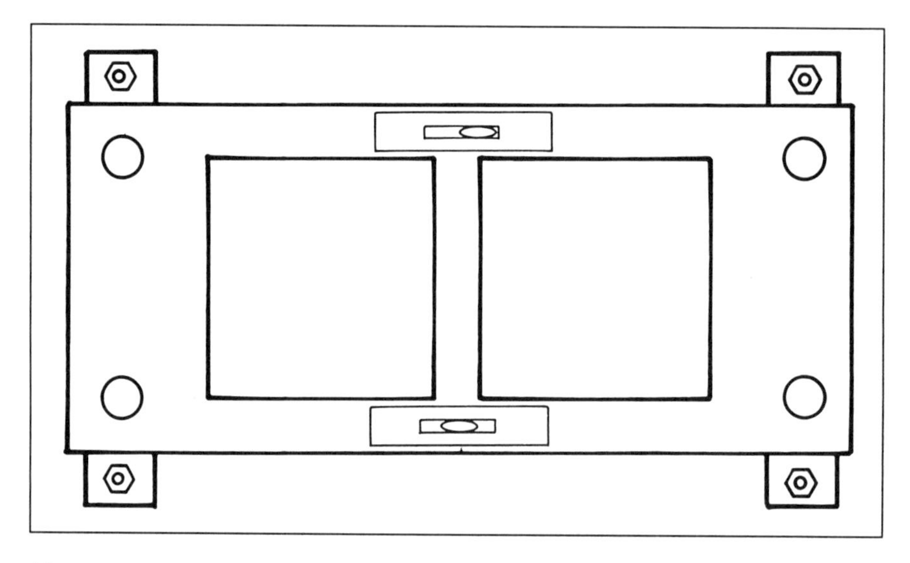

Figure 9-7. The level at the bottom (front) indicates that the bed is level left to right while the one at top (rear) is out-of-level indicating a skewed press bed condition. (Smith & Associates)

indicating a skewed condition. Figure 9-8 is a closeup view of the second level.

If the press is solidly mounted to its foundation, shims may be used to correct a skewed condition (Figure 9-9). The object is to remove any skew observed in either side of the press bed. The bed should be free from skew with equal weight resting on each support.

Some press beds may have a skew that cannot be corrected by shimming. The amount of inaccuracy should not exceed manufacturer's recommendations. For presses intended for close

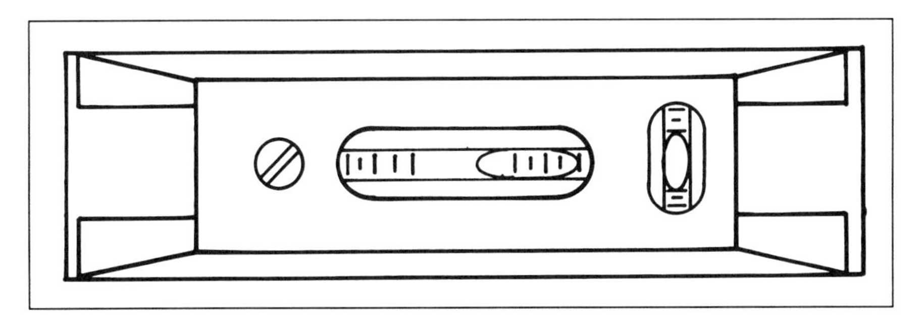

Figure 9-8. A close-up view of the top (rear) level in Figure 9-7: to correct the skew, shims must be added under the left rear press support. (Smith & Associates)

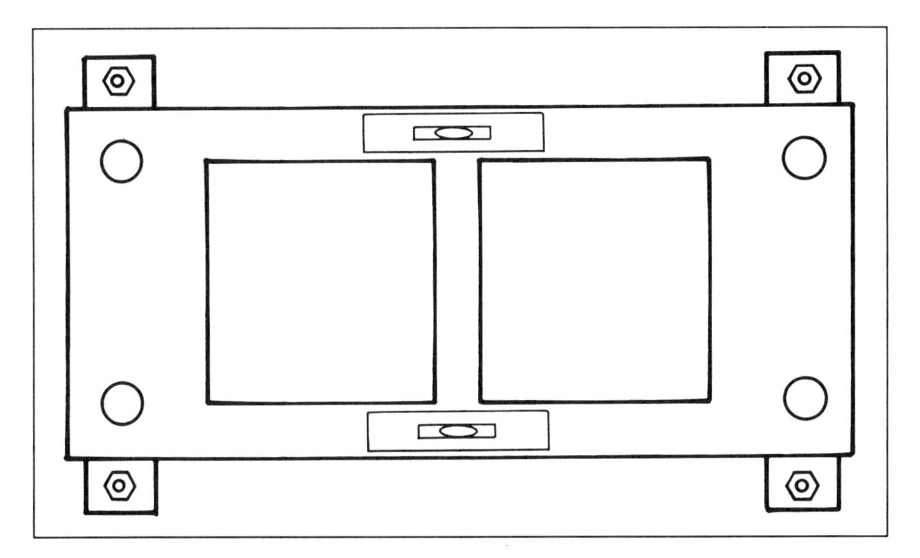

Figure 9-9. The press bed is level on both sides after the skew is corrected. (Smith & Associates)

work, the skew should not exceed 0.0005 inch per foot (0.04 mm per meter).

Large skews that cannot be corrected may have resulted from inaccurate machining, damage due to severe overloading, or stresses created during the fabrication process. New welded press components, as well as major welded repairs, require normalization to relieve any stresses prior to machining.

If steel shims are placed under a press foot, they tend to shift out of place as illustrated in Figure 9-10. To prevent this, anchor bolts

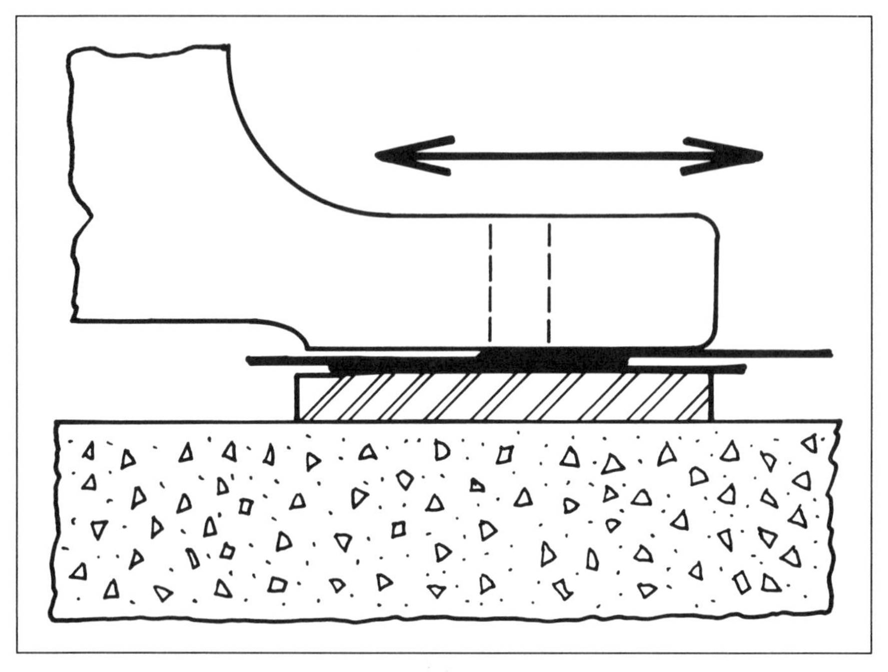

Figure 9-10. Steel shims placed under a press support point tend to shift out of position. (Smith & Associates)

grouted into the floor (Figure 9-11) should be used to secure both the press and shims.

Large presses with dedicated foundations may have steel plates in the concrete to create a firm level surface for mounting the press. This avoids extreme load concentrations on a small area of concrete and provides a smooth surface for placing leveling shims.

Adjustable Resilient Mounting Devices

There are some important advantages in using resilient adjustable mounts (Figure 9-12) for press installations. The resilient material reduces the transmission of shock and vibration to the mounting surface. The adjustment screw permits rapid leveling of machines weighing up to several hundred tons.

Jacks and jacking points are required for conventional shim placement. Jacks may also be used to permit the resilient mounting device screw to be adjusted on very heavy machines. Hydraulic lift cylinders are built into some of the larger mounts, a patented

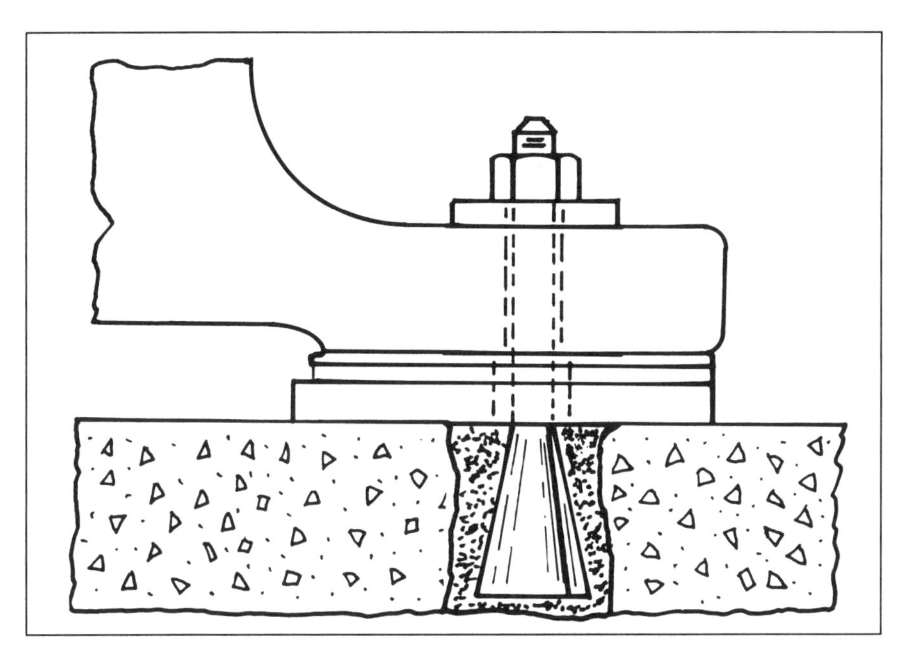

Figure 9-11. *Securing the press to the foundation with bolts grouted in-place serves to both anchor the press and hold the leveling shims in position. (Smith & Associates)*

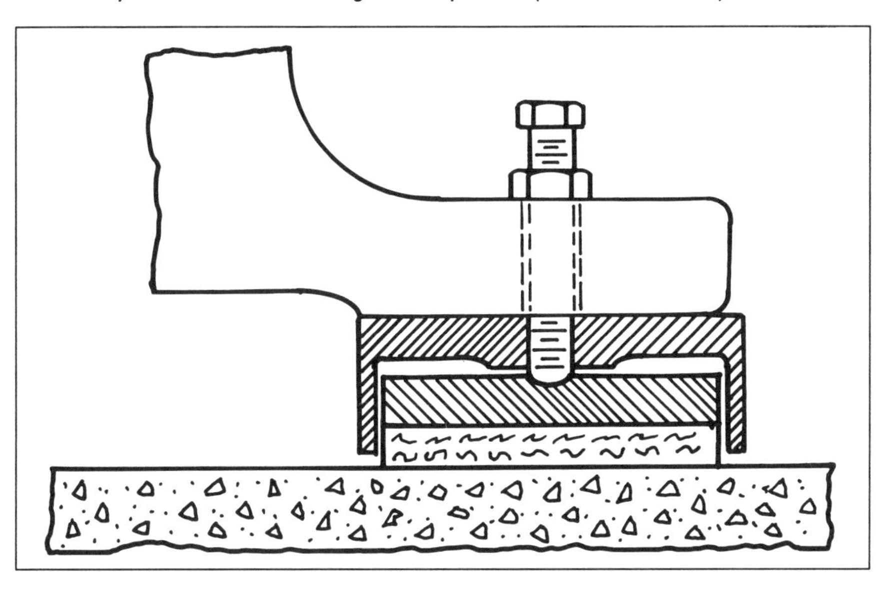

Figure 9-12. *Here is a section through a press mounting device housing under a press foot. The housing contains a resilient substance such as rubber or felt. A screw and lock-nut transmit the press weight to metal plate which spreads the load of the press foot over the resilient material resting on the press foundation. (Smith & Associates)*

feature that permits the screw to be easily adjusted when leveling very large machines.

Another patented feature is strain gages attached to the leveling screw to measure the actual weight supported with an external readout device. This system aids in adjusting the mounts for equal load sharing. The expense of this system can often be justified since it speeds press installation, especially in the case of very large machines.

A potential disadvantage is that large presses on flat concrete flooring are raised so high, operators need stands to reach the press controls. The height is also a problem in retrofitting existing tandem lines. Of course, new installations with press pits can easily accommodate the mount height called for in the planning stage.

PRESS FORCE MONITORS AND LOAD MEASURING SYSTEMS

Maintaining pressworking processes within the design capacity of the machine can be accomplished by careful analysis of the force requirements for each job by the tool engineer. For jobs that require the greatest force at bottom of stroke, operating loads can be computed from the amount that the shut height is reduced beyond initial contact, provided that the amount of press deflection per unit of force is known.

There are actually pressworking operations so precise that anyone setting up or operating the machine must have a mechanical engineering degree. However, the vast majority of presses are set up and run by persons with no formal engineering training, having learned the job by experience.

Electronic force monitoring is an excellent process setup and control tool, as well as an aid to achieving setup repeatability and troubleshooting the process. Training in theory and operation of the monitoring equipment is essential to achieve maximum utilization.

Force Measurement Historical Development

Measuring operating forces of presses is an old concept. Hydraulic press forces are easily measured directly from the fluid pressure applied to the cylinder. The gage usually indicates both the fluid pressure and the equivalent force based on the piston area.

A method for determining operating forces by measuring the displacement of a straightside press housing is illustrated in Figure 9-13. The housing or column, together with the crown and bed, is held in compression by prestressed tie rods. As the press is cycled

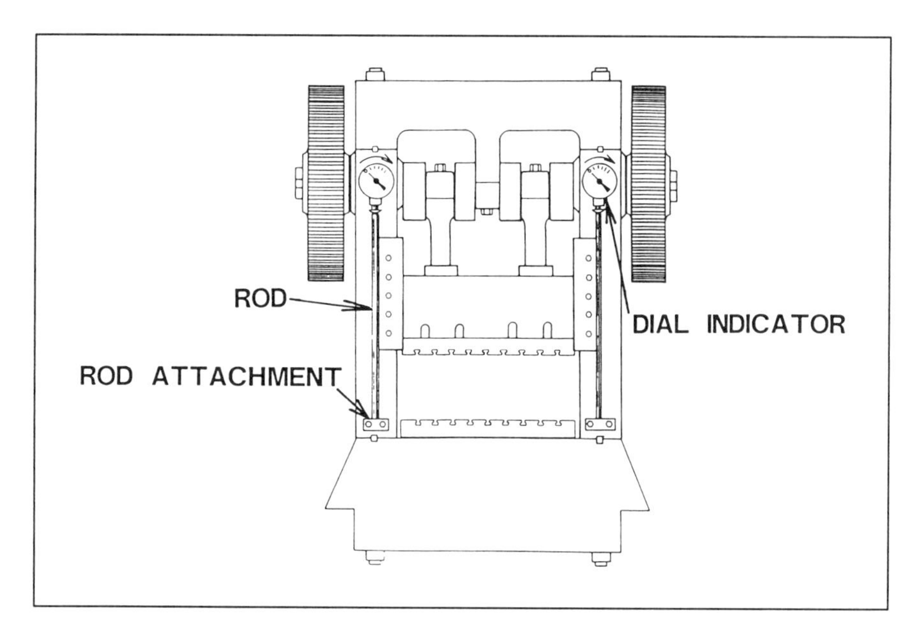

***Figure 9-13.** A dial indicator mounted on a press column used as a mechanical press force indicator. (Smith & Associates)*

through bottom, developing force, the preload on the columns is partially relieved.

Either a dial indicator with built-in mechanical amplification or a vernier scale is used to determine mechanical displacement of the press column. Practical dial indicator systems incorporate a second pointer to retain the peak reading. Vernier scales, provided with a mechanical decoupling apparatus to store the peak reading, can be used. This system was useful for press development and analysis of problem processes. It has been entirely supplanted by electronically amplified strain sensors.

Types of Electronic Strain Sensors

Several types of strain sensors are used in pressworking applications. They provide an electrical signal directly proportional to the mechanical displacement being sensed. The most common is the strain sensor or strain link used for tonnage meter applications.

The strain link has a rugged housing containing a calibrated assembly incorporating strain gages. Generally, tonnage meter installations require one strain link per channel of information.

They must be installed on the press at a location where the change in reading is proportional to the change in tonnage in the die-space. Strain sensors mount by means of screw attachments, and are generally interchangeable with another of the same type without need for recalibration. Several types of sensing elements are in use.

The modern foil strain gage evolved from the resistive wire strain sensor invented simultaneously by E. Simmons and A. Ruge in 1939. Development of the modern foil gage was done, in part, by the Baldwin Lima Hamilton (BLH) Locomotive Company. BLH management, which produced only locomotives, diversified its product line to include both large power presses and tiny strain gages.

Frank Tatnall, a key figure at BLH, joined The Budd Company in 1956 to add the strain gage to their emerging testing machine business. Today, The Budd Company continues to be a leader in applying instrumentation to control the sheet-metal forming process under Dr. Stuart Keeler and Michael Herderich.

The foil strain gage is one of the most widely used components for constructing strain sensors for press tonnage meter applications. The foil, generally a copper-nickel alloy known as constantan, is very thin. It is supported by a thin plastic substrate. The nominal resistance of most gages is 350 ohms. As the gage is stretched or compressed, resistance increases or decreases by a slight predictable amount, due to the change in length and cross-sectional area of the gage foil tracks.

The amount of resistance change per length change is predictable, and termed *gage factor.* Gage factor is defined as

$$GF = \frac{\Delta R/R}{\Delta L/L}$$

To provide a balanced output having high noise immunity and ease of connecting to instrumentation amplifiers, the usual practice is to use four gages connected as a Wheatstone bridge circuit. Two diametrically opposite gages are installed in line with the strain to be measured. The other two are installed at a 90-degree angle to the first set to measure the change in width. These gages are installed on a small metal structure which is housed in a rugged enclosure. This makes up the strain sensor or strain link.

The advantage of semiconductor strain sensors, compared to metallic foil types, is that much greater output (typically 10 or more times that of foil strain gages) is obtainable. Load cells using semiconductor sensors were pioneered in the early 1960s for the United States space program by Bytrex (now Data Instruments).

Strain links for press force monitoring employing silicon semiconductor strain sensors are available with built-in amplification. The high gage factor, together with built-in differential amplification, provides very high output, giving excellent immunity to electrical noise pickup.

A third type of strain link used for press tonnage meters employs a piezoelectric material, such as quartz. The strain link lends mechanical protection and electrical isolation for the piezoelectric material. Also provided is a mechanical mounting connection, to permit attachment to the machine. The output, a voltage proportional to load, is developed across the faces of the sensor.

This sensor requires no external power supply for operation. Its output is a large voltage proportional to strain — a high-level signal with excellent noise pickup immunity. Since the sensor can supply virtually no current, the tonnage meter incorporates a charge amplifier to drive the meter circuitry.

Measuring Press Strain to Determine Force

Accurate force monitoring requires that the strain gages or sensors be located on the press where the strain throughout the machine cycle is equal to the force being developed. In addition to a linear force-to-strain relationship, the location should exhibit large strains and be as free as possible from extraneous mechanical noise.

Long-term stability requires the use of high-quality adhesives. Depending on the part of the press being gaged, it may be necessary to remove the machine from service to allow overnight curing of the adhesive. For this reason, strain links, interchangeable bolt-on strain sensors containing strain-sensing elements, may be used.

Tapped mounting holes are placed at the point of attachment of the strain link. A second method is to attach prethreaded pads by welding. In either case, it is recommended that a drill or welding jig be used to obtain accurate hole spacing. In the event of misalignment, the strain sensor will be distorted, and give a false output. Should this occur, the force indicating meter may not be adjustable to zero output.

Another method is the application of strain gages that are prebonded to a metal carrier. These are attached with a 10 to 50 watt-second capacitive discharge spot welder. Attaching half the bridge circuit to either side of a pitman, or pull rod in the case of an underdriven press, provides immunity to signal errors due to bending or twisting.

The pitman, or eccentric strap, is an excellent location for strain sensor mounting. It is recommended that a strain gage or sensor be applied to each side of the pitman and the sum of the readings used. This will greatly lessen errors due to bending or twisting.

Figure 9-14 illustrates a strain sensor applied to either side of a press pitman. The sensors may be two Wheatstone bridge-type strain links connected in parallel. If the spot-weldable gages are used, two gages, each making up half the bridge, are applied to either side of the pitman.

The latter method is less costly and avoids tapping holes in, or placing weld-pads on, the pitman. Tapped holes and the heat of weld pad attachment are both objectionable because a stress-riser is created that may cause the pitman to fail.

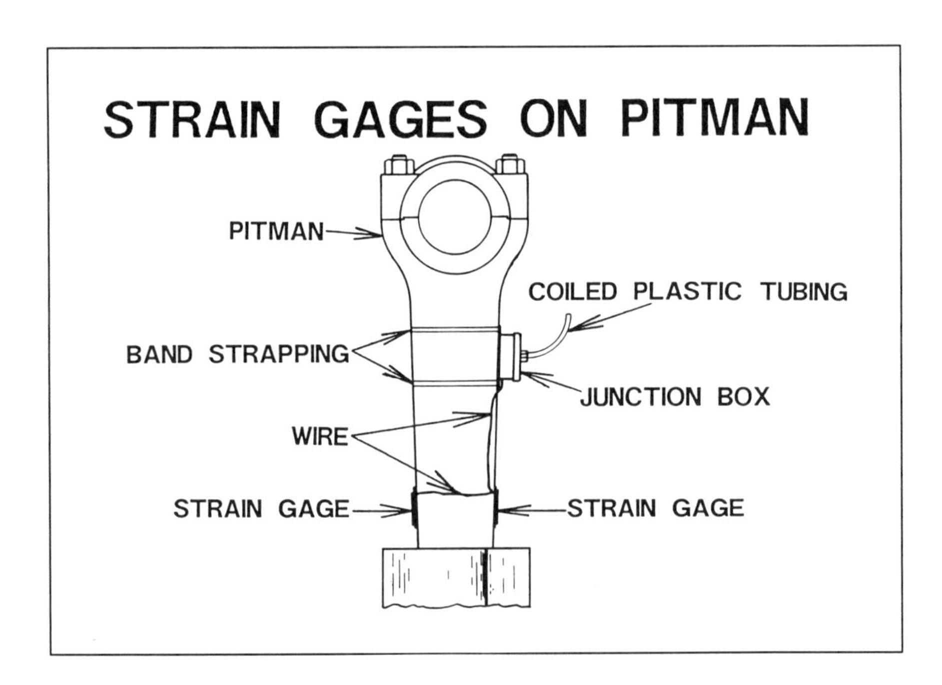

Figure 9-14. *A strain sensor applied to either side of a press pitman: The signal wiring is brought out through coiled plastic air hose, using an electrical junction box and pipe fittings.*

Properly installed, spot-welded gages can provide a robust installation on moving press members. A system used by Toledo Transducers brings the signal wiring out through coiled plastic air hose, using electrical junction boxes and ordinary pipe fittings to terminate the mechanical attachments. Such installations provide trouble-free service for many years. Good gage attachment and encapsulation practices must be followed.

The columns, or uprights, of straightside presses are popular locations for mounting tonnage meter strain links. Ease of installation is the main advantage over placing sensors on the moving press pitman.

For many years, some tonnage meter installers routinely mounted the sensors on the outside of the columns. While this location is convenient, with the sensor better protected from damage, it may not be the best for good process monitoring if the pitman is not accessible. Figure 9-15 illustrates why this location is often not preferred.

Where possible, strain sensors should be mounted on the inside of the columns in line with the center of the bed (L2) because the bed and crown deflection tend to drive the outside of the column into compression. This results in low sensor output. Inside of the column, near the crown, is another good location for sensor installation.

Controlled tests conducted on four straightside presses at Webster Industries involved moving all strain sensors from the outside to the inside of the columns. Carefully documented increases in sensitivity averaged from 236% to 694% per press.

The reason for moving the sensors was to obtain valid chart recorder waveform signatures needed for die-timing analysis. The data was successfully used in a snap-through reduction program.

Gap-frame Press Sensor Locations

Generally, the pitman is the preferred location for strain sensors on the gap-frame press. However, like the straightside press, ease of installation considerations have resulted in the frame being a more popular location.

Measuring the tensile strain on the side of a C-frame throat opening is done extensively. However, a more sensitive location is often found on the vertical frame at the back of the machine, in line with the center of the throat opening. Here, compressive strains are measured.

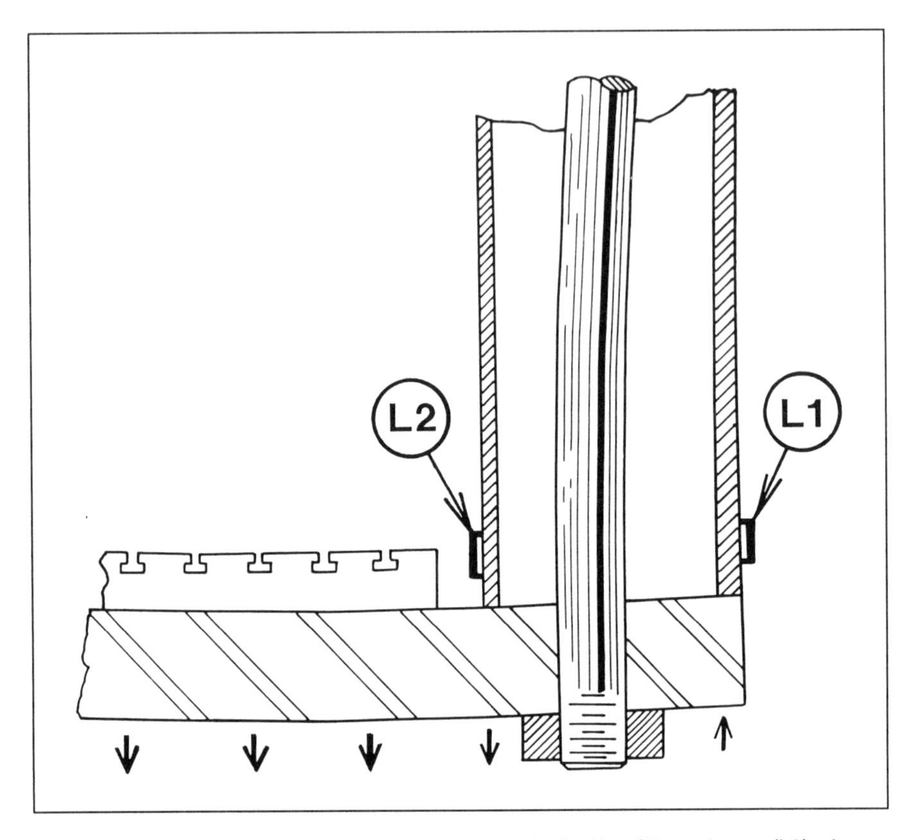

Figure 9-15. *Strain sensors should be mounted on the inside of the columns (L2) where possible because the bed and crown deflection tends to drive the outside of the column into compression, a factor that introduces waveform distortion and results in low output when mounted on the outside (L1).*

Cost-effective machine design takes advantage of the fact that iron and steel will withstand far higher loads in compression than in tension. Generally, the compressive strains at the rear of the machine are up to approximately two thirds greater than the tensile strains at the front of the press.

If possible, the sensor should be mounted on the center of the rear edge of the plate or casting making up the machine frame. Side-of-frame locations may sense strains attributable to buckling. Properly installed, the rear-of-frame location generally provides a good signal for process control.

Gap-frame Press Load Placement

Once the tonnage meter is calibrated, it is important to place the pressworking load directly under the center of the slide connection. Placement toward the rear of the throat opening will result in less strain on the frame and erroneous low meter readings.

Placing the die forward of the connection results in readings that err on the high side. In each case, the tonnage meter is reading actual strains in the frame. However, this is not the load seen by the pitman and crankshaft. This error can be avoided by installing the strain sensors on the pitmans of gap-frame presses wherever possible.

Measuring the strain on the pull rods of underdriven presses can be accomplished in the same way as pitmans are gaged on top-driven machines. In the author's opinion, the best installations make use of foil-backed spot-weldable strain gages. Attaching half the bridge circuit to 180-degree opposite sides of the pull rod, cancels out errors due to twisting or bending.

Strain gages provide very accurate direct readings of mechanical displacement. When applied to underdriven press pull rods of known material composition and dimensions, the true load for a given strain can be measured within ±2%.

Compressive Loads on Pull Rods

When analyzing tonnage meter and waveform signature data taken from underdriven presses, it is not unusual to observe the pull rod being driven into compression. This is normal in stretch draw operations that use large die cushions or nitrogen pressure systems for blankholder pressure.

The energy stored in the blankholder pressure system, less frictional losses, is being restored to the flywheel in this case. If a large compressive load is observed as the press passes through bottom dead center, it is important that the reverse loading of the press gear-train is not excessive when gear-tooth clearances are taken up in the reverse direction.

A correctly adjusted machine with proper die placement should show nearly equal loads on all the pull rods. Unequal pull rod load-sharing indicates a machine alignment problem.

In the case of four-point machines, low readings across diagonally opposite corners indicate an alignment problem. In severe cases, the low-reading pull rods may actually be driven into compression during a portion of the stroke, while the diagonally opposite pair may be overloaded. This may occur without a die in the press. In that case, the machine is running in a severe bind due to misalignment or broken press parts. The machine should be removed from service and inspected. The bed should be checked for a skewed condition as outlined earlier. The gear timing should be checked. Often, all that is wrong is that the drive timing to the slide adjusting screws is out of adjustment.

Double-action Presses

For drawing and stretch-forming operations, both top-driven and underdriven machines built with double actions are used. In each type, an outer slide or blankholder dwells on bottom of stroke to hold the edges of the blank, while the inner slide cycles through bottom dead center to draw or form the part.

The pull rod location is highly recommended for strain sensor installations on double-action underdriven machines. Installing dual or split-bridge sensors on opposite sides of the pitmans or eccentric straps of top-driven double-action presses is recommended.

Some installations of column-mounted sensors have been made on top-driven double-action presses. The column load change produced by the outer and inner slides must be separated by cam signals. Then, the outer slide reading must be subtracted from the inner slide load. In practice, this is difficult to accomplish accurately, and is not recommended.

A strain gage, or sensor, can only measure the actual strain where it is placed and in the direction of strain to which it is sensitive. Strain sensors mounted on the columns, slide, or press bed, do not, in the author's opinion, give as accurate information of press loads as the pull-rod or pitman locations.

Load Cells

Precision load cells for press testing and tonnage meter calibration have a metal structure to which strain sensors or strain gages are attached. The metal structure is compressed when subjected to load. The attached gages give an electrical output that is a linear function of applied load provided that the proportional limit of the metal structure is not exceeded. Usually, hardened tool steel is employed to increase the capacity of the supporting structure. Heat treatment increases the proportional limit of the steel, thus extending the cell's capacity. Strain gage based load cells are available in capacities ranging from a few grams through 1,500 tons (13,344 kN) or more.

Figure 9-16 illustrates a 1,000-ton (8896-kN) press calibration load cell having 32 strain gages installed on the inside and outside of a hollow steel cylinder. Both the inside and outside of the cell are gaged to increase accuracy, when placed on support surfaces not perfectly flat. Press calibration load cells, having gages installed in this way, are built by Toledo Transducers. Historically, the design of this type of load cell is similar to those developed by BLH for steel rolling mill slab thickness control. A comparable design was

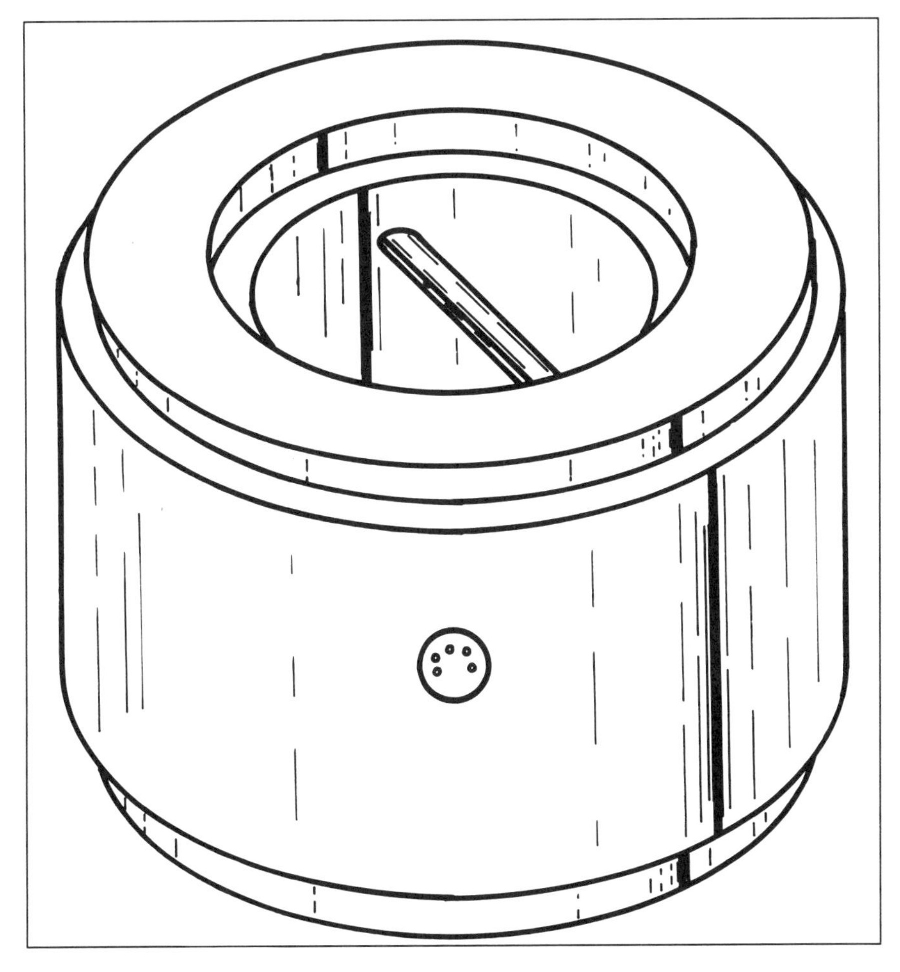

Figure 9-16. *A 1,000 ton (8896 kN) press calibration load cell having a total of 32 strain gages installed on the inside and outside of a hollow steel cylinder.*

produced with a capacity of 1,500 tons (13,344 kN) by BLH in 1954.

A second type of precision load cell often used for in-die force monitoring applications makes use of a piezoelectric material, such as quartz, directly compressed by the pressworking operation. The output is a voltage proportional to load developed across the faces of the sensor.

Precision in-die force monitoring may be accomplished by means of load cells placed under die members such as a punches, die buttons, staking anvils, or coining stations. Either a strain gage based or piezoelectric load cell may be employed. The output is an electrical signal directly proportional to force.

To obtain greater sensitivity in the strain gage based types, the metal load cell structure may be made of aluminum alloy rather than steel. The increase in sensitivity is inversely proportional to the difference in the modulus of elasticity, typically a factor of approximately three.

Force readings can be obtained by means of bore-hole strain probes. These are inserted into a drilled hole in a die shoe or tooling retainer and fastened by means of a drawbar or screw-actuated wedge. They are available in both strain-gage and piezoelectric types. The output is not easily calibrated and of little use for engineering calculations of tooling operating parameters. They are useful for alerting the operator to a change in force.

WAVEFORM SIGNATURE ANALYSIS

In pressworking, a waveform signature is a pictorial or graphical representation of relative movements or amplitude displayed as a rectangular coordinate line chart. The X or horizontal axis represents units of time, distance, displacement, or degrees of crankshaft rotation. The Y or vertical axis usually displays force or amplitude.

A familiar example is the stress versus strain curve or signature produced by a tensile testing machine. Figure 9-17 is a simplified curve of a typical stress-strain signature. A signature of this type is produced by tensile testing machines when test samples, or coupons, are subjected to increasing load until the sample is pulled apart.

Tensile test curve signatures provide important information about the formability and strength of the test coupon. The point

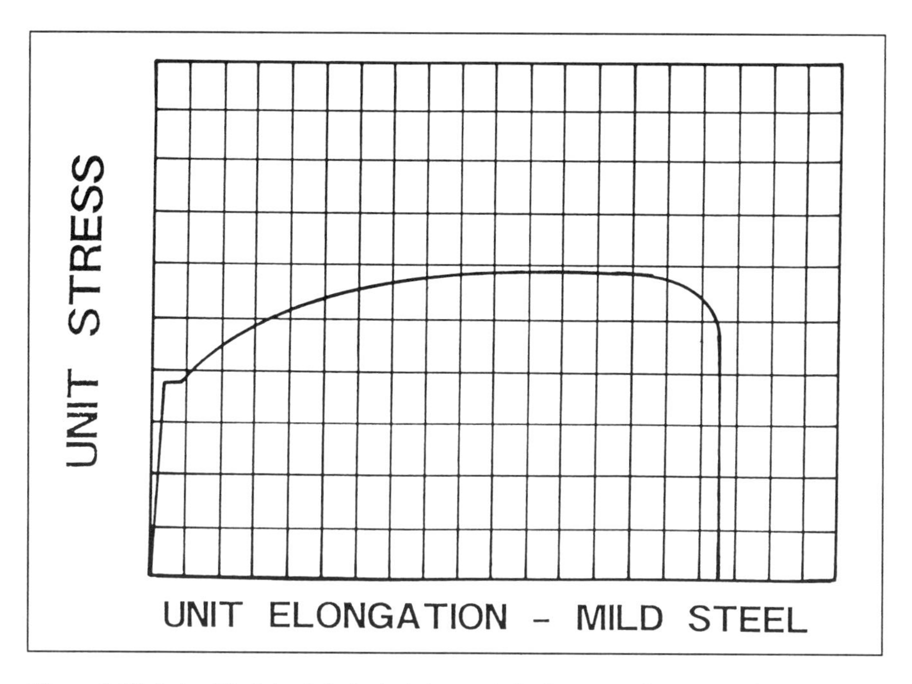

Figure 9-17. A simplified chart of a typical stress-strain signature of a test sample pulled apart in a tensile testing machine.

on the vertical axis where pronounced elongation or movement to the left occurs is the yield strength of the material. The peak unit stress or applied load is the ultimate tensile strength.

The amount of elongation that occurs before fracture is an important measure of material formability. The material yield and ultimate tensile strengths are useful to predict the force required to form the material. The stress-strain curves or signatures for various materials are available as an atlas, listing over 600 examples.

Waveform Signatures of Cutting Loads

Tensile testing machines have a load cell directly in line with the coupon being tested. The load cell output signal is proportional to the stress or force to which the sample is subjected at any moment.

Strain or displacement of the test coupon is measured by a sensor that provides a signal proportional to the distance the coupon is stretched. Zero force (stress) and distance (strain) correspond to the lower left-hand corner of the chart.

Tensile testing machines are of robust construction and operate smoothly. There is very little shock or machine vibration superimposed on the stress-strain waveform chart.

Punching Thick Carbon Steel, a Case Study

Except for superimposed noise signals, stress-strain relationships similar to those observed with a tensile testing machine occur when punching or shearing metal. In each case, the process results in a fracture or separation of the material.

Figure 9-18 illustrates the waveform signature of the stress-strain relationship when cutting off and piercing two holes in a chain side-bar. The material is AISI-SAE 1039 fine-grained steel 0.500 inch (12.7 mm) thick by 3.0 inches (76.2 mm) wide. The data was taken with a chart recorder paper speed of 200 mm (7.874 inches) per second.

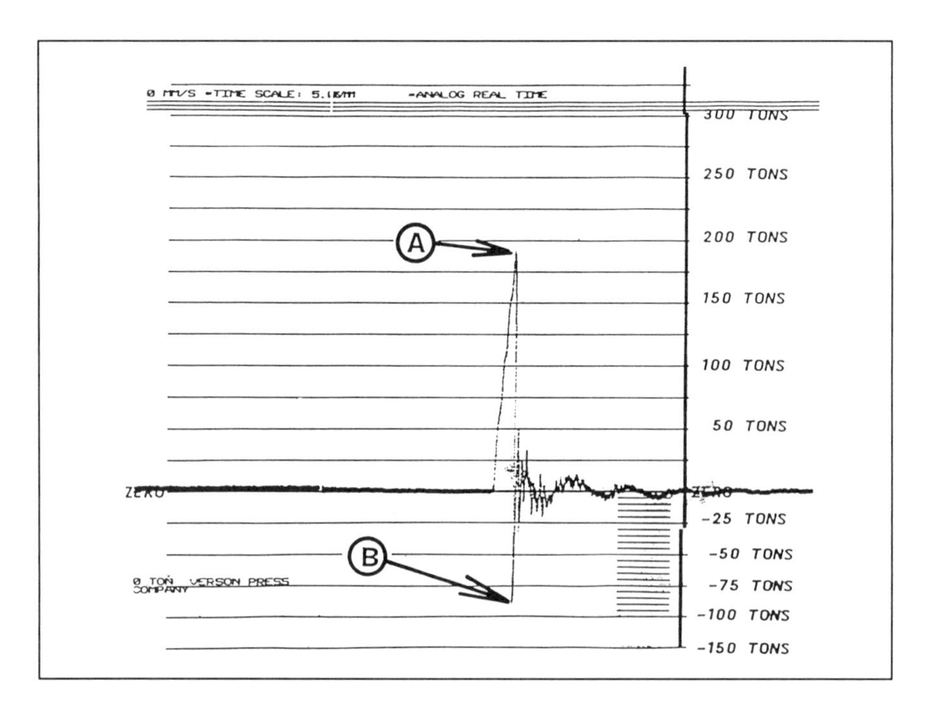

Figure 9-18. The actual waveform signature of a combined piercing and cut-off operation having excessive snap-through or reverse load. (Webster Industries)

Like the tensile test chart (Figure 9-17), the vertical axis indicates stress or force. However, the horizontal axis represents time rather than displacement.

The press speed is 60 strokes per minute (SPM). Even at 200 mm per second, the waveform trace distance is very short from initial contact of the punch on the work until it breaks through. The portion of the waveform from initial punch contact to breakthrough occurs in 5 mm (0.20 inches) or 0.025 seconds.

The punching waveform exhibits a sharp negative spike below the zero trace at breakthrough. This is due to a sudden release of the energy stored in the press and die in the form of strains and/or deflection of the various parts.

The magnitude of the actual energy released increases as the square of actual tonnage developed at the moment of final breakthrough. The actual energy is given by:

$$E = \frac{F \times D}{2}$$

In this equation:

F = Pressure at moment of breakthrough in short tons. (lbf × 2,000)

D = Amount of total deflection in inches.

E × 166.7 = Energy in foot-pounds

or:

F = Pressure at moment of breakthrough in metric tons (kgf × 1000).

D = Amount of total deflection in millimeters.

E × 9.807 = Energy in joules or watt seconds

Note: (1 foot-pound = 1.356 joules or watt seconds)
(1 joule or watt second = 0.7375 foot-pounds)
(1 inch ton = 166.7 foot-pounds)

In timing punch entry or die shear, care must be taken for a gradual release of the force developed. With the exception of high-speed applications, a shock load is not normally generated by the impact of the punch on the stock. In fact, when the punch first contacts the stock, the initial work may be done by the kinetic energy of the slide.

To complete the work, energy must be supplied by the flywheel. An analysis of the quantity of energy involved will show why a gradual reduction in cutting pressure prior to snap-through is very important. A general rule for the amount of snap-through or reverse load that a press can withstand without sustaining damage, is 10% of rated force. Reverse loads significantly higher than this may damage the machine. An especially critical part is the connection, which may fail and allow the slide to fall.

For example, using English units: if 400 tons resulted in 0.080 inch total deflection to cut through a thick steel blank, the energy released at snap-through, from the formula, is 2667 foot pounds.

Careful timing of the cutting sequence results in reducing tonnage at the moment of snap-through to 200 tons. A dramatic reduction in shock and noise occurs because half the tonnage produces only half as much deflection or 0.040 inch. The resultant snap-through energy is only 667 foot pounds, or one fourth the former value.

Timing shear and punch entry sequences to provide a gradual release of force prior to snap-through is a straightforward way to reduce the shock and noise associated with this problem. The simplified analysis of the square-law relationship can be applied to our case study.

A 300-ton (2,669 kN) straightside press was used for this operation. The allowable reverse load is 30 tons (267 kN). Point (A) on Figure 9-18 illustrates a peak load of 191 tons (1,699 kN), well within press capacity.

The reverse load (B) is 87 tons (774 kN), nearly three times the allowable amount. The die was immediately taken to the repair bench and one punch shortened 0.312 inch (7.92 mm). Balanced angular shear was ground on the punches, with balanced shear also ground on the parting punch.

Figure 9-19 illustrates the improvement achieved by modifying the tool. The peak tonnage was reduced to 82.8 tons (737 kN), less than half the initial value. The reverse load was reduced to 22 tons (196 kN), about one-fourth the former value.

This and the documented results of many other tests show snap-through reductions conforming closely with the square law formula. To get chart recordings sufficiently free of noise to give good results, the sensors were all installed as illustrated in Figure 9-15.

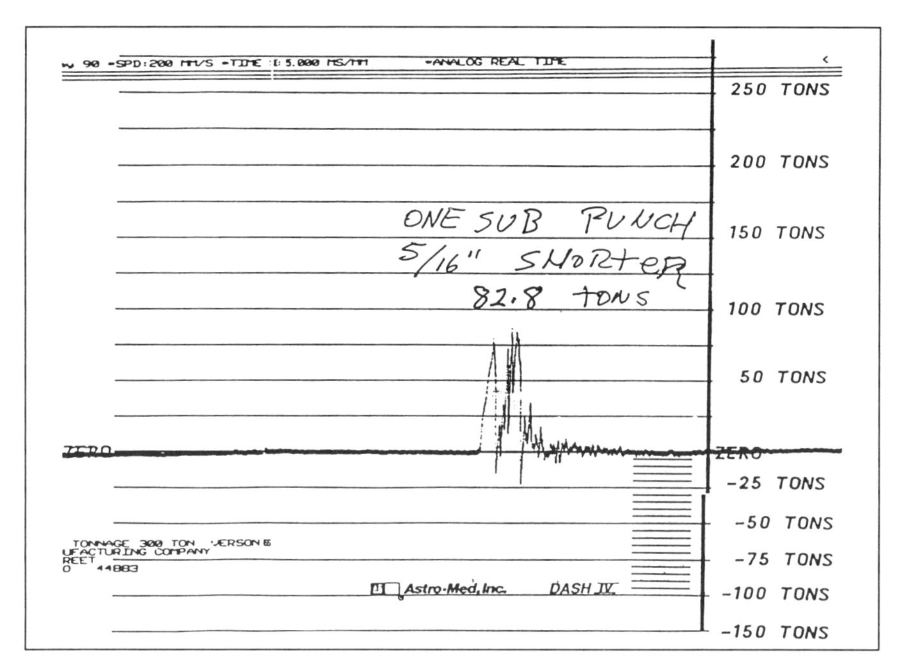

__Figure 9-19.__ Waveform signature of the operation illustrated in Figure 9-18 after modifying the die by adding timing and balanced shear. (Webster Industries)

Waveform Signature of Combined Cutting and Bending

Figure 9-20 illustrates the waveform signature of a combined punching, cutting-off, and joggle bending operation. Here, AISI-SAE 1039 steel 0.500 inch (12.7 mm) thick by 2.000 inches (50.8 mm) wide has two holes punched, and a 0.562 inch (14.27 mm) joggle formed. The part, an engineering-class chain side-bar, is also cut off in this combined operation.

Webster Industries, located in Tiffin, Ohio uses a portable chart recorder to check every setup. This is done to ensure that optimal forces are being used. It is especially important not to exceed 10% of press capacity as snap-through load.

The battery-powered recorder is mounted on a cart having all needed supplies, and plugged into the press electronic force meter to obtain readings. To speed up the analysis, laser-printed plastic overlays are used to interpret the readings. The waveforms illustrated in Figures 9-18, 9-19, and 9-20, each have the appropriate overlay for the press tonnage in place.

The operation is performed in a 500-ton (4448-kN) Minster straightside press. While the machine is quite robust to avoid

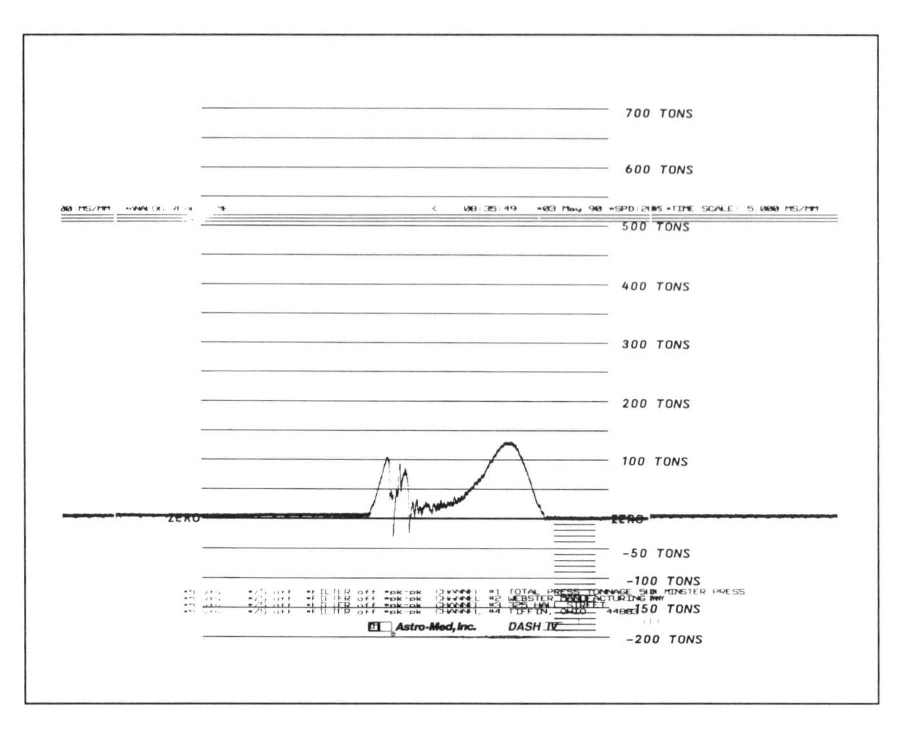

Figure 9-20. *The waveform signature of a combined punching, cut-off and joggle bending operation: the part, an engineering-class chain side-bar, is also cut off in this combined operation. (Webster Industries)*

maintenance problems, the reverse or snap-through load is maintained within 10% of capacity, 50 tons (444.8 kN) or less.

Reading the chart from left to right, the first peaks, which have sharp excursions below the zero reference line, result from the cutting loads. A combination of balanced shear and staggered breakthrough sequence results in a peak reverse load of 45 tons (400 kN), which is within the 10% limit. Better optimization of shear angles and timing may further reduce the energy release.

Both the chart and force meters indicate a peak load of 122 tons (1085 kN). The chart shows that the peak force is the joggle-bending operation. This portion of the waveform signature is a somewhat symmetrical bell-shaped curve. The longer slope on the left-hand side of the peak is where the bending action takes place.

The trace on the right-hand side of the forming peak is half of a bell-shaped curve. This portion of the force signature is normal for coining loads, in simple crankshaft and eccentric driven presses, having sinusoidal slide motion.

Diagnosing Hydraulic Overload Problems

Process variation problems may be encountered in presses equipped with a hydraulic overload system from time to time. Waveform signature analysis is an excellent tool to help pinpoint the cause. Waveform signatures taken directly from load cells placed on strong supports in the die space is the procedure used in the case study. However, charting successive hits from the analog voltage output of a dedicated press force meter can also show variation from stroke to stroke.

Here, the complaint was a serious variation of the depth of embossed features in automotive inner doors. The machine is a Danly 1000-ton (8896-kN) 120 inch (3048 mm) wide triple-action toggle press.

Both the outer slide, or blankholder, and inner slide have toggle mechanisms to provide dwell at bottom of stroke and hydraulic overload protection. In the event of an overload, hydraulic cylinders in the slide of this underdriven machine will dump, furnishing approximately 0.750 inch (19.05 mm) of additional shut height. Figure 9-21 illustrates a waveform signature taken from one of four 250 ton (2224 kN) load cells used to test the inner slide.

The toggle mechanism cycles the slide dwells on bottom by means of idle points in the driving linkage. A double hump-like variation in pressure of equal amplitude during the stroke is normal for this type of slide actuation method.

Clearly, the force variation is caused by the hydraulic overload problem. This in turn is responsible for the inconsistent depth of part embossment. A number of factors have contributed to this problem. Among them: The press is several decades old. Any maintenance is generally in response to a breakdown. The lube oil, which is also used to charge the hydraulic overload cylinders, is not changed unless the press is torn down to repair a broken part. The recommended antifoaming additives are not used. Contamination with water-based drawing lubricants is a factor. The cylinder packing used is the least costly obtainable, not what the manufacturer recommends.

The most serious flaw was a jumper that defeated the limit switches that signal the press to stop if an overload dumps. This can cause unsafe press overloading as well as machine and die damage. This incident resulted in thousands of dollars in scrap parts. Here, portable instrumentation was used to highlight a problem. The overall root cause, however, was poorly organized maintenance management practices and no systematic employee training in basic pressworking machine maintenance skills.

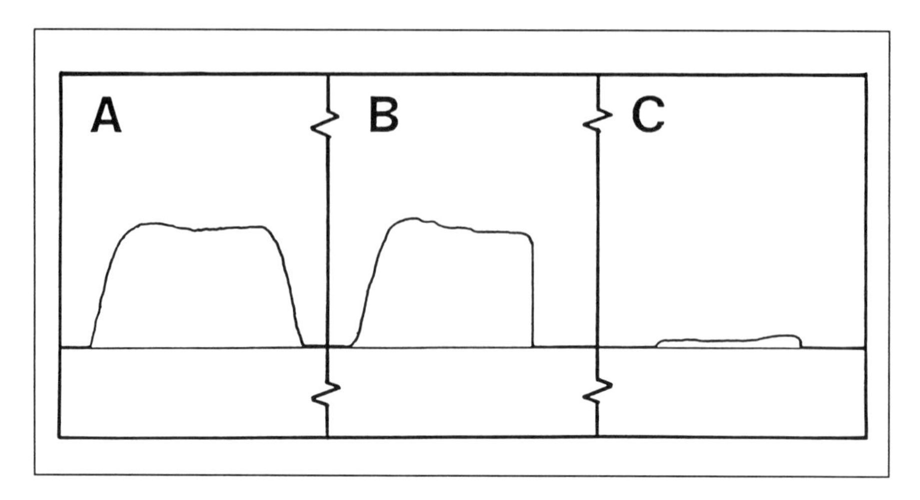

***Figure 9-21.** A waveform signature illustrating a partial dumping of a hydraulic overload system on a toggle press during three successive cycles: the result is serious product variation. The waveform is abnormal. Reading the illustration from left to right: during the first waveform (A), the cylinder is leaking down under load, resulting in a loss of force during the dwell portion of the cycle. Near the end of the second cycle (B), the relief valve dumps suddenly. During the third cycle (C), the cylinder is being recharged. The pressure can be seen increasing during the dwell on bottom.*

Monitoring Double Action Press Tonnage Curves

Large conventional double-action drawing presses used to stamp automotive body panels are designed to develop full force a short distance from bottom of stroke. The blankholder is typically designed to exert full tonnage only 0.250 to 0.500 inch (6.35 to 12.7 mm) above bottom.

Many processes now employ stretch-forming dies in these older presses designed for conventional draw dies. There are advantages, especially better part rigidity by optimizing the amount of biaxial stretch. Draw rings employing up to 200 tons (1779 kN) or more, of die cushion or nitrogen pressure are used. Draw ring travel up to 4.000 inches (101.6 mm) or greater is used to form the part. Most of the pressure is needed upon initial contact in order to form and hold locking draw beads as the metal is stretched.

Figure 9-22 illustrates the waveform of an inverted stretch-form die running in conventional double-action press. Both the inner and outer slide signatures are shown, with the press tonnage curve limits superimposed.

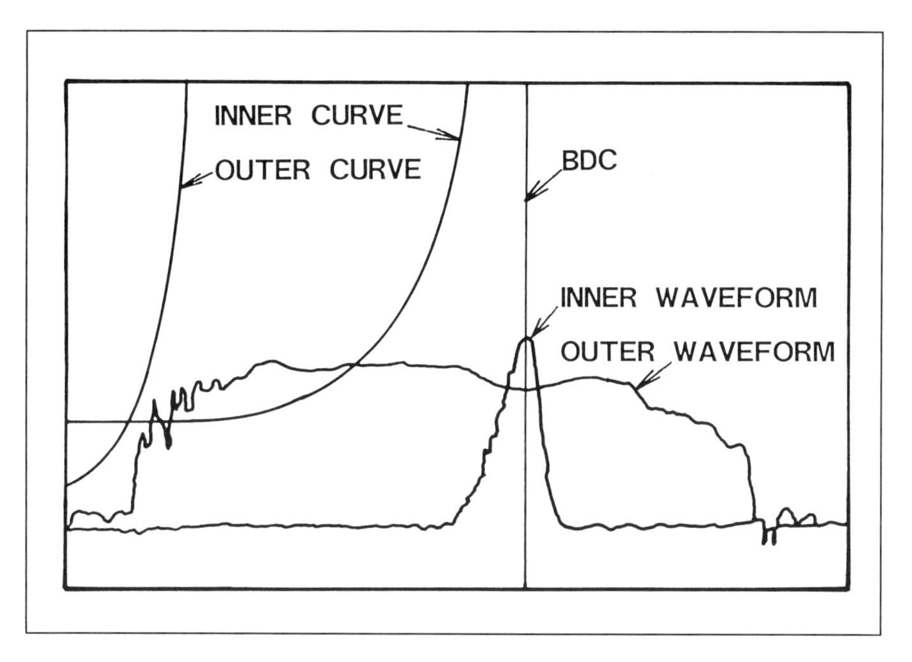

Figure 9-22. *Double-action press waveform of a inverted stretch-form die. Both the inner and outer slide signatures are illustrated with the press tonnage curve limits superimposed. (Toledo Transducers, Inc.)*

The press tonnage curve limits of both the inner and outer sides are compared to the force of each slide at many points during the press cycle. This is accomplished as follows:

- The tonnage curve data supplied by the press manufacturer, in terms of maximum force versus distance from bottom dead center (BDC), is programmed into nonvolatile digital memory in the press tonnage or force meter.
- A rotary resolver, connected to the press crankshaft, sends angular position data to a specially adapted press tonnage or force meter.
- Force data taken from strain gages or sensors on the press pull rods, pitmans, or eccentric straps is converted to an accurate force value for both the inner and outer slides.
- If the force throughout the critical die closure portion of the cycle should violate the maximum value specified by the press manufacturer, an alarm signal stops the press.

- In the event that a force or tonnage curve overload condition is detected, waveform analysis is used. A display, like that illustrated in Figure 9-22, is acquired and shown on the screen of a portable digital computer. This data can be stored on a magnetic disk, for detailed analysis, or as hard copy generated at the press on a portable printer with graphics capability.

In the event a tonnage curve violation is found, one or more of the following corrective measures should be taken:

- The engagement of nitrogen cylinders may be stepped to lessen the initial force.
- If the press is equipped with a Eaton Dynamatic® constant energy drive, the press may be slowed down upon initial draw ring contact to lessen problems with bounce or rebound.
- Die modification to lower the draw ring should be carried out if it has more travel than necessary.
- A press with a better force versus distance from bottom rating may be used if available. Hydraulic presses are ideal.
- The press may be modified, in some cases, by installing stronger parts to increase the tonnage curve capability.

Press Monitors With Waveform Signature Readout

Figure 9-23 illustrates the operator terminal of a press control system having an integrated force meter. In addition to press and auxiliary equipment control functions, the force meter portion of the system has useful capabilities, including:

- A built-in screen with waveform signature capability.
- Selectable high limits at a number of crankshaft angular positions in the press cycle.
- Presettable reverse load alarm monitoring.

A system of this type is especially useful for progressive die and other combined operations where cutting, snap-through, and forming forces are present. The alarm window set points can be stored by job number to speed up diesetting operations.

Pressroom-wide Monitoring Systems

Press force meters having remote monitoring capability are available from several manufacturers. Force data and, in some

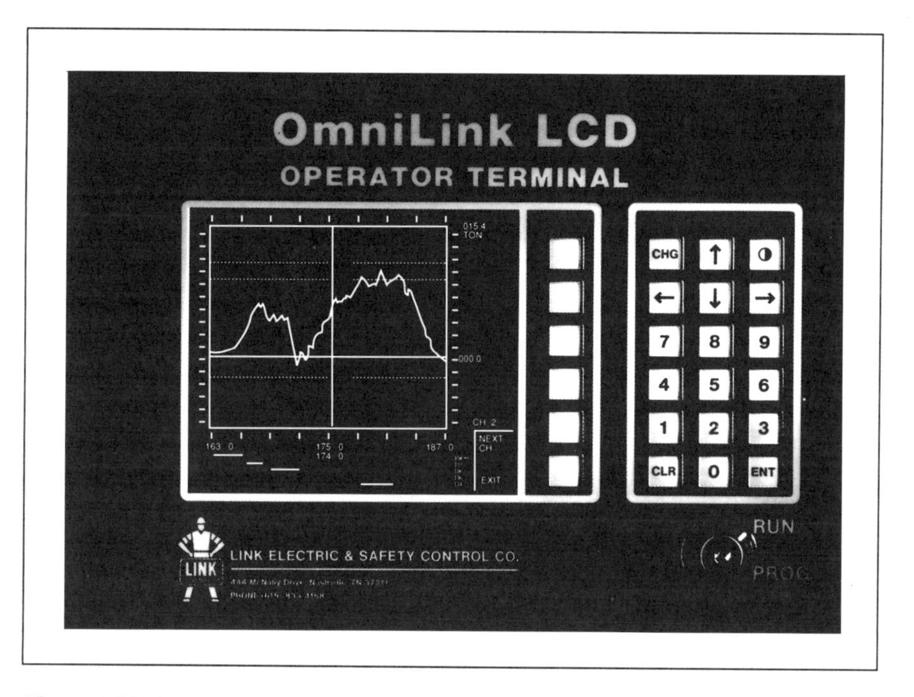

Figure 9-23. The operator terminal of a press control system having an integrated force meter. A built-in flat panel screen provides waveform signature display and alarm capability. (Link Systems, Inc.)

cases, waveform signatures can be exported to a remote monitoring location. While the technology is quite feasible, the desirability and cost effectiveness of gathering such data in a central command center, are not universally agreed upon.

Some managers believe that force-metering equipment is a management tool to prevent press overloading damage. Experience has shown that efforts to use the meter for this purpose alone usually fail, with the equipment becoming disabled.

Properly used, the tonnage, or force meter is a tool for the pressroom work force to set up, troubleshoot, and control the stamping process. Case studies show that training, trust, and teamwork, are the key factors in achieving the goal of improving the efficiency of stamping operations.

In the author's experience, training all pressroom personnel in the fundamentals of pressworking operations is a part of improving efficiency. This training must teach how presses, dies, and auxiliary equipment work. In most cases, it is far better economy to train employees to control the process at the press than to transmit and analyze data at a remote location. The latter approach often is a waste because it does not add value to the product.

Die and Press Protection System Limitations

Several types of die and press protection systems have been used many years. In addition to force meters, press hydraulic overload systems, stretch links, and shear collars are employed.

Problems involving misfeeds and mislocated stock in progressive die and transfer press applications can cause tool damage in the process of tripping the overload set point on a force meter. In many operations, just one bad hit can damage the die. The result is an increase in product variation and a need for prompt bench repair. Double hits involving very heavy stock can result in catastrophic press and die damage.

Hydraulic overload systems are designed mainly to protect the press. The trip points are often adjusted to loads of 50% or more of press capacity. Such forces may cause expensive damage to delicate tooling.

The value of hydraulic overload systems is undisputed. Indeed, such systems should be specified wherever possible for gap-frame and solid-frame straightside presses, that are liable to become stuck on bottom dead center.

Both straightside and gap-frame presses may incorporate a replaceable steel shear ring. The ring has a stepped diameter. One is placed under each connection, with one side machined out to provide an area that shears under a press overload condition, causing the ring to collapse.

Underdriven presses may incorporate a stretch link in line with each pull rod. This device is shaped much like a dumbbell used for calisthenics. It is retained by split collars, which are designed to facilitate replacement.

Both shear collars and stretch links provide a degree of protection for the press. However, because shear collars and stretch links are subjected to large cyclical loads, the force required to deform the device changes with time.

There is a serious problem with this form of protection on presses with multiple connections. If only one device yields, the slide will be tipped out of alignment severely. This may not be noticed until severe scoring of the gibbing — and perhaps die damage — has occurred.

PART SENSING SYSTEMS

Press force meters are of great value as a process set-up and control tool. Trends away from correct operating forces can be detected and the process stopped. Force variation may indicate a change in tooling conditions such as dulling of cutting edges or a slug buildup. Often problems can be detected before tool damage or scrap parts result.

To avoid mishits, a positive means must be provided to assure that the work being processed is in the correct position. If an out-of-position condition is detected, in most cases, the press can be stopped before damage occurs through the use of electrical or electronic part detection.

Electrical Limit Switches

Limit switches are a simple means to interrupt the press run circuit in a misfeed. The limit switch with suitable wiring is usually interfaced directly with the 110-volt AC press control wiring.

Typical applications include possessive die pitch notch switches, part-in-place detectors, and many similar functions. Often the switch is timed to look for a part at specific degree of crankshaft rotation. This may require manually setting a rotary cam limit switch.

Before the advent of electronic cam switches, this type of protection was a popular means of process protection. In the author's opinion, it is unwise to run the 110-volt press control wiring outside of the control panel using drop cord.

Such wiring is subject to damage, which may compromise the control system safety circuit integrity. This concern also applies to 110-volt optical and electromagnetic proximity sensors used for in-die monitoring.

Low-voltage Contact Sensors

Contact sensors, metallic probes, and low-voltage snap-action limit switches are popular for die misfeed protection. Typically 12 to 24 volts DC is used. There are a number of electronic die protection systems that accept contact sensor inputs and provide the required signal conditioning to assure reliable operation.

Neither mechanical switches nor metallic contact sensors can be depended upon to make and maintain perfect electrical continuity. Contact bounce, current leakage, and electrical noise pickup may all contribute to erratic action.

When metallic contact is established in a switch or contact sensor, the contacts may bounce, sending an intermittent signal. Signal conditioning must be built into electronic die protection equipment to avoid false triggering and erratic output.

Electronic die protection equipment gives the low-voltage DC current for contact sensing. Excessive current leakage to ground may cause false signals. One popular brand of equipment detects a resistance to ground of 2,000 ohms or less as a valid signal. Here, good insulation resistance is required, especially in the presence of water-based lubricants.

Immunity to electromagnetic interference (EMI) must also be provided. Pressrooms usually have many sources of EMI, including press motor contactors, control relays, welding operations, and induction heaters.

Filtering and antibounce logic are needed to prove signal conditioning and noise immunity. This delays output response time of the electronic protection unit.

Pitch-notch sensors are actuated by the forward advancement of the progression strip. A conventional pitch-notch stop, or French stop, as it is also called, is intended to hit a solid stop. A special type of pitch-notch stop is illustrated in Figure 9-24. Item 10 in this chapter's bibliography is a video, animated to clarify the concept.

First, a French cut (1), where metal is trimmed away from the edge is made. Next, the forward motion of the strip is arrested by the stop. In this case, the stop (2) is spring-loaded and able to pivot several degrees to permit actuation of a limit switch (3) or sensor.

In this die design, the strip is also gutted (4), leaving only a center carrier. Thus, this operation has the picturesque name of a French gut. The combination of a French stop on each side of the strip (not shown) and gutting operation can have important quick-die-change and productivity advantages. Stock with a pronounced cambered condition can be successfully run, without trial-and-error adjustments. This is because of the trimming of the strip edges and flexibility provided by the center carrier.

The homemade sensing system, illustrated in Figure 9-25, is easy to adjust, very rugged, size-for-size compatible, and costs much less than the 110-volt limit switch it is designed to replace.

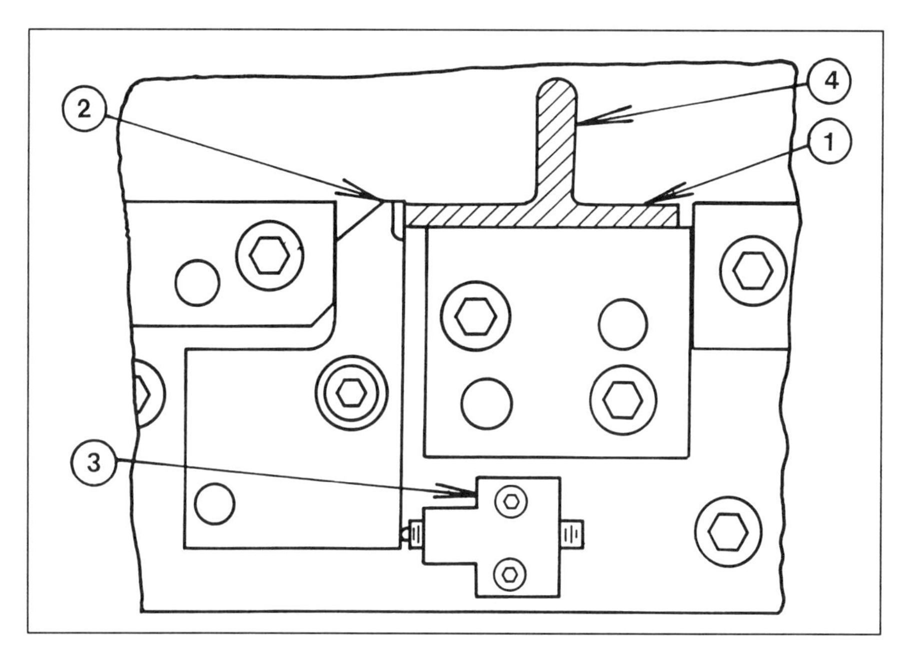

Figure 9-24. Operation of a pitch-notch sensor. A pitch notch is made (1) trimming metal away from the edge; the forward motion of the strip is arrested by the stop (2), which is spring-loaded and able to pivot permitting actuation of a sensor (3); the strip is also gutted (4), leaving only a center carrier. (Smith & Associates)

The sensor is housed in a thermoplastic body. The contact itself (1) is a commercially available threaded spring plunger. The jam screw (2) is a set screw drilled out with a number 36 drill to accommodate a standard banana plug.

Safety is a big advantage of low-voltage contact sensing. However, should electrolysis occur, carbide tooling may be damaged by stray currents. Ground current leakage may also cause false signals. If any of these problems arise, low-voltage fiber optic or proximity sensors may be a better choice.

Electronic Sensors

An advantage of properly installed electronic sensors is relative freedom from mechanical wear. The response speed of electronic sensors may be faster than mechanical switches. However, the limiting factor in safeguarding high-speed stamping operations usually is the time it takes for the presses to stop, rather than the reaction time of the sensor.

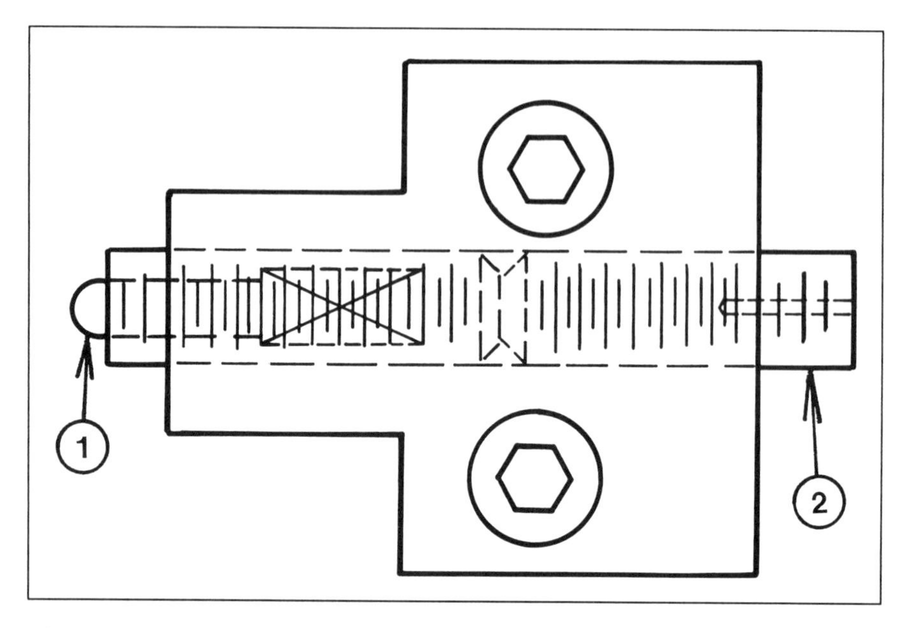

Figure 9-25. A homemade sensor in a thermoplastic body: the contact (1), is a commercially available threaded spring plunger; a set-screw (2), is drilled out to accommodate a standard banana plug. (Smith & Associates)

Inductive Proximity Sensors

Of the many types of electronic sensors available for die protection, inductive proximity sensors are the most popular. Self-contained, inductive proximity sensors are available in several configurations. For die protection, the most common type is the cylindrical style.

An electromagnetic coil in the sensing face is excited by an oscillator built into the housing. The electromagnetic field is damped or weakened by both ferrous and nonferrous metals. When oscillation strength is weakened to a preset level, determined by a detector circuit in the unit, an output signal indicates that the target is in proximity with the sensor.

The detector circuit provides snap action through an a electrical hysteresis switch. The hysteresis of the sensor is the amount of reverse motion of the target required to switch the sensor output. A light emitting diode (LED), which indicates target detection, is a desirable feature. The exact alignment of sensor and the target should be established on the bench before modifications are made to the tooling.

Adjustable Proximity Sensors

Adjustable proximity sensors offer the advantage of an externally variable change-of-state set point. This feature provides flexibility in applications where fine adjustments are needed due to tooling wear and material variations.

The electronics are housed in an external package, Only the coil is contained in the sensor housing. The pickup coil can be very small.

The remote electronics package may feature multiple adjustable set points with LED status indicators. Some units provide a 4 to 20 milliampere current loop output. This standard instrumentation output is easily interfaced with external data acquisition and process control equipment.

Inductive proximity sensors are used in many specialized measuring applications. These include:

- Double blank detectors on blank feeders;
- Inductive ring sensors to detect part ejection;
- Analog distance measurement; and
- Special purpose types built into automation components.

Photoelectric Sensors

Photoelectric sensors are quite versatile. Their main advantages over inductive proximity sensors are the ability to sense nonmetallic objects and a very long sensing range.

Optical sensors are available in several basic types. The opposed type makes use of a separate transmitter (light source) and photoelectric detector (receiver).

Retroreflective types sense objects by detecting the reflected light. They may also sense the interruption of a light beam between the sensor and a remote reflector, resulting from the presence of an object.

The versatility of optical sensors has been greatly increased by development of optical fiber light conduits and miniature optical systems. This has resulted in a wide range of in-die sensing methods.

However, there are disadvantages. One problem is great variations in the sensing range resulting from contamination of optical surfaces by oil and grime. Color or light reflectance variations of the object to be sensed can also be a problem.

These sensors are widely used to control stock decoilers and maintain the correct strip loops in blanking and progressive die applications.

Miscellaneous Stamping Sensor Applications

Inductive proximity, optical, laser, capacitive, and ultrasonic sensors monitor and control the stamping process. Both monitoring and, in some cases, active feedback control of the stamping process are feasible. Applications include:

- Strip thickness and width measurements;
- Stock buckle detection;
- Monitor part geometry;
- Misfeed detection; and
- Feed length sensing.

Designing Dies to Include Sensors

The best time to decide where and how to place sensors in a die is at the die design stage. Retrofitting existing dies with electronic sensors can be rather costly. Even so, the cost of retrofitting an existing die is usually more than justified in terms of damage avoidance.

BIBLIOGRAPHY

1. S. Young, *Press Isolators: Their Function and Effectiveness, Metal Stamping*, The Precision Metalforming Association, Richmond Heights, Ohio, February, 1980.
2. W. Whittaker, "Preventing Machine Installation Problems," *Manufacturing Engineering*, Society of Manufacturing Engineers, Dearborn, Michigan, April, 1980.
3. H. Nielsen, Jr. *From Locomotives to Strain Gages*, Vantage Press, New York, 1985.
4. M. Herderich, *Experimental Determination of the Blankholder Forces Needed for Stretch Draw Design*, SAE Paper 900281, Society of Automotive Engineers, Warrendale, Pennsylvania, 1990.
5. B. Mettert, *Load Sensor Placement and Tonnage Data for Underdrive Presses*, SME Technical Paper MS90-384, Society of Manufacturing Engineers, Dearborn, Michigan, 1990.

6. Published proceedings, educational and reference materials of the Society for Experimental Mechanics, Bethel, Connecticut.
7. Hands-on workshops, and extensive technical information on strain gage installation and environmental protection, including a full line of strain gages and installation supplies, are available from the Measurements Group, Inc., Raleigh, North Carolina.
8. D. Smith, *Using Waveform Signature Analysis to Reduce Snap-Through Energy,* SME Technical Report MF90-11, Society of Manufacturing Engineers, Dearborn, Michigan, 1990.
9. D. Smith, *Quick Die Change,* Chapter 29, "Control the Process with Waveform Signature Analysis," Society of Manufacturing Engineers, Dearborn, Michigan, 1991.
10. D. Smith, Video, *Quick Die Change,* divided into 26 training sessions; referenced material includes workbooks and facilitator's guide, Society of Manufacturing Engineers, Dearborn, Michigan, 1992.
11. H. Boyer, *Atlas of Stress-Strain Curves,* ASM, International, Materials Park, Ohio, 1986.
12. D. Smith, *Die Design Handbook,* Section 4, "Shear Action in Metal Cutting," Society of Manufacturing Engineers, Dearborn, Michigan, 1990. Anthony Rante, PE., Manager of Mechanical Engineering, Danly Machine, Chicago, Illinois, is thanked for reviewing the Editor's formulas and examples.
13. D. Smith, *How to Improve Hit-to-hit Time With a Tonnage Monitor,* SME Technical Paper TE88-780, Society of Manufacturing Engineers, Dearborn, Michigan, 1988.
14. David A. Smith, *Quick Die Change,* Chapter 28, "Instituting a Tonnage Meter Program," Society of Manufacturing Engineers, Dearborn, Michigan, 1991.
15. D. Smith, "Visual Indicators in the Workplace," *The Fabricator,* The Fabricators and Manufacturers Association, International, Rockford, Illinois, January-February 1993.

CHAPTER 10

PRESS INSPECTION AND MAINTENANCE

Inspecting the condition of stamping presses can range in complexity from checking for irregular action or sounds while running production, to a detailed inspection of the individual parts when the machine is disassembled. Pinpointing problems requires skill, experience, and training in what to look and listen for.

Inspection and record-keeping requirements vary. In the United States, current OSHA law requires the employer to establish and follow a program of regular periodic inspections of the press. This ensures the machines — including the auxiliary equipment and safeguards — are in safe condition and properly adjusted.

OSHA also requires a maintenance file for each press. The file includes a record of all inspection reports and maintenance performed. A backlog of open maintenance items must not be allowed to develop, and safety-related items must be corrected at once.

SCOPE OF INSPECTIONS

Performing press inspections as part of the preventive maintenance program allows orderly scheduling of repairs, and permits the correction of problems before they cause downtime or danger to personnel. Formal press inspections must be made at regular intervals based on the manufacturer's recommendations. In addition, a routine inspection must be made at the start of each shift, when operators are changed, and whenever dies are changed. Most importantly, the engaging means should be checked to make sure they are functioning properly, and the safeguarding of the point of operation should be checked for correct installation.

Many things can be noted by a visual and listening check. It is especially important to note any new vibration or noises, and to look for loose machine parts, evidence of machine overloading, and lubrication problems.

Management's Responsibility

Management has the primary responsibility to ensure safe press functioning. However, the operator and/or diesetter should be part of the process. In many shops, they are assigned the task of making the informal inspection. They are usually most familiar with the machine's operation and can easily detect any new sound or change in operation.

A simple set procedure can simplify the routine inspection task. Management must investigate any change at once, and schedule maintenance. The machine should always be removed from service immediately to correct a problem. Ignoring the situation creates a risk of personnel injury or serious machine damage.

The Formal Inspection

All machine components should be inspected at regular intervals by competent personnel. The results of the inspection together with any corrective action taken or needed should be entered into a formal record-keeping system. Often the record keeping is aided by a computerized maintenance management system.

Any safety-related items such as improper functioning of clutch, brake, counterbalance, and control systems must be corrected before the machine is returned to service. Also, proper power lockout procedures are to be followed when required during the inspection procedure.

Follow a Set Procedure

Just as an aircraft flight crew follows an established checkoff procedure before takeoff, maintenance technicians are required to inspect the press systematically. To aid this procedure, many shops develop a checksheet based on the press manufacturer's recommendations and applicable law.

Figure 10-1 illustrates a generic example of such a sheet. It is reproduced for purposes of discussion only. Readers may find it helpful as a starting point for developing a standardized form to suit their shops' requirements.

All Parts Tight

The term *all parts tight* encompasses every part. A loose part may cause an expensive machine failure and, should a part fall, there is danger of injury to personnel. In some components, covers or safety cables prevent a broken or detached part from falling.

PRESS INSPECTION RECORD

REFER TO MANUFACTURER'S RECOMMENDATION

PRESS #	DATE•
INSPECTION ITEM	REPORT CONDITION - OK - Y/N
ALL PARTS TIGHT	
LUBRICATION SYSTEM	
BEARING CLEARANCES	
GIB CLEARANCES	
DRIVE GEARS AND KEYS	
CLUTCH & BRAKE	
DRIVE ''V'' BELTS	
GAUGES ACCURATE	
AIR LEAKS	
PUSHBUTTONS & WIRING	
LIMIT/ CAM SWITCHES	
OPERATOR SAFETY	
STOPPING TIME	
COUNTERBALANCE	
DIE CUSHION(S)	
INSPECTED BY•	DATE•
CORRECTED BY•	DATE•
PRODUCTION APPROVAL BY•	DATE•

Figure 10-1. *An example of a formal inspection form. The items on the list are not all-inclusive; immediate action must be taken to correct any safety-related items.*

Figure 10-2 illustrates a safety cable ensuring a drive motor cannot fall off if the motor mount should fail.

Screws, studs, nuts, and other fasteners used to hold bearing caps, gears, gibbing, and other parts should be manually checked for correct tightness. Safety covers and guarding devices intended to retain broken machine components in place should also be checked for proper attachment.

***Figure 10-2.** Example of a safety cable used to ensure that a press drive motor cannot fall off if a motor mount fails.*

Press Lubrication

The type of lubrication system varies with the type, size, and operating speed of the press. Small OBI and OBS presses often have a manual lubrication system. If grease is used, it may be applied with a manual grease gun on a scheduled basis.

It is important not to overlubricate the machine. A checklist should be used to make sure fittings are serviced as needed. An important part of machine lubrication is manually wiping up lubricant seepage with rags or cotton waste. Keeping the machine clean makes leakage problems easy to find.

Lost Oil and Grease Systems

Presses where the lubricant is not recirculated are termed *lost lubricant* systems. The main advantage is lower initial cost of the press. Both oil and grease are used in such systems.

Grease has important advantages for some presses, especially older machines with open gearing. The advantages include:

- Grease tends to stay where applied;
- Grease can withstand shock loads;
- Generally fewer applications are needed than required for oil; and
- More than one type of lubricant can be applied, permitting optimization for each bearing.

Major disadvantages of lost oil and grease systems include:

- Essentially all lubrication results in a one-time use of the lubricant;
- Residue of waste oil and grease accumulating on and around the press must be cleaned up;
- Lost lubricants may require disposal as hazardous waste; and
- Housekeeping problems and potential fire hazards may result.

Both lost grease and lost oil lubrication systems may use intermittently actuated pumps to apply lubricant from a central point. A simple system may use a hand pump operated as needed. Automatic systems meter the lubricant by press-driven mechanical actuating devices, or a signal from a timer or stroke counter.

Lubricant Distribution Systems

Both metered and recirculating lubricant systems usually require lubricant distribution points, each point having a series of small pistons contained in a valve block. Lubricant under pressure is applied to the valve block inlet. When pressure is applied, each piston displaces a fixed amount of lubricant metered to a series of outlets.

Typically, the system shoots lubricant to each bearing in a fixed sequence as long as pressure is applied to the inlet. A blocked line can stall the entire system. To avoid this failure in manually pumped systems, the last outlet can be connected to a return line that actuates a signal when the lubrication sequence is completed. Automatic systems often use either electrical limit or inductive proximity switches with the lubrication distribution devices to remotely signal distribution piston positions.

Figure 10-3 illustrates a centralized lubrication system having pins that indicate when each piston has shifted to deliver lubricant to the respective outlet. Also illustrated are broken line indicators used to signal a loss of lubrication to critical points. These detectors are useful to safeguard critical bearings not easily inspected visually. During normal operation, lube pressure is maintained in the line at all times. This is done by a check valve in the broken line indicator and a special pressure relief valve, called a *simulator*, at the bearing. Loss of pressure actuates an extendable pin on the indicator, which may actuate a sensor to signal a loss of lubrication.

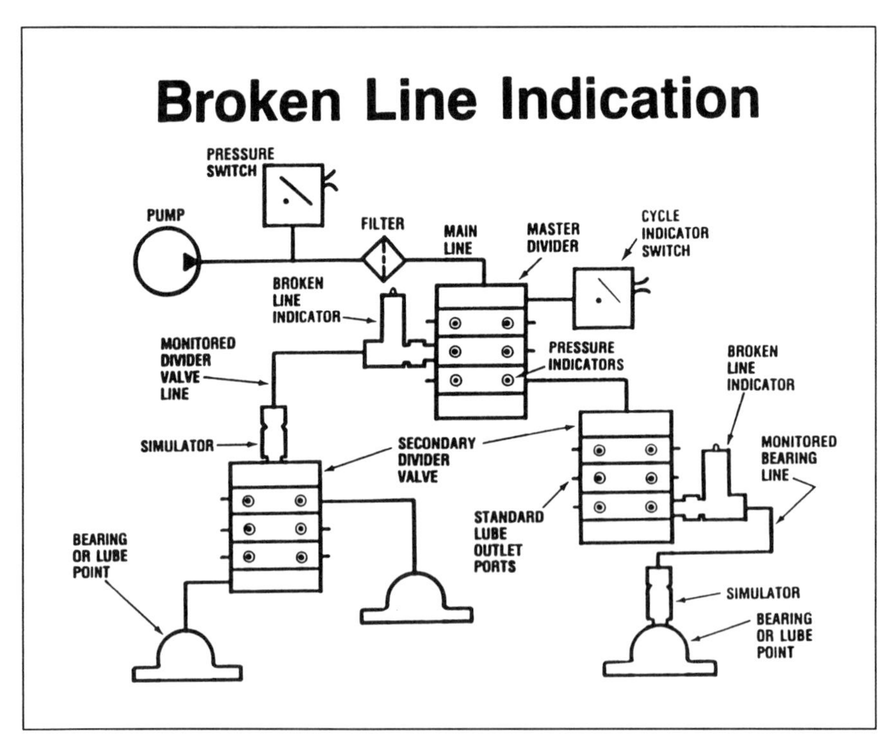

Figure 10-3. *A diagram of a centralized lubrication system having a number of features that signal lubrication problems. (Lubriquip, Inc., A unit of IDEX Corporation)*

Cost of Cleaning

Any lost lubricant must be cleaned up with mops and/or oil absorbents. In some cases, press lubricant may contaminate the parts being produced. Increasingly, mixed waste lubricants are treated as hazardous waste and the cost of disposal by a licensed contractor can exceed the cost of new lubricant.

The best economy of machine life and overall lubricant cost is realized if the correct type and amount of lubricant are applied as needed. Over-lubrication is wasteful and can actually harm the machine by creating excessive pressures that may damage seals. Excessive lubricant can contaminate parts and seep into electrical control systems. The potential for a fire must always be considered, especially if welding operations are nearby.

Recirculating Lubricant Systems

Larger presses generally have recirculating lubrication systems. This is also the case with many smaller presses intended for high-speed and heavily utilized applications.

The three most important benefits of a recirculating lubricant system are:

- The lubricant is continuously filtered and recycled;
- The oil flow cools bearing surfaces; and
- Any particulate matter is flushed from critical bearing surfaces.

Figure 10-4 illustrates a high tonnage geared press having a recirculating cascade lubrication system. The solid lines in the illustration distribute the lubricant under pressure. Reservoir return lines are identified with dashes. The oil reservoir is located in the press bed. The oil, pumped from this reservoir, is filtered and distributed to all points needing constant lubrication.

Usually the reservoir is contained in the bed of the press as illustrated in Figure 10-4. Some very large presses requiring pits may have the reservoir, filtration, and pumping system in a self-contained unit. It is important that the reservoir is protected from the entry of contaminants and designed for easy cleaning when the lubricant is periodically changed.

Motor driven press lubricant pumps are usually of the positive-displacement gear or rotary vane type. A wire-mesh strainer is provided on the inlet to keep out foreign objects that could damage the pump. Depending on the requirements of the press, sizes

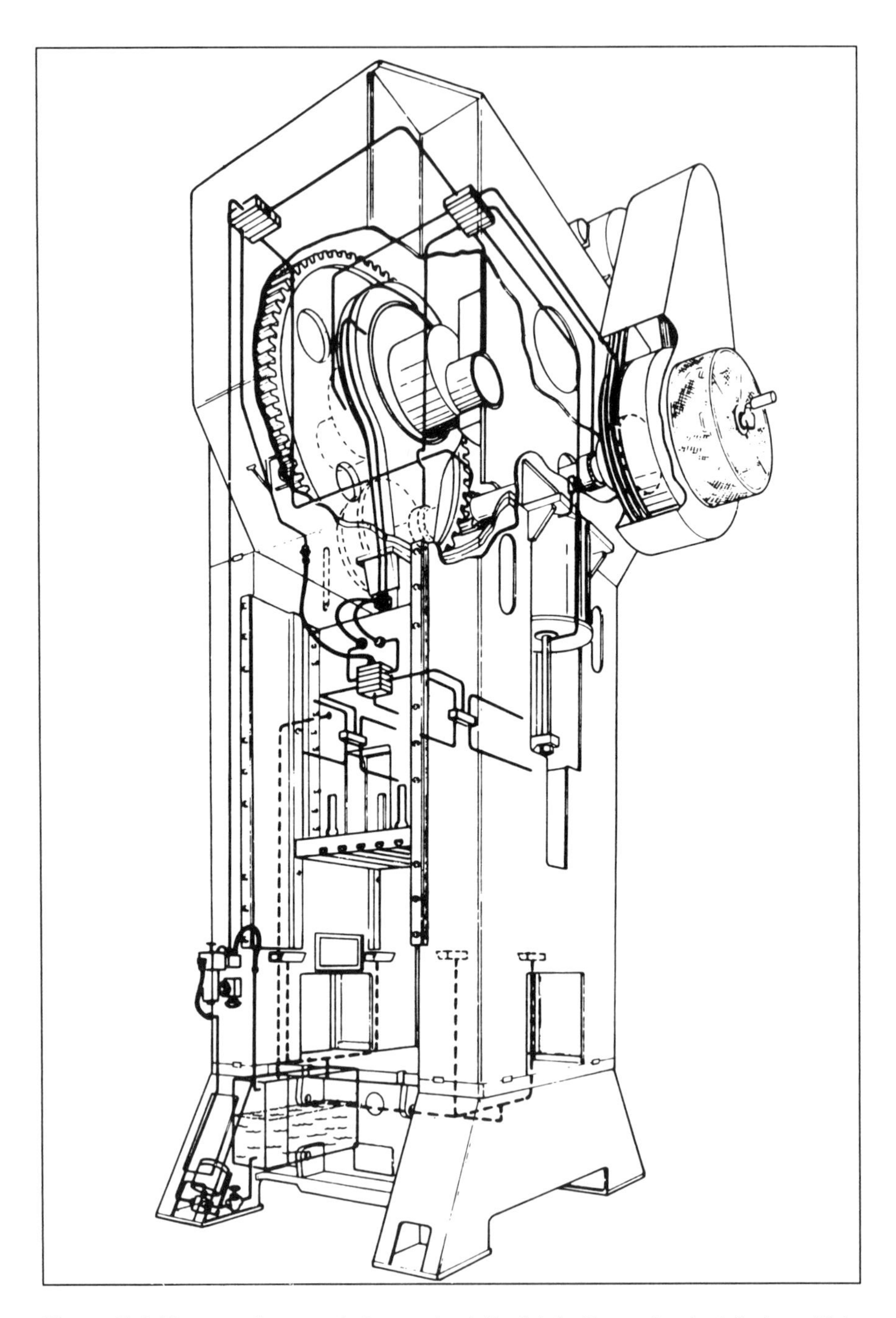

Figure 10-4. *Diagram of a cascade-type recirculating lubrication system installed on a high-tonnage geared press. (Verson Corporation)*

range from one quarter (0.186 KW) to five (3.73 KW) or more horsepower.

Pressurized lubricant is filtered. The filter, usually a replaceable cartridge type, can be easily changed. A relief valve maintains the correct pressure. Any excess lubricant is returned to the reservoir.

Lubricant Lines

Steel tubing or pipe is an excellent material for lubricant distribution lines. Copper tubing is also used, although it is not as strong as steel and is subject to work-hardening from vibration. Plastic lubrication lines are successfully used on small presses and multislide forming machines. Lubricant distribution lines should be routed away from areas exposing them to physical damage.

Lubricant lines to pitman bearings and other moving parts must withstand flexure. Flexible hydraulic hose is a good choice.

The lubricant supplied to reciprocating bearings tends to be thrown out and must be replenished continually. Rapid failure of such critical bearings can occur if they run dry. Providing two lubricant lines supplied from separate distribution device ports ensures additional lubricant, and is a good way to prevent failure in case one line should fail.

Figure 10-4 illustrates lubricant return lines with dashes. All oil returns, or *cascades*, to the reservoir by the force of gravity. Contaminants, such as metal-forming lubricants, solvents, welding debris, and grinding grit must not be permitted to mix with the lubricating oil.

Oil supplied to the gears and bearings in the crown drains down the inside of the columns or uprights. Oil from the pitman and connection bearings drains into the slide. A pipe, which connects to the side of the slide, returns oil to the inside of a column. The return oil from the gibs drains into a small trough under each gib. These troughs each have screens, which must be kept clean to keep debris from stopping up the return lines or entering the reservoir.

Scheduling Filter and Lubricant Changes

Frequency of filter maintenance may be determined in several ways. The press manufacturer's recommendations are usually based on days of elapsed time or hours of operation since the last filter change. Several filter changes may be recommended before a complete lubricant change is specified.

Some filter housings have a gage indicating the pressure differential across the filter. As the filter becomes clogged, the difference in pressure between the inlet and outlet increases, and the amount of pressure difference is indicated on the gage. In addition to the dial face being calibrated in units of pressure, color coding may be used. For example, green may indicate a safe or low pressure difference, yellow an intermediate condition, and red an excessive pressure drop.

Most press manufacturers recommend a specific grade of medium-to-high viscosity mineral oil for recirculating systems. The oil itself never wears out. However, lubricant changes are required because of contamination and loss of additive effectiveness. While scheduled intervals may be followed, the lubricant should be promptly changed whenever contaminants, such as metal-forming lubricants, accidentally enter the system.

Lubricant Requirements for Recirculating Systems

Recirculating systems employ the same lubricant to perform tasks ranging from lubricating gearing, where extreme pressures may be encountered, to supplying pressure to a hydraulic overload system. Flat sliding bearings, such as gibbing and cushion liners, as well as counterbalance packing, may use the same lube.

The choice of lubricant may be a compromise to meet all requirements. Extreme pressure, antioxidant, and antifoaming additives are often used to improve the lubricant properties for such multiple applications.

Some presses have oil reservoirs with thermostatically controlled electrical heaters. Heated oil is easier to pump and distribute throughout the system. Temperature control of the recirculating lubricant helps keep the press at a constant temperature. Supplying lubricant at a constant temperature lessens the amount of component expansion caused by frictional heat buildup. This is an important means of reducing process variability, especially in precision high-speed applications.

Inspecting and Troubleshooting Lubrication Systems

Proper functioning of the press lubrication system is essential for ensuring long machine life. A thorough check is an important part of the formal inspection procedure. In addition, a visual check should be part of the routine inspection performed several times a day. Manually applied lube procedures may be followed at this time, if needed.

Visual inspection of the gibs for proper lubrication is easily done each shift by a trained press operator or diesetter. If the pitmans and other critical bearings are visible from floor level, they should also be checked visually to be sure they are wet with lube, which indicates proper flow. The reservoir level and pressure gage readings should be checked frequently for any change.

The source of any oil leak reported by the operator should be investigated at once. While some oil seepage is normal, a pronounced drip or flow often indicates serious problems. In many cases, a broken line in the crown or slide will result in an oil leak, especially if lube is being thrown from the slide or crown. Another common, and easy to remedy, source of leaks, is blockage of a return line. For example, if a rag or glove is carelessly left in the slide or crown, it can stop up an oil return pipe.

Reporting an oil leak is not a labor complaint. Erecting a pasteboard or plastic tent to shield the operator is foolish. The cause of the problem must be found and corrected. Failure to repair oil leaks wastes lubricant and creates a combined fire hazard and housekeeping problem. If a broken line is the cause, an expensive press failure is very likely.

Foaming of lube oil can result in poor lubrication. For example, should any foaming or air entrapment occur in the lubricant, it will enter the hydraulic overload system and cause spongy action. This difficulty can result in severe product variation problems that are difficult to pinpoint.

Foaming of the lube oil can result from several causes. These include:

- Contamination by cleaning solutions and metal-forming lubricants;
- An air leak on the suction side of the recirculating pump; and
- A lack of antifoaming additives where needed.

Lube Inspection and Chemical Analysis

Providing a sample valve at the outlet of the recirculating pump is a simple way to draw lube samples. Visual inspection of the sample in a clear glass container should focus on:

- Cloudiness;
- Contamination;
- Particulate matter; and
- Air entrapment.

Visual inspection of a lube sample may be made a part of the periodic inspection procedure.

Chemical analysis, including a spectrographic test for small amounts of bearing materials such as copper, tin, and chromium, can reveal wear materials unseen in visual inspections. The results of these periodic tests can be used to both pinpoint press problems and determine the optimum frequency of lubricant changes.

Whenever filter housings and lube reservoirs are drained, any sediment should be thoroughly cleaned out. The sediment should be examined for metallic particles and debris. Metallic particles such as bronze and cast iron indicate a specific type of machine wear. Welding slag and welding rod stubs indicate careless maintenance practices. Food and smoking material remains suggests a need to review present housekeeping practices.

PRESS ALIGNMENT AND BEARING CLEARANCES

Figure 10-5 illustrates a maintenance technician checking a straightside press for parallelism of the slide with the bed at bottom dead center (BDC) of stroke. To perform this test, the bolster has been removed. A cast iron die parallel is used to provide a movable reference surface for the dial indicator. A precision machinist's level is also placed on the parallel, which should be done before the rest of the machine is checked for alignment accuracy.

When making formal press inspections, several approaches may be used to check and correct alignment. This discussion illustrates a straightside press having two connections.

Press Identification

The press used for purposes of illustration has a die-space (and bolster) that is 60 inches (1.524 m) long and 42 inches (1.067 m) wide. A typical tonnage for a press of this size is 300 tons (2,669 kN). Standards adopted by the Joint Industry Conference (JIC) of press manufacturers and users provide for a uniform system of press identification. This press would be conspicuously identified in a central position on the front of the press as follows:

S2-300-60-42

The "S" indicates straightside top-driven construction. If the press were underdriven, the letter "U" would be used. The 2 following the "S" indicates the number of driving points or

Figure 10-5. A maintenance technician checking a straightside press for parallelism of the slide and the bed with a dial indicator. (Midway Products Corporation)

connections. The figure 300 indicates the tonnage. The next two numbers give the bolster size in inches, starting with the left-to-right dimension.

STANDARD MEASURING PROCEDURES

There are several approaches to measuring press alignment and wear. The traditional approach to a comprehensive inspection involves making a number of measurements with a dial indicator, straightedge, and square. One or more jacks capable of lifting the slide and attached linkage are also needed.

The inspection can be performed by press repair contractors on an as-needed basis. However, it is highly recommended that in-house skills be developed to conduct the tests as part of the regular maintenance program. Training in press maintenance skills is available from several sources. The traditional testing procedure includes:

- Checking and aligning the bed to correct any skew;
- Checking the ram, bed, and bolster surfaces for flatness;

- Measuring slide to bed parallelism at 90, 180, and 270 degrees;
- Inspecting the bearings for proper clearances;
- Checking slide perpendicularity (tracking) with a square and dial indicator;
- Checking the gears for correct clearance, amount of wear, and any damage;
- Checking the gib clearances and wear condition;
- Checking the clutch and brake for wear and proper actuation; and
- Examining all parts for cracks and other damage.

Figure 10-6 illustrates an exaggerated view of a precision straightedge used to check a press bolster for flatness. Feeler gages are used to locate and measure any low spots. A low spot in the bolster may result from wear resulting from many die changes. Quick die change rollers installed in the bolster will greatly lessen the wear due to metal-to-metal sliding friction.

***Figure 10-6.** Exaggerated view of a precision straightedge used to check a press bolster for flatness. Feeler gages are used to locate and measure any low spots. (Smith & Associates)*

If the low spot comes from an extreme overload, the bolster's bottom will have a convex bow as well. The bed of the press also may be damaged by such a large overload. In the same way, a straightedge is used to check the face of the press ram and bed for low spots. A low spot in the center of a press bolster due to wear can be corrected by remachining. The cost and downtime are comparatively low. If the bed or ram requires remachining, the cause is probably an extreme overload condition. The machine should be disassembled for such work. These repairs are costly.

Figure 10-7 illustrates a dial indicator having one inch (25.4 mm) travel equipped with a magnetic base that can be turned on and off with a knob. It is used with accessories such as swivel adapters and extension rods. The magnetic base can also serve as a movable surface gage when the magnetism is switched off.

The dial indicator, including the magnetic base and other accessories, is an essential tool for measuring press alignment and bearing clearances.

Figure 10-7. *A dial indicator having one inch (25.4 mm) travel equipped with a magnetic base. When used with swivel adapters and extension rods, it is useful for measuring press bed to slide alignment and bearing clearances. (Smith & Associates)*

Ram to Bed Parallelism

The most common test method, ram to bed parallelism, measures parallelism of the slide to bed or bolster at BDC as was shown in Figure 10-5. The counterbalance air is raised above normal empty press values to draw up as much bearing clearance as possible. This test is easy to perform and provides useful information, especially when compared with previous readings over time.

Measuring Bearing Clearances

A comprehensive formal machine examination normally involves measuring the clearance in each bearing. Here, the magnetic-base dial indicator is also used. To assist with drawing up the bearing clearance to permit measurement, one or more hydraulic jacks capable of lifting the slide and attached linkage are used. This method is illustrated in Figure 10-8.

Complexity of the procedure needed to check and adjust press alignment varies according to the type of machine. Single-point machines are relatively simple to check and adjust. Measuring and correcting multiple-point press alignment can be fairly complex. The maintenance technician must thoroughly understand the mechanical principles involved to safely and correctly check and align any machine.

For a single-point machine having only one connection, the press to bed parallelism is determined mainly by the condition and adjustment of the gibs. If the wear surfaces of the slide and gibs are in good condition, the side-to-side and front-to-back alignment of the ram to bed can be set within close limits by adjusting the gibbing.

A two-point press, as illustrated in Figure 10-8, depends upon the gibbing to maintain correct front-to-back alignment. However the left-to-right parallelism can be affected by bearing wear, adjusting screw synchronization, gear timing, and a number of other factors, including gibbing adjustment.

Machines having four connections in many cases are self-guided by the driving mechanism. This self-guiding effect occurs in both the front-to-back and left-to-right directions. This is especially true of those machines having guided plunger-type connections.

However, correct adjustment of the gibbing is necessary to increase the rigidity of the machine. The gibbing greatly increases the machine's ability to resist lateral forces.

Four-point machines are more expensive than the less complex two-point machines. However, the extra expense is justified: if the tooling used requires excellent guiding by the press, if the press is

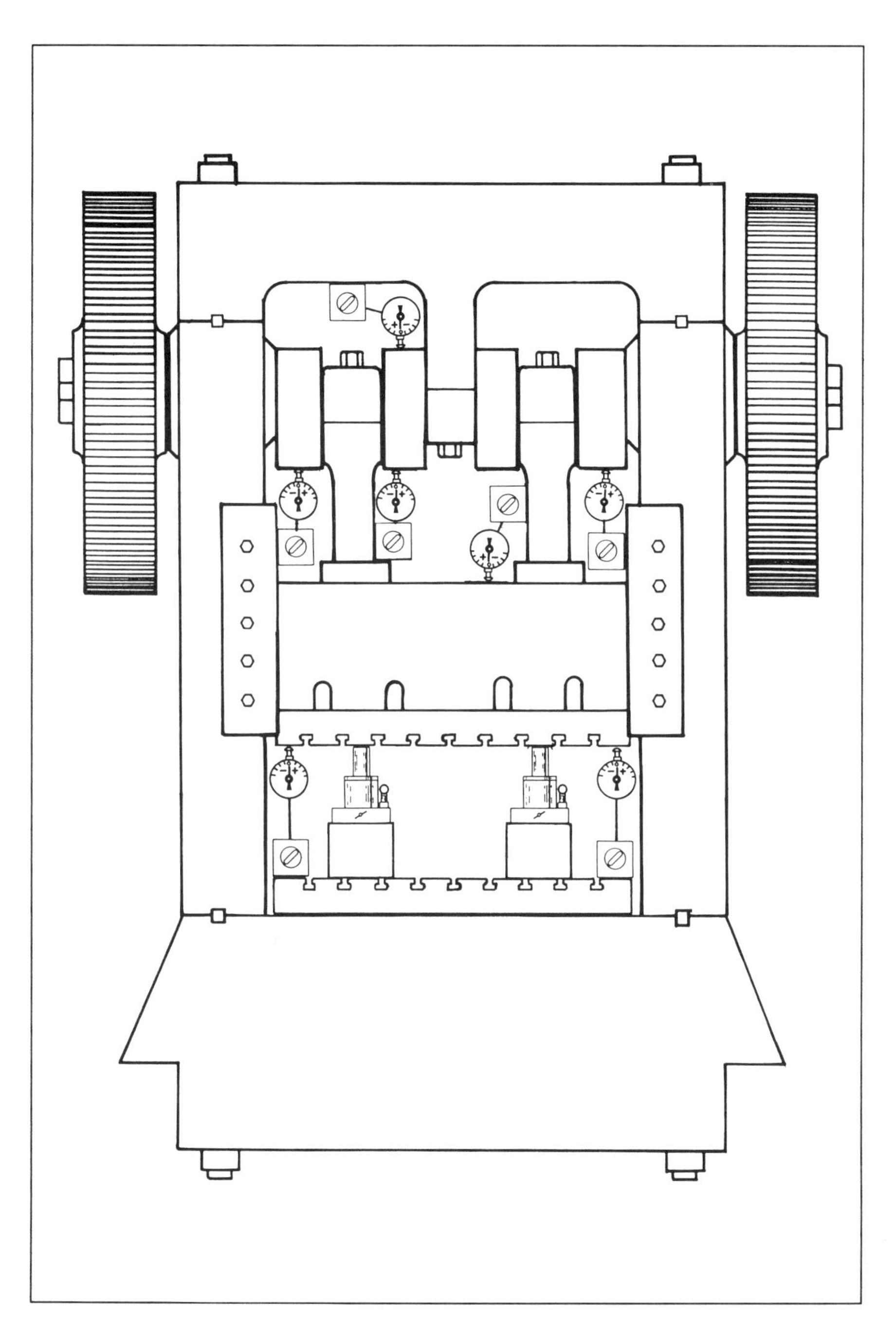

Figure 10-8. *Examples of magnetic-base dial indicator and hydraulic jack placement on a straightside press to measure alignment and bearing clearances. (Smith & Associates)*

very large, or if the tooling used in the process tends to tip the ram due to uneven loads.

The four-point and, to a lesser extent, the two-point press have redundant guiding systems. There is a very real possibility of the machine guiding systems being adjusted incorrectly, causing the machine to run in a bind. If this happens, the press and die may be damaged, and poor quality work can result.

A Quick Static Test

The test for ram to bed or bolster parallelism performed at BDC of stroke illustrated in Figure 10-5 should be done on a regular basis to track the health of the press. The test is easy to perform. Only a dial indicator with a movable base and extension rods are required.

To take up as much bearing clearance as possible, the counter-balance pressure is raised to its maximum value once the press is inched onto BDC. The readings should be recorded on a standard inspection report form and filed for reference.

The maximum out-of-parallel condition considered acceptable for good-quality presswork is 0.001 inch (0.0254 mm) per foot (304.8 mm). For high-speed work with tooling having precise clearance requirements, including electrical lamination and computer terminal work, 0.0005 inch (0.0127 mm) per foot (304.8 mm) is the maximum allowable out-of-parallel condition.

Attaining and exceeding the press alignment tolerance in all classes of pressworking operations is a worthwhile goal. The extra effort spent in maintaining press alignment will greatly reduce machine wear, and close clearance cutting operations will produce more parts before the dies need to be sharpened again.

In addition to measuring the press bed or bolster to ram parallelism at bottom of stroke, a thorough check involves taking measurements at top dead center (TDC), one-quarter down (90 degrees), and one-quarter up (270 degrees) on the up-stroke.

It is important to check ram to bed parallelism at these three other points in the stroke periodically. A record should be entered on a standard reporting form, and the information maintained, to identify any changes over time.

Examples of Readings

When working on two- and four-point machines each of the following areas has an effect on parallelism:

- Adjustment screw drive;

- Drive train gear timing;
- Differences in stroke lengths of crankshafts and eccentrics;
- Slide centering with the gibbing; and
- Proper tie rod prestressing.

The following example covers an analysis of the readings obtained from an inspection of the type illustrated in Figure 10-8.

	LEFT	RIGHT
TDC	0	+
1/4 DOWN	0	+
BDC	0	+
1/4 UP	0	+

Interpreting the Readings

From our readings, the right side of the machine remains higher than the left side throughout the stroke. The adjusting screw drive was probably not adjusted correctly.

This problem is normally corrected by disconnecting the coupling between the left-hand and right-hand adjustment assemblies and lowering the high side to bring the machine back into parallel. The gibbing clearance should be checked and adjusted as needed after realigning the screw drive.

Misalignment due to incorrect drive train timing or a twisted crankshaft will cause the machine to run in a bind and greatly accelerate wear on the press. The following is an example of the readings taken from a machine with a timing problem.

	LEFT	RIGHT
TDC	0	0
1/4 DOWN	0	+
BDC	0	0
1/4 UP	+	0

In this case, the slide is level at both BDC and TDC. The right side is high at the quarter down or 90-degree position. Likewise, the left side is high at the quarter up or 270-degree position.

On many geared machines, the problem can be corrected by leveling the slide. A driving key is removed and the gear or pinion rotated to the point where the gear clearance is evenly distributed throughout the machine. An offset key is machined and fitted in the keyway.

Provided the problem is not a twisted crankshaft, this procedure will often correct timing problems at very little expense. Of course, if a large correction is needed, the key offset should be distributed through two or more keys to avoid any one key being offset so much that it cannot transmit the required torque without danger of shearing.

If a twisted crankshaft is the source of the bad alignment, probably an extreme overload caused the problem. Here, the correction will require removing the crankshaft for repair or replacement.

A press can exert a force several times its rated capacity when severely overloaded. Typical causes are diesetting errors, part ejection failures, and foreign object damage. Such extreme overloads may subject some parts, normally stressed within safe limits, beyond the yield point. Whenever a component fails due to overload, a survey for other damaged parts is required.

Machining Errors

Never assume a new or recently rebuilt press is in perfect condition. Like automobiles, presses can have incorrectly designed, manufactured, or installed parts that affect the performance of the machine.

In the case of the following readings, the slide is out of parallel at both TDC and BDC. However the slide is parallel at both the quarter down (90-degree) and quarter up (270-degree) positions.

	LEFT	RIGHT
TDC	0	+
1/4 DOWN	0	0
BDC	+	0
1/4 UP	0	0

Here, the problem to look for is a difference in the stroke of the two throws of the crankshaft. In the case of a machine driven with eccentrics, the problem can also exist if they are not all machined identically, but this is not common. If this problem should be found when the machine is new or returned from rebuilding, the only solution is to replace or rework the defective parts.

Slide Centering in Front to Back Shaft Presses

On presses that have the crankshaft(s) or eccentric shaft(s) running from front to back, the slide and driving mechanism must be aligned with the center line of the machine to run properly.

	LEFT	RIGHT
TDC	0	0
1/4 DOWN	0	+
BDC	0	0
1/4 UP	+	0

The resultant readings exemplify alignment errors caused by improper centering of machine parts. The cause often is due to improper centering of the slide. Here, the gibbing must be adjusted to center the slide again in the machine.

Another common cause is excessive wear on the thrust flanges of the drive mechanism. The solution here is to rework or replace the worn thrust surfaces.

How the Bolster Affects Alignment

On most presses, the bolster plate is attached to the press bed with relatively large bolts that must remain tight. These bolts are normally designed to withstand large stripping or upward forces as a safety factor in the event of an excessive stripping load. A typical cause of high stripping load is mislocated stock or double hits in deep drawing operations. Should the bolster attaching bolts fail, the bolster may lift up with the die. Broken bolster bolts and machine parts may become airborne, endangering pressroom personnel. Severe press and die damage can also result, especially if one end of the bolster or die lifts first.

The bolts also serve to hold the bolster in proper contact with the press bed. A feeler gage check should reveal no detectable clearance. Look for the following problem areas:

- Dirt, slugs, and debris between the bed and bolster plate;
- Tightness of the bolster tie-down bolts; and
- Proper locating device function and clamp actuation in the case of presses with self-moving bolsters.

The bolster and press bed should be inspected for burrs and low spots as was illustrated in Figure 10-6. A file or oilstone will reveal any burrs on these surfaces. High spots may be removed by filing, stoning, or careful hand grinding.

MEASURING BEARING CLEARANCES

Figure 10-8 illustrated a number of locations on a two-point straightside press where bearing clearances can be measured. The equipment requirements are simple. A dial indicator with a magnetic base such as was shown in Figure 10-7 is used. One or more hydraulic jacks capable of lifting the weight of the slide and attached linkage are also needed.

Figure 10-9 illustrates a procedure for measuring bolster to ram parallelism at BDC of stroke. This test setup also provides a rapid means to check the total clearance in the press bearings.

As was shown in Figure 10-5, a quick check for parallelism of the press ram to bed can be performed BDC with just a dial indicator and movable base. The counterbalance air is raised to the maximum allowable value to remove as much bearing clearance as possible.

The test illustrated in Figure 10-9 uses large hydraulic jacks to lift the slide. After inching the press onto BDC and following lockout procedure, the counterbalance air is drained with the jacks in place, but not touching the slide.

To speed up the work and assure the most accurate results, it is best to place a jack under each connection and a separate dial indicator on each side of the machine. The dial indicator plungers are placed against the slide, and the dials set to zero. The jacks are then pumped to lift the slide, with the dial indicators showing the amount of upward movement. As much clearance as possible is taken up, and the readings noted. These readings, while subject to several sources of error, indicate the approximate total bearing clearance on each side of the press.

If a connection or crankshaft bearing has an extreme amount of wear, and the gibbing is correctly adjusted to guide the slide with minimal sideplay, the jacks may not take up the full clearance. In such a case, the gibbing may be loosened for this and other bearing clearance tests.

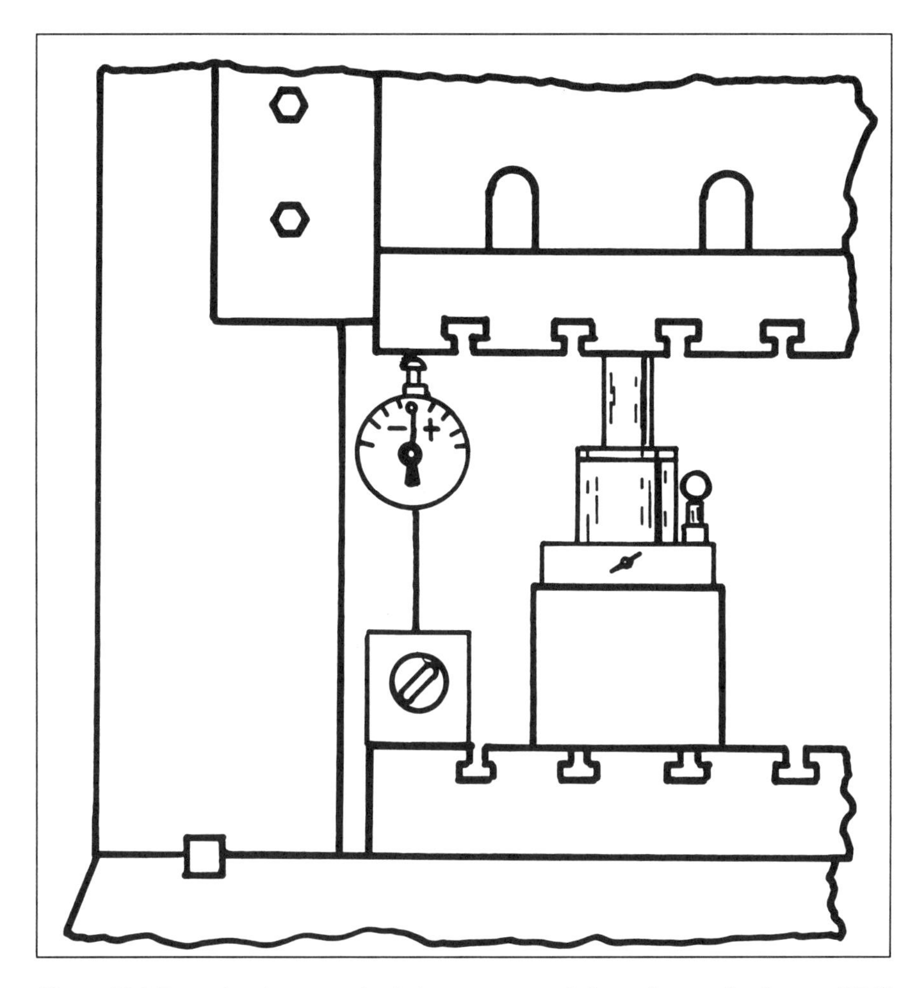

Figure 10-9. Procedure for measuring bolster to ram parallelism at bottom dead center (BDC) of stroke. A quick check of the total clearance in the bearings can also be made. (Smith & Associates)

Measuring Main Bearing Clearance

Figure 10-10 illustrates how to place the dial indicator to measure the clearance between the crankshaft end main bearing and press housing. Here the procedure is to measure play with the slide landing without counterbalance support, and with the slide and attached parts raised with the jacks.

Maintenance technicians must not enter the space between the top of the slide and crown if the counterbalance air is turned on to

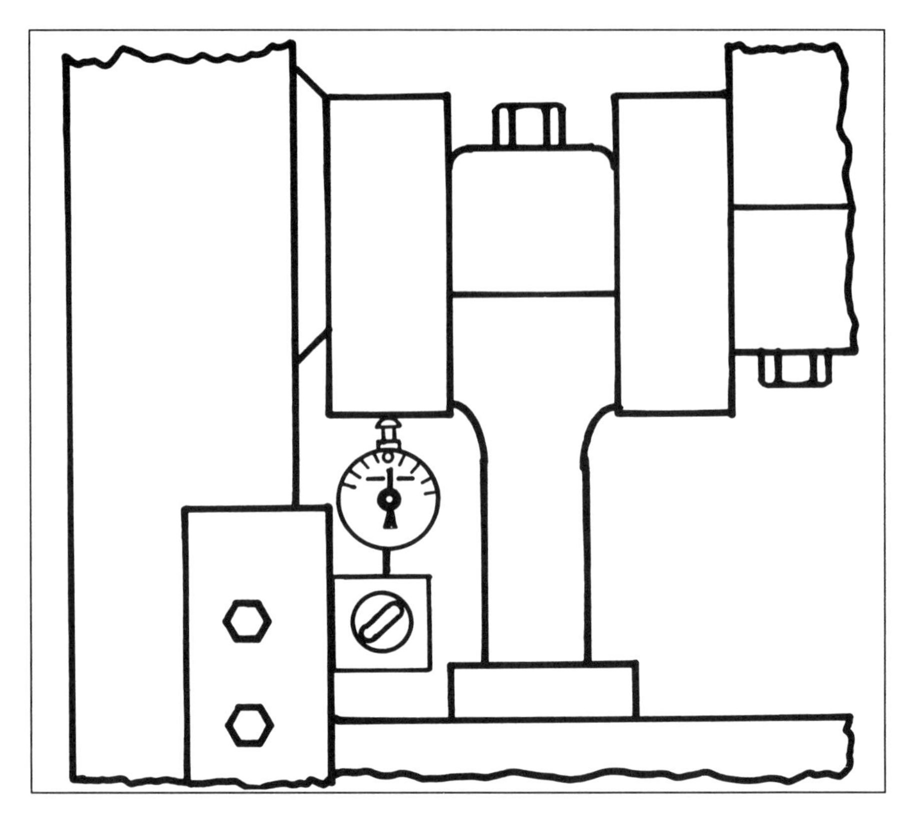

Figure 10-10. Dial indicator for measuring the clearance between the crankshaft end main bearing and press housing. (Smith & Associates)

assist in taking up as much play as possible when measuring bearing clearances. The reading must be made by remote observation. Should the brake release, the slide may be forcefully drawn up to the top.

Moderate amounts of wear in an individual crankshaft bearing can be accurately determined. However, if extreme wear is present in the bearing, the crankshaft may not bend enough under the jacking force to show total clearance.

One cause of large amounts of bearing wear is lubrication failure. The maintenance technician should check for this when making the inspection. A sign of such a failure is the presence of bronze particles around the bearing.

Measuring Pitman Bearing Clearance

Figure 10-11 shows the dial indicator placement to measure crankshaft throw to pitman bearing clearance. The gibbing should be set loose enough to permit accurate measurement if a large clearance is found.

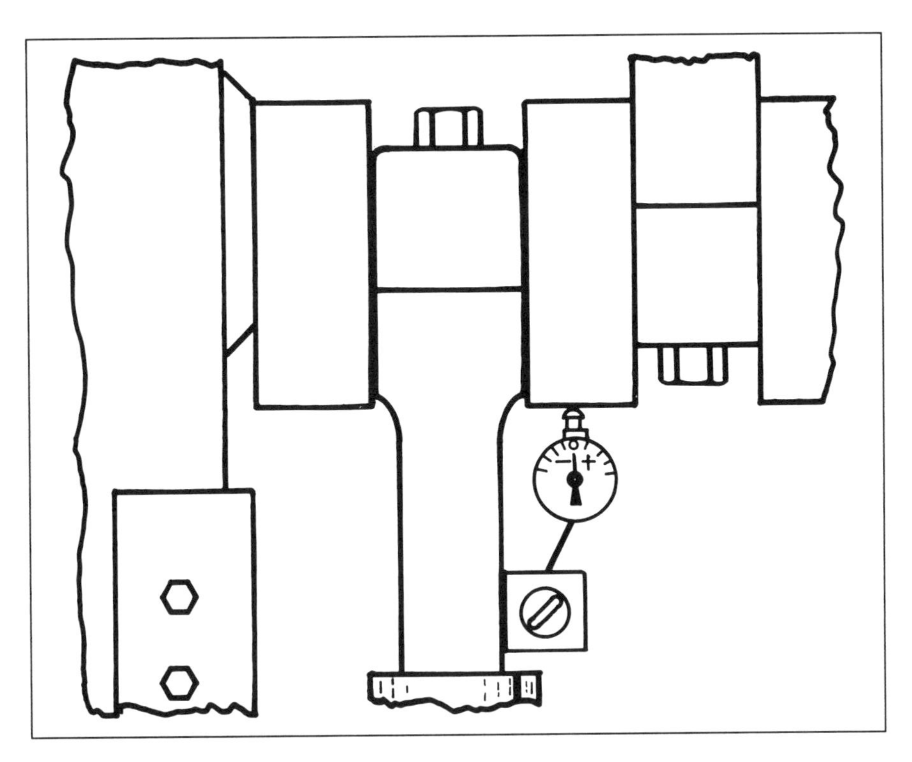

Figure 10-11. Dial indicator for measuring crankshaft throw to pitman bearing clearance. (Smith & Associates)

Usually, a single high-capacity jack placed under the connection on the side of the press being checked is sufficient. Unless extreme amounts of wear are present in both the throw bearing and the connection, this method provides accurate results.

Figure 10-12 illustrates the dial indicator placement to measure crankshaft center bearing clearance. Depending on the width of the machine, a press of the type shown in Figure 10-8 may have two center main bearings. Like the end main bearings, if extreme wear is present, the total clearance may not be indicated.

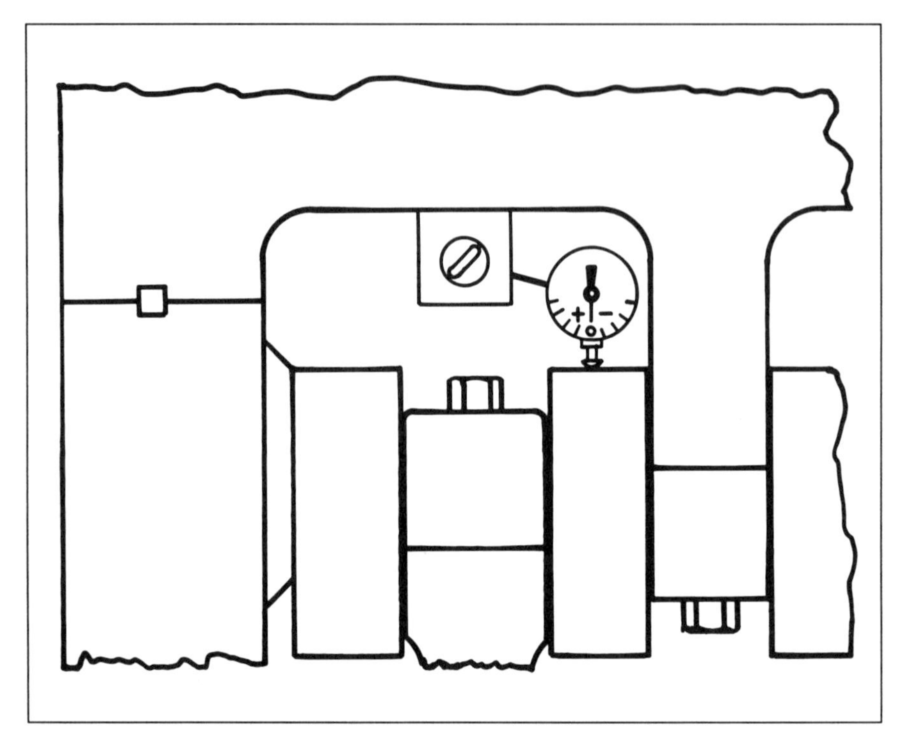

Figure 10-12. Dial indicator for measuring crankshaft center main bearing clearance. (Smith & Associates)

Measuring Connection Bearing Clearance

Figure 10-13 shows the dial indicator placement to measure the connection bearing clearance. The procedure for obtaining a true reading is essentially the same as that for the crankshaft throw to pitman bearing.

The connection is a weak point of the press in terms of stripping loads. If the press is used in heavy punching applications involving considerable snap-through energy release, there is a good possibility the connection retaining means will be damaged or fail.

The connection should be carefully inspected, especially if a large clearance is measured. The condition of the connection is an important safety consideration. The connection(s) and counterbalance system are all that prevents the static weight of the slide from falling.

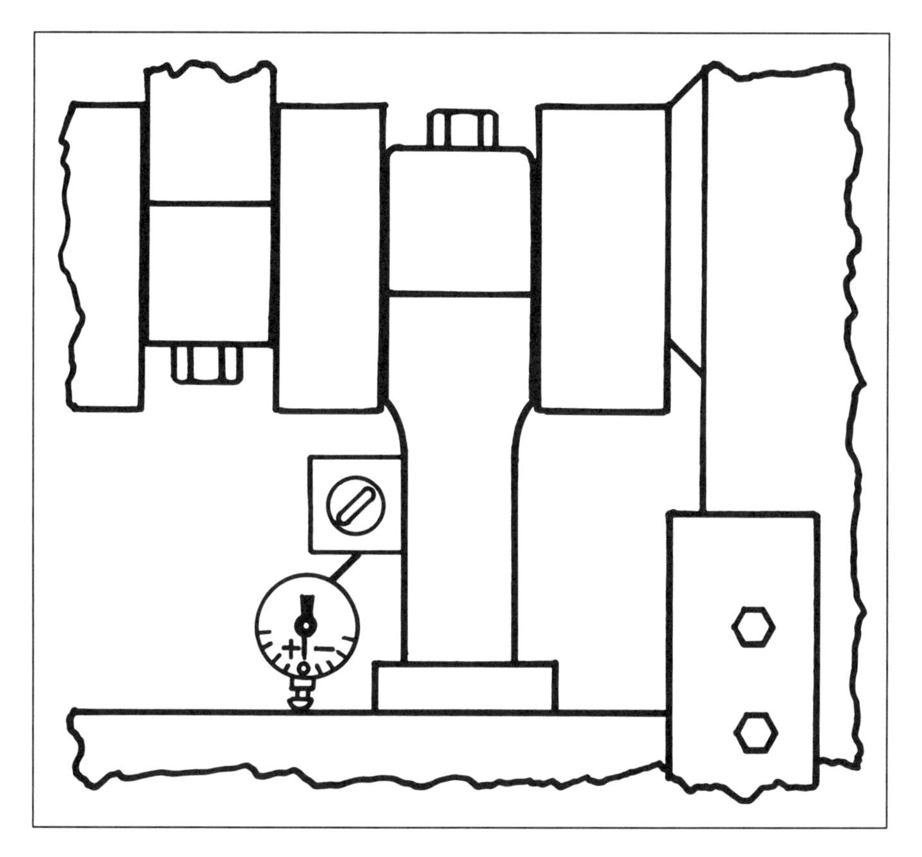

Figure 10-13. Dial indicator for measuring connection clearance. (Smith & Associates)

Typical Bearing Clearance Allowances

A rule of thumb for round bearing clearances in mechanical power presses is one unit of clearance for each 1000 units of journal diameter, or 0.10%. Thus, for each 1.000 inch (25.4 mm) of shaft diameter, 0.001 inch (0.0254 mm) of bearing clearance is provided.

The actual amount of clearance may vary depending upon the type of lubricant specified and the class of work for which the press is designed. A slow machine using grease lubrication will have a greater percentage of clearance than a high-speed precision machine using heated temperature-controlled lightweight mineral oil. Manufacturers' recommendations and good engineering practice should be followed.

Wear will ultimately increase the clearance beyond acceptable limits for the class of work being performed. When the press is in motion, a film of lubricant separates the shaft or journal from the bronze bearing liner by means of hydrodynamic action. Very little wear occurs, especially in machines supplied with filtered recirculating oil under pressure. However, once the clearances become excessive, the hydrodynamic action keeping the surfaces separated is no longer effective. The bearing surfaces will come into metal-to-metal contact, accelerating wear. Timely adjustment or replacement of the bearings is highly recommended. If maintenance is delayed, not only will the quality of work suffer, but the shaft or journal surface may be worn out of round, creating expensive rework and extended downtime.

Example of Total Clearance Calculation

The following example of total machine clearances is based on typical clearance specifications for the type of press illustrated in Figure 10-8. Assuming a 6.000 inch (152.4 mm) diameter crankshaft and crankshaft throw diameter, the three main and two throw bearings would each have 0.006 inch (0.1524 mm) clearance. The connection clearance is somewhat greater, typically 0.008 inch (0.2032 mm).

The adjusting screw also has clearance. However, on many presses, the clearance is taken-up with a locking plug. Some precision presses automatically apply hydraulic pressure to the elevating screw connection to take up all play and avoid wear during operation.

Example of Clearance Calculation

Bearing	Total Inches	Total mm
Crankshaft Main	0.006	0.1524
Crankshaft Throw	0.006	0.1524
Connection	0.008	0.2032
Total	0.020	0.5080

Checking the Slide for Perpendicularity

Figure 10-14 illustrates the procedure for checking the slide motion path for perpendicularity with the press bolster. Before this check is performed, any skew should be removed from the press bed by following the leveling procedure, and high spots and burrs should be removed from the press bed and bolster.

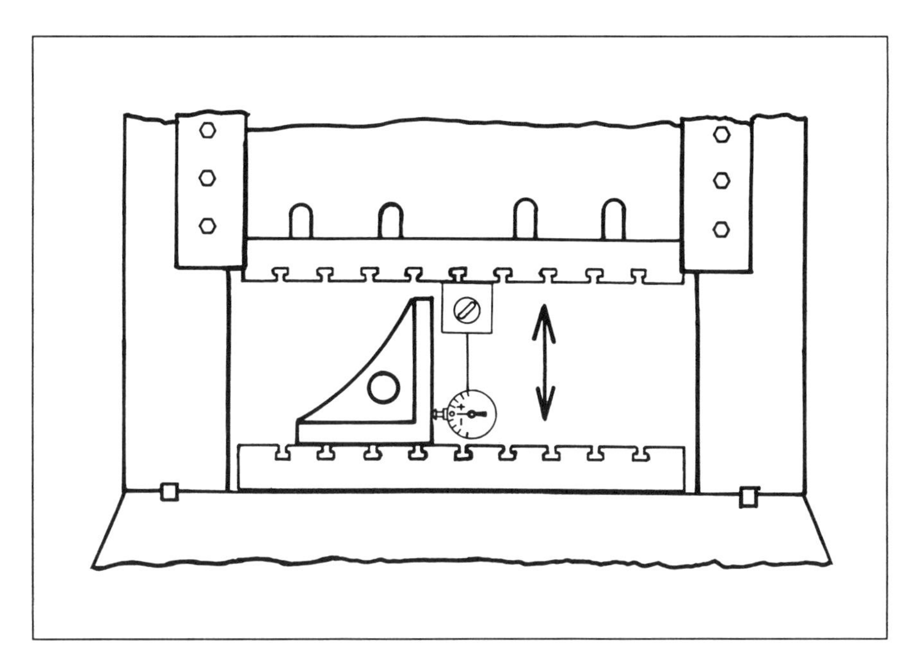

Figure 10-14. Checking the slide motion path for perpendicularity with the press bolster. (Smith & Associates)

The perpendicularity test, known as "tracking the slide," determines the path of the slide as it travels through the stroke. Equipment required consists of a dial indicator with a magnetic base and a precision square. The square used may be a precision machinist's square, an angle plate, a tooling inspection cube, or a cylinder square precision ground for the test.

As with any pressworking operation, all applicable safety rules must be followed and constant caution exercised. The press should be inched onto BDC before the square and dial indicator are placed in the press opening. There should be enough flat vertical surface on the square to permit inching the press through its entire stroke while maintaining contact with the dial indicator plunger. Properly done, this test is a fast reliable method for determining machine perpendicularity throughout the stroke.

Run the press completely through the stroke several times while noting the readings. The test is performed in both the front-to-back and left-to-right positions.

Acceptable results depend on the size and precision of the machine. For our example, the total indicated reading in both planes should not exceed 0.005 inch (0.127 mm). If ram to bolster parallelism and accurate vertical tracking cannot be obtained simultaneously by adjusting the machine, it is likely that the gib surfaces on the slide are not square with the face of the slide.

TIE-ROD PRESTRESSING

The structure of nearly every top-driven straightside press is normally held in compression by four tie rods. The amount of compressive preload used to hold the bed, columns, and crown together is determined by the amount of prestress or shrink applied to the tie rods when the press is assembled.

Most press service manuals explain the correct procedure to correctly prestress tie rods. The prestressing force is sufficient to maintain a compressive preload on the press housing components at full press capacity plus a safety factor. Once the safety factor is exceeded, the tie rods stretch, allowing the press crown to lift from the top of the columns.

Once the crown lifts due to an overload, the tie rods stretch easily. This serves much like a safety valve to protect the other press parts. The amount of force developed by the machine in a catastrophic overload is limited to the yield strength of the tie rods. The tie rods will require prestressing if they are not damaged completely.

The tie-rod nuts are installed and manually snugged up. Except for very small machines, it is not feasible to develop enough torque to manually prestress the tie rods sufficiently. Unless the press is equipped with hydraulic tie-rod nuts, the tie rods must be heated, the nuts tightened the correct amount, and the correct prestress tension is attained upon cooling.

Amount of Tie-rod Prestressing

Proper prestressing of the press tie rods is essential if a straight-side press is to function properly. It is very important to equally tighten all tie rods the correct amount.

An accepted amount of tie-rod prestress used for many years is 0.0007 inch (0.018 mm) per inch (25.4 mm). This equates to approximately 0.0084 inches (0.213 mm) per foot (302.8 mm), which is about 700 parts per million or 700 microstrain. The tensile stress developed in a tie rod made of medium carbon steel at this strain level is approximately 20 ksi or 137,880 kPa. The actual stress is somewhat less once the tie rod has cooled, due to the elasticity of the members held in compression. An accurate way to achieve the 700 microstrain tie-rod prestress when the manufacturer's recommendations are not available, may be calculated using the following formula:

$$Deg = \frac{L \times T}{4}$$

$$Deg = \frac{200'' \times 4\ TPI}{4}$$

$$Deg = 200$$

Where:

DEG = Degrees of rotation to tighten the tie-rod nut after the rod is heated.

L = Length between tie-rod nuts in the same units as T.

T = Threads per unit measure in the same units as L.

4 = Constant used in formula irrespective of units used.

Example of Calculation

In the example seen in Figure 10-15, inch units are used. The length between tie-rod nuts is measured and found to be 200 inches. The thread pitch is four threads per inch. The tie-rod length, 200 is multiplied by four, which equals 800. Dividing 800 by the formula constant four equals 200; the number of degrees that the tie-rod nut must be turned for proper prestressing.

Figure 10-16 illustrates how the tie-rod nut is marked with chalk prior to heating. Use a protractor to ensure accuracy.

Mark the nuts prior to applying the heat, because when the tie rods are heated with an oxygen fuel-gas torch they tend to bow toward the side being heated. This movement can cause a properly snugged-up nut to walk or loosen.

If the bottom tie-rod nut is easily accessible, it can be tightened instead of the top nut. Here a simple procedure is to heat the tie rod until the correct space between the lower nut and bottom of the bed is attained.

The spacing is measured with a thickness gage. In the case of the press illustrated in Figure 10-15, a preload of 0.0007 inch (0.0178 mm) per one inch (25.4 mm) of tie-rod length is required. By multiplying the required preload per inch of length (0.0007 inch)

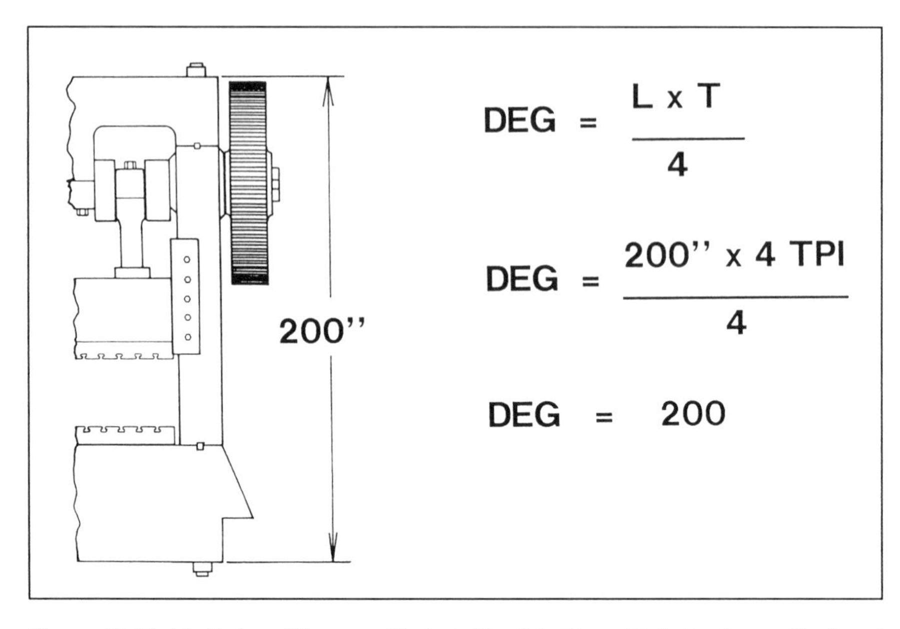

Figure 10-15. *A half-view of the press illustrated back in Figure 10-8 showing application of the tie-rod shrinkage formula.*

times the overall tie-rod length of 200 inches (5080 mm), the result is 0.140 inch (3.556 mm).

The four tie rods are snugged up as tightly as possible. All tie rods are gradually heated until a 0.140 inch thick feeler gage can be inserted into the space caused by the tie-rod expansion. When this space is created, the lower tie-rod nuts are snugly tightened.

Proper Application of Torch Heat

Presses designed for tie-rod prestressing with a torch have one or more access holes in the uprights or columns through which the flame is applied to the tie rod. Prior to heating, it is very important to thoroughly clean any oil, grease, and debris out of the column housing and press area. Portable fire extinguishers should be on hand and a fire watchman should be posted to observe the basement or press pit for any problems during and, for a reasonable period of time, after the heating work is completed.

Mineral fiber insulation is placed between the tie rods and sides of the column to keep the uprights as cool as possible. This will decrease the amount of heat needed to expand the tie rods and

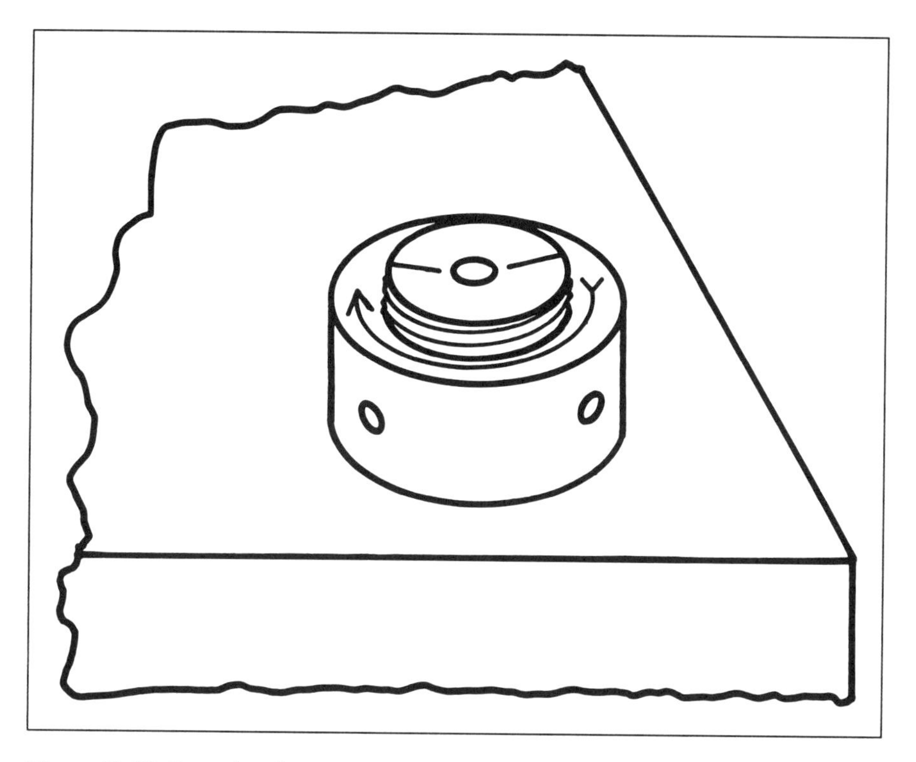

Figure 10-16. *Procedure for marking the tie-rod nut with chalk and illustration of how it is turned after heating.*

avoid damage to oil lines and wiring. If wiring is too close to the area to be heated, it should be temporarily disconnected and pulled out of the way. Both serious accidents and expensive damage have resulted from control wiring ruined by careless tie-rod heating.

All four tie rods are heated gradually, moving the flame from one to the others in turn. Time is required for the concentrated heat to be conducted along a substantial length of the rod. More than one torch may be needed for large work. Care must be taken not to overheat the metal. If a small area of the tie rod is heated excessively, the metal may yield as the tie rod shrinks, resulting in less than the desired amount of prestressing.

The final nut adjustments should be completed at the same time to ensure an even preload on the frame of the machine. Finally, the press is allowed to cool and the alignment checked as outlined previously.

Electrical Resistance Tie-rod Heaters

Many presses built over the last several decades have holes drilled several feet (a meter or more) into the top end of the tie rods. These holes are intended to hold electrical heating elements designed especially for prestressing tie rods. The larger sizes are designed for connection to a three-phase current source.

Four heaters in a matched set are used simultaneously, assuring uniform heating. Once the proper expansion is attained, the tie-rod nuts are tightened.

The use of electrical heaters is much easier and the results more certain than flame heating. However, a plan should be worked out in advance, especially if an emergency, such as a press stuck on bottom, should occur.

The heaters draw a large amount of current so a safe electrical supply of the required capacity, voltage, and phase must be available. Other common sense precautions include making sure that no liquids or debris are in the holes before inserting the heaters, which must be fully inserted into the hole. If the heater surface projects above the tie rod, it will overheat, perhaps ruining the unit.

Hydraulic Tie-rod Nuts

Tie rods with hydraulic pistons to permit cold tensioning greatly speed up press maintenance work. They are available as standard items on many new presses and can be retrofitted to older machines.

Figure 10-17 illustrates a sectional view through a hydraulic tie-rod nut. The nut engages the threads on the tie rod, and has been tightened manually to remove all clearance between the top and bottom nuts and the mating surfaces on the press bed and crown. The piston is shown in the withdrawn or unpressurized position.

The hydraulic power source may be either a hand-actuated pump or a small high-pressure power-driven unit. Relatively low volumes of fluid at high pressure are required. Typical peak operating pressure is 10,000 psi (68,940 KPa). Even with such high operating pressures, the effective piston area must be several times larger than the cross-sectional area of the tie rod. The hydraulic force must be capable of tensioning the tie rod to well over 20,000 psi (138,880 KPa), the normal working load of the tie rod. Additional force is required to install the nut and to assist in getting a press stuck on bottom of stroke freed-up in an emergency.

Figure 10-18 illustrates a hydraulic tie-rod nut with the pressure applied. Here the piston is shown in the fully extended position.

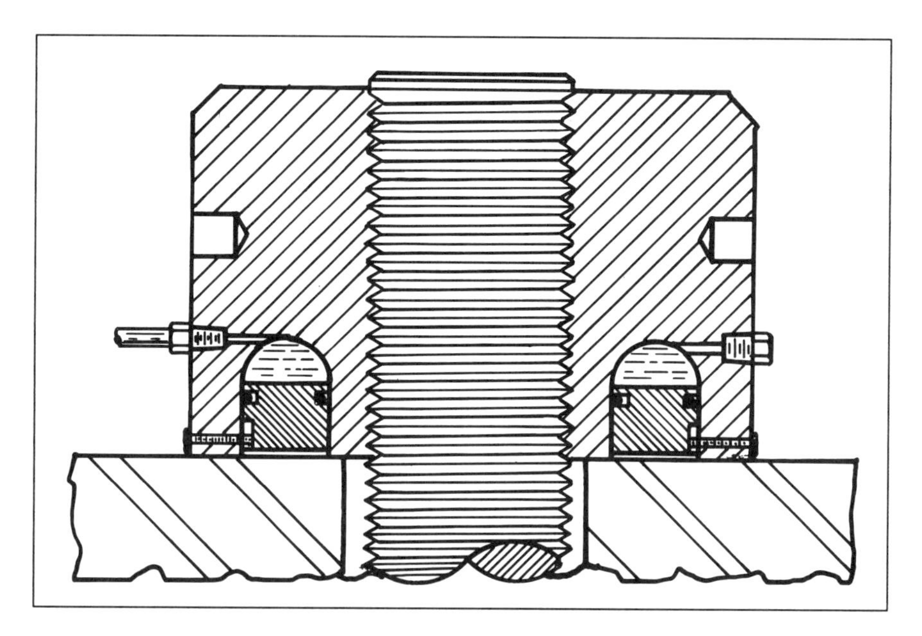

Figure 10-17. *A sectional view through a hydraulic tie-rod nut, which has been manually tightened prior to prestressing. The piston is shown in the withdrawn or unpressurized position. (Verson Corporation)*

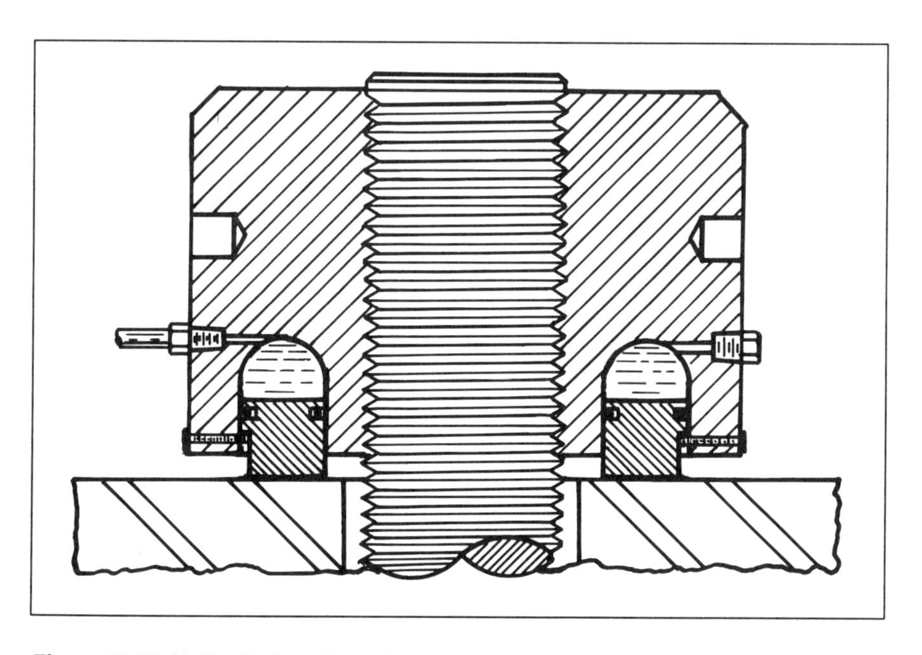

Figure 10-18. *Hydraulic tie-rod nut with pressure applied. (Verson Corporation)*

The piston is a donut-shaped ring fitted with hydraulic packing. The matching piston bore is machined into the tie-rod nut body. Note the pressure inlet connection and air bleed valve.

Once the nut is pressurized, segmented steel shims are inserted around the outer edge of the nut. Figure 10-19 illustrates the shims inserted prior to releasing the hydraulic pressure. The thickness of the shims, which are supplied by the press manufacturer, provides the correct value of tie-rod pretensioning for the press.

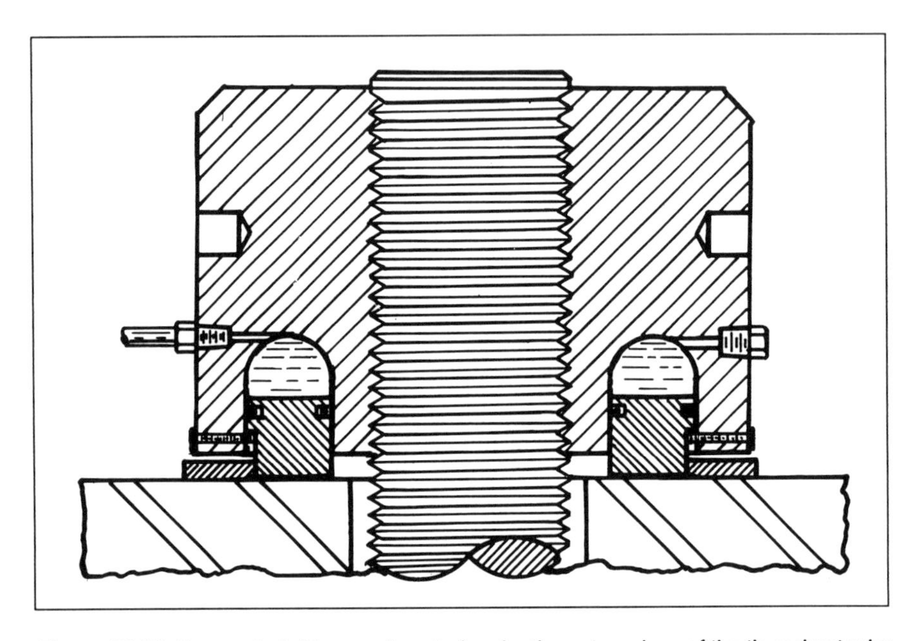

Figure 10-19. Segmented shims are inserted under the outer edges of the tie-rod nut prior to releasing the hydraulic pressure. (Verson Corporation)

Finally, the hydraulic pressure is released, as illustrated in Figure 10-20. Here, the segmented shims are tightly clamped in place. The press frame is uniformly held in compression.

Figure 10-21 illustrates a view of how a hydraulic tie-rod nut piping system connects to four nuts. A centrally located hand-operated pump is used to pressurize the system. All four nuts are properly snugged up, as was seen in Figure 10-17. The correct segmented shims supplied by the manufacturer should be used. If they are missing, shims of the correct thickness can be fabricated. It is advisable to make them of alloy steel and heat-treat them for toughness. The thickness to achieve a nominal prestress of 20,000

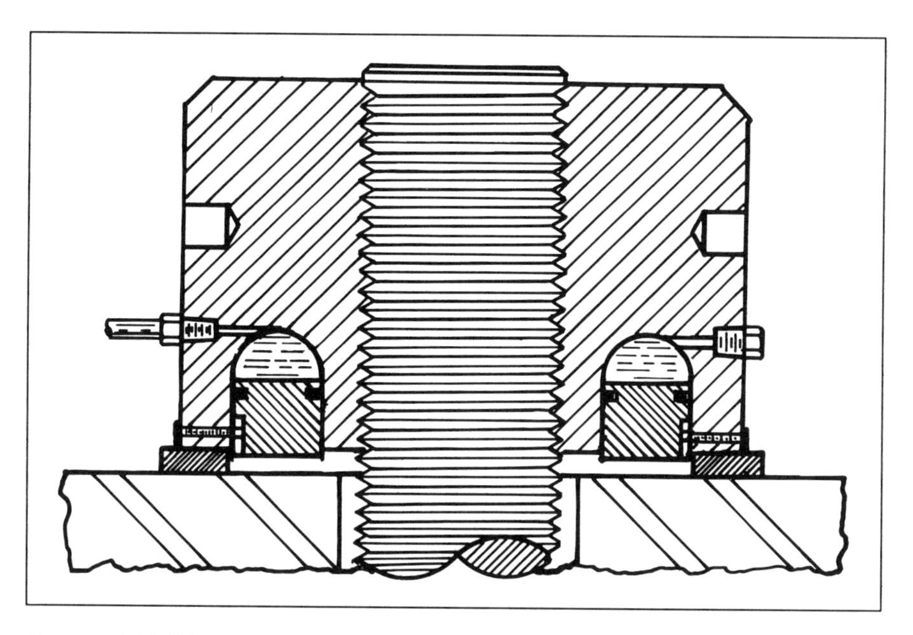

Figure 10-20. When the hydraulic pressure is released, the segmented shims are tightly clamped in place, holding the press frame in compression. (Verson Corporation)

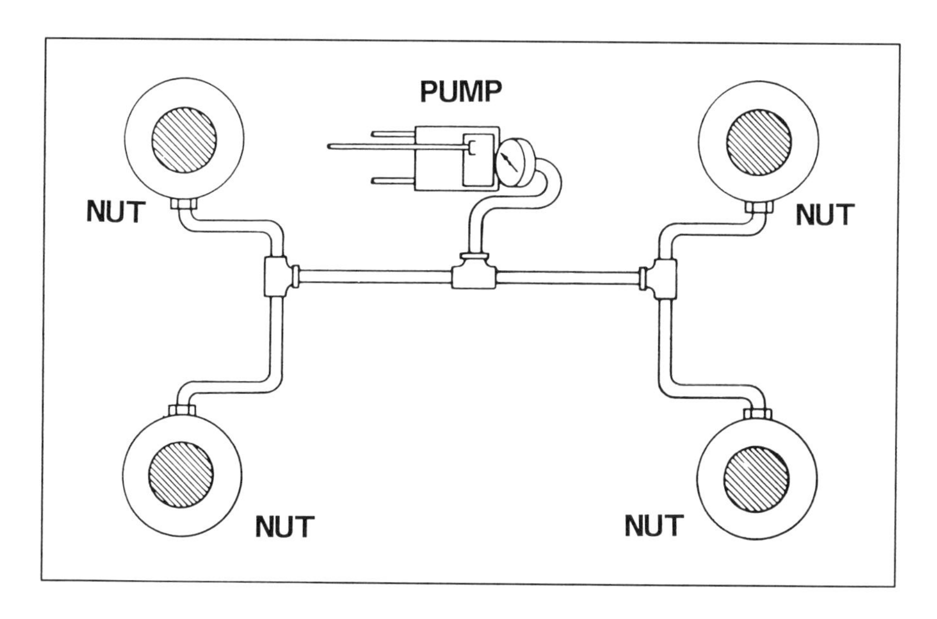

Figure 10-21. Top of a press crown illustrating hydraulic tie-rod piping and hand-actuated pump. (Verson Corporation)

psi (137,880 KPa) tension in steel tie rods can be determined from the formulas presented earlier.

Some presses have heat-treated washers used under the tie-rod nuts to spread the load over the crown and bed surface. They also serve to avoid wear to these surfaces. It is very important that they be used if specified by the manufacturer.

Assembling Straightside Presses

With few exceptions, the frames of most straightside presses are held in compression by four tie rods. Figure 10-8 illustrated how the frame members in an assembled press fit together. The process of assembling the press is termed stacking the machine or stacking the press.

The bed, uprights, and crown may first be stacked together. The tie rods may be inserted last if there is sufficient overhead clearance. An eyebolt hole is usually provided for lifting. If not, a cable sling must be used.

An alternative procedure is to insert the tie rods before stacking the crown and, in some cases, the uprights. Some very large machines require a cylindrical pit or well under each tie rod location. After placing and leveling the bed in position, the tie rods are lowered through the bed into the wells. Once the uprights and crown are stacked in place, the tie rods are lifted into position with an eyebolt and cable.

Machined mating surfaces such as the bed, uprights, and crown should have any burrs removed and these parts should be checked for flatness. The keyways should be in good condition to provide a proper locational fit.

If a lack of flatness is found, the machine possibly was subjected to an extreme overload condition. If such an overload produced a large lateral force, mating surfaces on the bed, uprights, and crown may be warped. The only good solution is to remachine or replace all machine components as needed. If proper performance is expected, all fits and tolerances must be restored to the manufacturer's specifications.

Responding to a Large Overload

Common causes of large overloads are diesetting errors and foreign object damage. The machine may have damage in such an overload, making it dangerous to operate. The press should be removed from service and thoroughly inspected for damage.

After a conventional inspection with dial indicators and inside micrometers, load cell testing in conjunction with waveform signature analysis is an especially good procedure. Load cell tests provide an accurate means to gradually increase the force of the machine to full capacity.

This test permits observation for cracked members as evidenced by breathing and irregular operation. The waveform signature can be used to detect loose (stretched) tie rods, as well as pinpointing timing errors caused by partially sheared keys, twisted shafting, and similar problems.

If no damage is apparent after performing the tests described earlier in this chapter, a note of the incident should be made in the press maintenance records. Even though no damage may be detected, latent damage — such as the initiation of a stress crack — may occur which could cause a failure. In the event of future irregular operation, the proper procedure is to remove the machine from service and run a thorough inspection.

LOAD CELL TESTS AND CALIBRATION

Many stampers use load cells to verify press condition. This is excellent practice. Good machine alignment is a basic requirement for quality presswork with minimal die wear. It is important that the press maintain its alignment under full load and load cells provide an efficient and speedy method of making sure that the press remains in alignment when developing full rated force at the bottom of stroke.

Load Cells

The load cells used for press testing are patterned after designs originally developed for high force and precision weighing applications for the aerospace and transportation industry. Baldwin Locomotive Works pioneered the development of the modern foil strain gage. In 1954, Baldwin was marketing a variety of load cells, having capacities of 3,000,000 pounds (1,361,000 KGm), or more.

Four cells are recommended for press testing. Load cells specifically designed for press testing are commercially available in matched sets of four with individual capacities of 100, 250, and 1,000 short tons (889.6, 2,224, and 8,896 KN) each. Other sizes and configurations are available on special order. Figure 10-22 illustrates load cell placement for such testing.

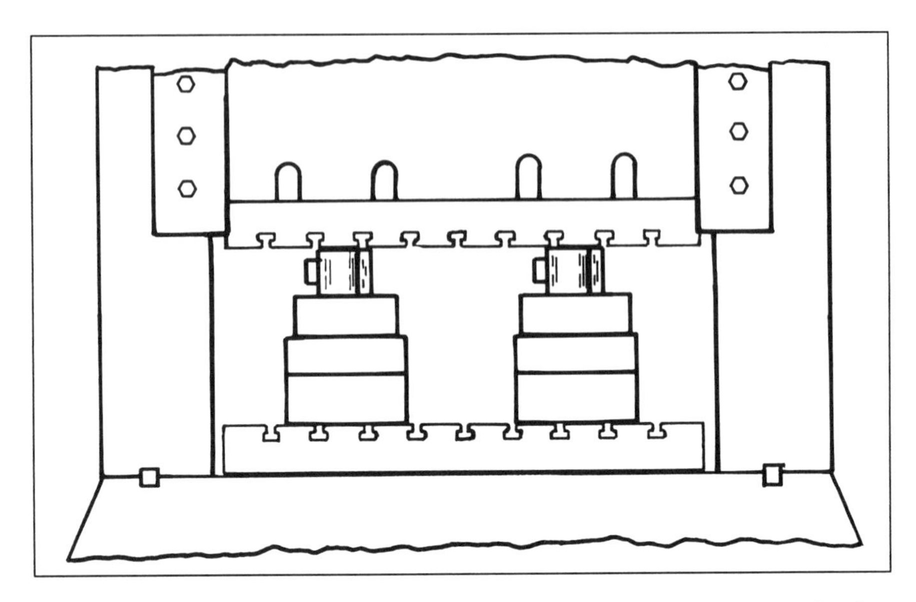

Figure 10-22. *Load cells placed on solid steel support blocks for press testing and calibration. If the load cells do not have built-in readout devices, an external readout is used. (Smith & Associates)*

In performing the load cell test, as with any other press operation, it is extremely important to exercise constant caution. Before the supports and load cells are placed, the press must be inched onto BDC. To be sure of accurate repeatable results, it's important to make sure that the load cells are placed at the same relative location on the press bolster for every test.

The slide motor limit switches must never be bypassed or jumpered out to decrease the press shut height to less than the press rating to make contact with the load cells. Not only is there danger of mechanical interference, there may not be enough adjusting screw thread engagement to withstand full press tonnage.

Load Cell and Support Placement

One factor in deciding the number and placement of load cells is that the cells will cover only a small fraction of the bed and slide area in most presses. Generally, the press can be tested to one-quarter slide capacity per corner by using four load cells. Care must be exercised to ensure that localized forces do not damage the press.

On presses having four connections, the cells are usually placed directly under the connections. The support provided by the bolster and bed must be adequate. With large presses, where testing at full capacity is desired, heavy-steel plate may be required under the load cell supports to distribute the pressure over a sufficiently large area of the bed.

A good location for placing the cells is directly in line with the vertical plate making up the box-like structure of the press bed and slide. The press manufacturer should be consulted for information on safe load cell placement if there is any question as to the procedure's safety.

Once the load cells are correctly placed, slide and shim adjustments may continue until full press capacity is attained. While making the test, position observers around the machine to look for problems such as cracks breathing in the bed, crown, and slide. Listen for any unusual sounds that could indicate mechanical problems. Immediately terminate the test if improper press operation is noted as loading is increased.

Proper Load Cell Supports

Do not put a person in charge of load cell testing who does not understand the forces involved and how they are transmitted through the machine, the load cell supports, and the cells themselves. For example: at capacity, the average pressure on the working face of a 250 ton (2224 KN) load cell having a diameter of five inches (127 mm) is 25,478 PSI (175,645 KPa). A sufficient quantity of solid steel or cast iron load cell support blocks must be available to provide proper support within the press shut height rating.

The pressures involved are sufficient to slightly mark the freshly machined surfaces of a new press. To prevent marking the surfaces of the slide, bolster, or load cells, sheet aluminum or thin fiberboard can be used to even out slight irregularities in the mating surfaces.

Supports for load cells should be made of solid material slightly larger than the diameter of the cell itself. Avoid using supports made of hollow mechanical tubing. Even if the strength of a hollow support is sufficient to avoid permanent deformation, the nonuniform support of the cell face and resulting uneven strain in the cell structure may contribute to inaccurate readings.

The care that must be exercised when adjusting the slide to strike load cells is no different than that required when placing a die in the press. A tape measure should be used to verify that there is enough shut height to accommodate the load cells and supports, with approximately 0.250 inch (6.35 mm) extra press shut height as a safety factor.

When cycling and adjusting the press slide, exercise extreme caution at all times to avoid damage to the press, the load cells, and other equipment due to overload conditions. The press should be adjusted downward a little at a time while observing the load cell readout.

To avoid sticking the press on bottom, the press should be rapidly cycled through the bottom of the stroke. Instruct all persons involved in the test to be careful not to trip press safety devices, such as light curtains, while the press is being cycled as this may result in sticking the press on bottom.

If a misadjusted or damaged press member is discovered, the test should be interrupted and the cause corrected. Always rely on traditional observation skills rather than instrument readings when lowering the ram because the readout device may be incorrectly calibrated.

The load cells are normally connected to a portable tonnage meter, which is calibrated to provide readings in either actual force or percentage of load cell capacity. To simplify the work, load cells with built-in digital readout capability are available.

The readouts indicate the load on each of the cells. Then, to equalize the tonnage on each cell, shims are added to the load cells with the lowest indicated tonnage. The thickness of the shims indicates the amount of press misalignment at the location of each load cell.

The load cell test enables fast and accurate calibration of the tonnage meter for the press. It is a simple matter of adjusting the dedicated meter on the press to the same readings as those of the portable meter.

In-house Capability

Many companies have a policy of doing as much in-house mechanical and electrical work as possible. Tasks range from installing simple safety equipment to state-of-the-art die, press, and personnel protection equipment.

Installing tonnage meters and other electronic equipment in house can save money. Even if electronic technicians who specialize in installing and calibrating tonnage meters are hired, local support personnel are still required to mount the equipment, do welding, run power wiring, and often provide load cell supports.

Total cost of a set of load cells, portable tonnage meter, chart recorder, and support blocks, can be recovered in saving the installation fees of a number of tonnage meters. Also, there is the added bonus of having the equipment available, for the periodic press preventive maintenance inspection program.

There are several reasons why this is a good practice. Once the maintenance personnel are trained to install the equipment properly, they also understand how to maintain it. Developing this in-house capability is a way to maximize human resources and employee pride in accomplishment.

Load Measurement on Underdriven Presses

The best measuring system for underdriven single-action and toggle presses is direct application of strain gages to the pull rods. The strain gages are used in a Wheatstone bridge configuration with the two halves of the bridge applied to opposite sides of the pull rod. This method cancels any errors due to bending. The wires are brought out through coiled plastic air hose. Such installations give years of dependable service. An added feature of tonnage meters used with this system is a means to compare crankshaft degree readings with the instantaneous load to provide an alarm if the tonnage curve of the press is exceeded.

ANALYSIS OF MACHINE FAILURES

Figure 10-23 illustrates the characteristic appearance of a failed crankshaft due to gradual crack propagation. A sudden overload may have caused the crack. The crack growth leading to failure may have occurred over several years of operation. The failure itself often occurs quite unexpectedly under moderate loads.

Most of the fracture surface displays "oyster shell" marks typical of crack growth bands. This crack growth pattern is often caused by slight changes in the direction of crack propagation due to changes in load levels.

Often, there is also considerable evidence of fretting corrosion, especially in the old part of the fracture. Here the opposite sides of the crack were rubbing under conditions of high pressure for some time.

The final failure occurs when there is no longer enough area of sound metal to transmit the required torque. That area has a rough appearance, indicating that the metal was torn apart. The final part

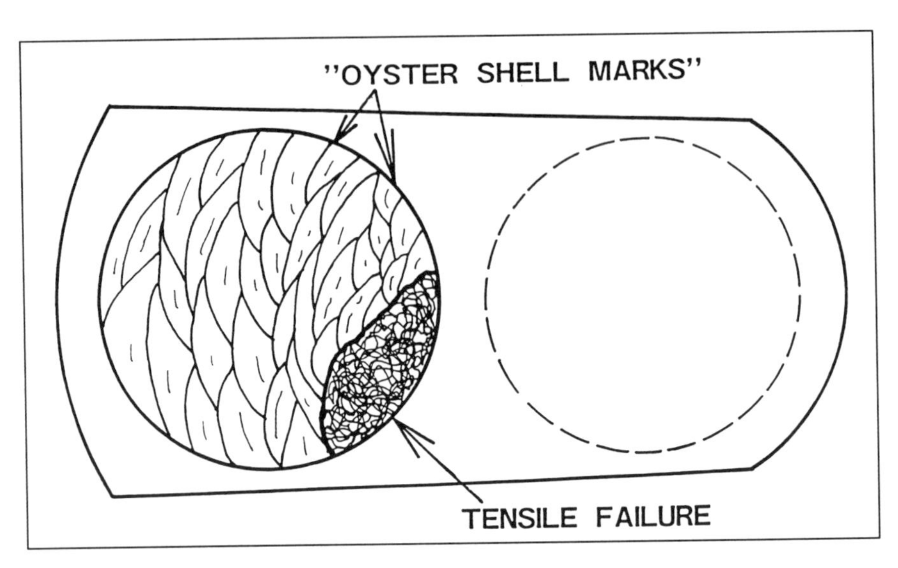

Figure 10-23. *The characteristic appearance of a failed crankshaft due to gradual crack propagation. Most of the fracture surface displays "oyster shell" marks typical of crack growth bands. The final tensile failure occurs when there is no longer enough area of sound metal to transmit the required torque. (Smith & Associates)*

of the fracture is essentially a tensile failure of the remaining sound metal.

Gradual failures can have many contributing causes. While a sudden, long-forgotten overload may have initiated the crack, there may be other causes. Repeated cycling at loads too high for the machine design may initiate the crack. The ultimate fracture often occurs unexpectedly under moderate loads.

Sharp corners will create a stress riser and should be avoided in the design of crankshafts and other cyclically loaded machine parts.

Small cracks may be difficult to spot. If the machine is disassembled for any reason, it is wise to carefully inspect the crankshaft, gears, pitmans, and other highly stressed components with an appropriate nondestructive testing method.

Sudden Catastrophic Failure

The majority of press component failures exhibit the evidence of crack propitiation illustrated in Figure 10-23. However a very large overload, typically three or more times press capacity, may result in complete fracture of a component.

Figure 10-24 illustrates a crankshaft failure resulting from a single extreme overload. A typical cause is extreme lack of shut height when diesetting, or mechanical interference due to hitting multiple blanks in a drawing or reforming operation.

When such a failure occurs, it is likely that several other machine components may have been damaged as well. Careful inspection and alignment of the entire machine is required.

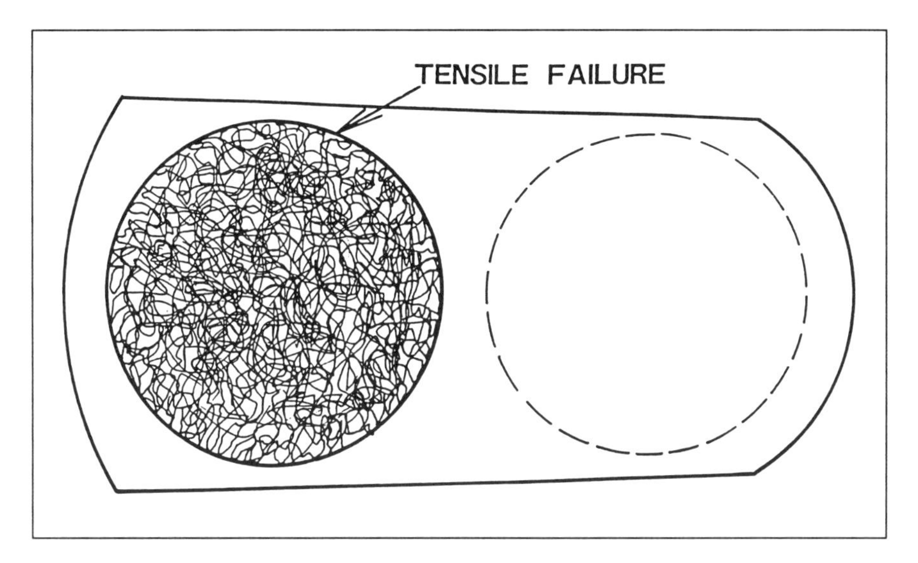

Figure 10-24. *The appearance of a failed crankshaft due to a catastrophic overload. Entire fracture surface displays the rough torn metal typical of a tensile failure. (Smith & Associates)*

Inspect Driving Keys for Tightness

All driving keys should be checked for tightness and signs of any play. Even a slight amount of play can permit the key to work loose, resulting in increasingly severe movement and wear. If not corrected, the key may fall out or experience shear failure due to repeated shock loading. Tensile stress concentration at the corners of the keyway may result in crack propagation and component failure. This is often a contributing cause when gear hubs fail.

REPAIR VS REPLACE CONSIDERATIONS

A machine is simply an assembly of parts that work together to accomplish the intended function. When the parts become worn or broken, the machine cannot function as intended. Repairing or replacing the defective parts will restore the machine to like-new performance.

A press of good design is capable of several decades of service without major maintenance requirements, provided it is not abused. All that is generally required is strict periodic maintenance that includes a carefully followed lubrication schedule.

If a press fails to give long service before major work is required, the root causes usually fall into one or more of several categories of abuse and neglect. These are:

- Overloading;
- Lack of lubrication; and
- Poor alignment.

Cost Factors

Press repairs caused by normal wear usually are cost effective, provided the machine serves the needs of the business. Such items include rebuilding the clutch and brake, replacing worn bearings, and updating electrical controls. These costs can be predicted by developing a scheduled maintenance program that incorporates the normal wear of press parts. However, if the press structure is cracked or broken, the components in question either need to be replaced or expensive welded repairs carried out, followed by remachining. Any welding must be done by a shop capable of exercising the strict procedural control needed to ensure like-new strength and durability.

Clutch and Brake Repairs

Clutch and brake parts usually are not repaired by welding. Metallic failures in these components indicate chronic overloading problems. Any error in the welding procedure can compromise the safety of pressroom personnel.

If clutch or brake components fail due to metallic fracture, the possibility of press overloading should be investigated. Overloading failures come from repeated shock and overheating. If tooling or process modifications cannot correct the overloading problem, a higher capacity press is needed.

Overloading a press to perform work beyond its rated capacity is not only illegal, it is dangerous and almost certain to damage the machine. If only a small volume of work exceeds available press capacity, work should be sourced to a shop having a machine with the needed tonnage available. If this option is not acceptable, a larger press must be purchased.

Gear Repairs

Gears are expensive to rework or replace. It is often possible to turn or rekey gears to use surfaces subjected to little wear in high-wear areas. This essentially doubles the life of the gearing. Presses having worn gearing can often be rebuilt without the expense of gear replacement.

Should new gears be required, this factor, together with other required repairs, may favor finding a replacement for the press.

Suitability of the Press

If the press is not ideally suited for the type of work performed, one may assume that reduced productivity and/or tooling life will result. An example is a long-stroke triple back-geared deep-drawing press used to run a progressive die only because the machine is available, and has the needed bed size.

BIBLIOGRAPHY

1. C. Wick, J. T. Benedict, R. F. Veilleux, *Tool and Manufacturing Engineers Handbook,* Volume 2, Fourth Edition — Forming, Society of Manufacturing Engineers, Dearborn, Michigan, 1984.
2. J. Fredline, *Practical Solutions for Troubleshooting Metal Stamping Presses*, SME Technical Paper MS88-200, Society of Manufacturing Engineers, Dearborn, Michigan, 1988.
3. D. Smith, *Quick Die Change,* Chapter 27, "Selecting Maintenance Software," (courtesy of P. Stang, Vice President of Professionals for Technology), Society of Manufacturing Engineers, Dearborn, Michigan, 1991.
4. J. Ivaska. *Press Lubrication, A Predictive Maintenance Approach,* Presented at "How to Increase your Stamping Profitability," Society of Manufacturing Engineers, Dearborn, Michigan, October 27-28, 1992.
5. J. R. Pugh, *Advances in Automatic Lube Systems for Large Stamping Presses,* Presented at SME-FMA Presstech Conference, Detroit, Michigan, May, 1990.
6. *Bliss Power Press Handbook,* E. W. Bliss Company, Hastings, Michigan, 1950.
7. H. Daniels, *Mechanical Press Handbook,* Third Edition, Cahners Publishing Company, Boston, 1969.

8. H. Nielsen, Jr., *From Locomotives to Strain Gages*, Vantage Press, New York, 1985.
9. D. Falcone. Information on self-contained load-cells having direct electronic readout is found in the published data of Toledo Transducers, Inc., Toledo, Ohio, 1992.
10. D. Smith, *Adjusting Dies to a Common Shut Height*, SME/FMA Technical Paper TE89-565, Society of Manufacturing Engineers, Dearborn, Michigan, 1989.
11. D. Smith, *Die Design Handbook*, Section 27, "Press Data," Society of Manufacturing Engineers, Dearborn, Michigan, 1990.
12. D. Smith, *Quick Die Change*, Chapter 28, "Instituting a Tonnage Meter Program," Society of Manufacturing Engineers, Dearborn, Michigan, 1991.
13. D. Smith, Video, *Quick Die Change*, Society of Manufacturing Engineers, Dearborn, Michigan, 1992. Illustrates load cell testing, waveform analysis, proper sensor placement, and tonnage meter installation.
14. Bradley K. Mettert, *Load Sensor Placement and Tonnage Data for Underdrive Presses*, SME Technical Paper MS90-384, Society of Manufacturing Engineers, Dearborn, Michigan, 1990.

CHAPTER 11

PRESS SAFETY AND AUXILIARY EQUIPMENT

Edward V. Crane, in his classic reference work, *Plastic Working of Metals and Non-Metallic Materials in Presses,* first published in 1932, cites the advantages of automatic press operation. In addition to much greater productivity, Crane cited avoidance of operator mental fatigue and amputations as a benefit of automatic operation. Automatic operation avoided the need for the operator to place his or her hands in the "danger zone" which is also termed "a pinch point" or "the point of operation." Crane also explained the cost effectiveness of electrical and electronic die protection systems, and illustrated examples.

Today, the same advantages of automatic operation obviously exist. Even so, many pressworking operations continue to be hand-fed in a manner that exposes the operator to serious injury should an equipment malfunction occur. It is not this book's intent or purpose to give advice on making specific operations safe. Instituting proper provisions for personnel safety in the workplace is the duty of the employer.

Voluntary standards organizations also call on industry experts to formulate guidelines for safe operation of power equipment. In some political jurisdictions, government authorities make laws defining minimum requirements for safe operation of power presses. Often, these laws embody the recommendations of voluntary standards organizations.

OVERVIEW OF PRESSWORKING SAFETY

For liability reasons, neither SME nor the author represents the information on press safety systems contained in this work to reflect current law, or to meet the exact safety requirements of any company or governmental regulatory body. Rather, this is a partial discussion of some of the risks historically associated with pressworking, together with some safety measures that minimize the danger to personnel.

Any time a hand or any part of the body is in a point of operation, i.e. pinch point or crush area, in or around a power press, there is a statistical probability that the press or ancillary device, such as a feeder, may actuate unexpectedly and cause an injury. For example, the clutch and flywheel system operate with close spacing of frictional surfaces and bearing clearances when not engaged.

Should a mechanical seizure of the clutch occur, the press repeats through several cycles, even though the clutch was signaled to deactivate. The brake slips if a mechanical failure locks the clutch in solid engagement. Because of this danger, power presses are equipped with clutch/brake control reliability monitors to provide warning of clutch or brake deterioration some time before a catastrophic failure.

Power Lockout

It is a requirement that a power lockout procedure be followed whenever any part of the body is in the point of operation during repair or maintenance operations. The essence of power lockout is to lock out all power sources and dissipate stored energy that could cause unexpected movement of the press or auxiliary equipment.

If maintenance work is to be done with the press in the open position, special props called *die safety blocks* are used to block the slide from moving. The blocks must support the static weight of the slide, attached linkage, and maximum upper die weight. While a safety factor is included, the blocks are not designed to withstand full press tonnage load under power.

The block must be a snug fit in the press opening to prevent any slide movement. An aluminum block having a large screw for length adjustment and an attached interlock plug is illustrated in Figure 11-1. The interlock prevents the press from accidentally cycling with the block in place, which could cause catastrophic press and die damage. It could also result in serious injury to personnel from the flying safety blocks and machine components.

Avoiding Operator Injury

Many company safety rules do not permit hand-in-die operation. For low-volume production jobs, the use of safety tongs, hand vacuum tools, magnetic lifters, and simple gravity slides serve to accomplish the desired production, while avoiding employee exposure to the point of operation. However, such tools and production aids are not a substitute for proper point of operation guarding.

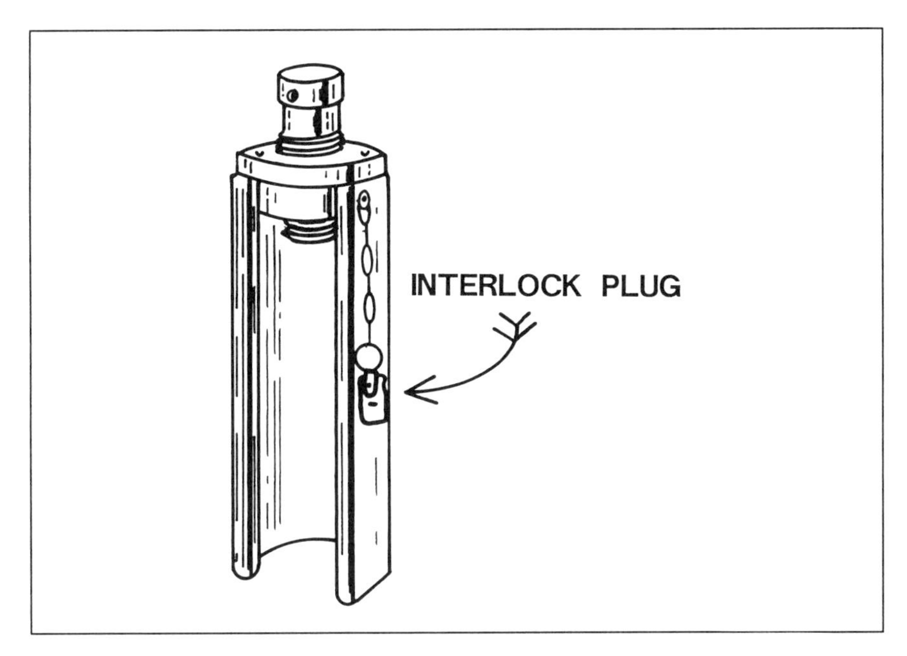

Figure 11-1. Aluminum die safety block equipped with a captive adjusting screw to provide a snug fit; note the attached chain and interlock plug. (Rockford Systems, Inc.)

Frequent amputation injuries result in high industrial compensation costs and fines. Depending upon various circumstances, employers and equipment builders may be required to pay the injured employee civil damages.

Holdout or Restraint Devices

Holdout (restraint) devices must prevent the operator from inadvertently reaching into the point of operation at all times. Figure 11-2 illustrates a holdout device used in conjunction with hand loading tongs. Attachments (wristlets) are provided for each of the operator's hands. These attachments must be securely anchored and adjusted in such a way that the operator is restrained from ever reaching into the point of operation.

Pullout Devices

Pullout (pullback) devices must withdraw the operator's hands should they be inadvertently located in the point of operation as the dies close. Pullout devices must have attachments provided for

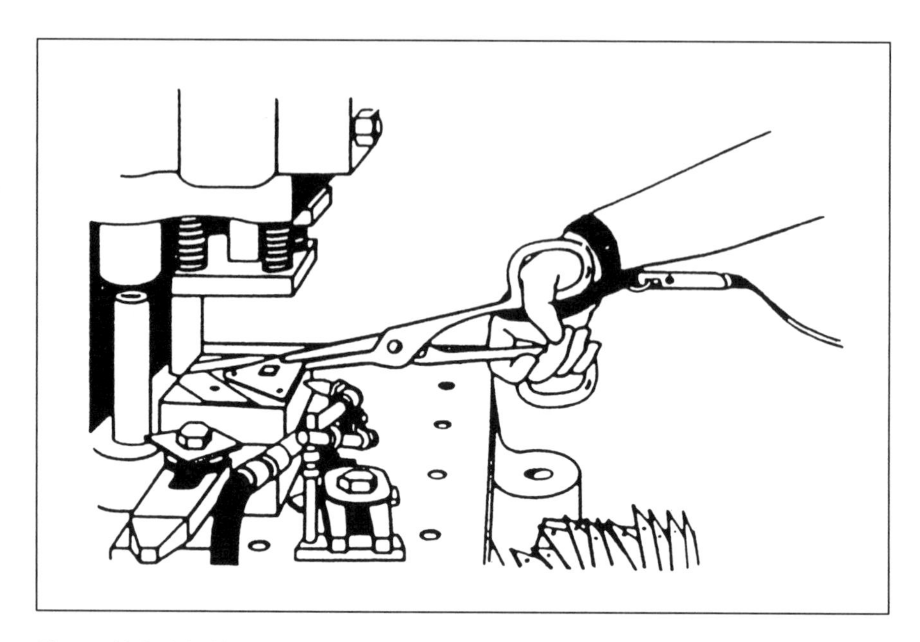

Figure 11-2. A holdout device used in conjunction with hand loading tongs. Attachments (wristlets) are provided for each of the operator's hands. (Rockford Systems, Inc.)

each of the operator's hands. These hand attachments are connected to, and actuated by the motion of the press slide. The mechanism withdrawing the operator's hands by means of slide motion is supported by a rigid framework which must be securely attached to the press. Figure 11-3 illustrates a pullout device in use.

Precautions for Using Holdout and Pullout Devices

For low-volume production, these methods are effective, provided proper safety precautions are taken. A key factor in achieving operator safety with both the holdout and pullout systems is strict supervisory oversight and inspection to ensure proper adjustment and maintenance of the equipment.

The entire pullout system should be carefully maintained, inspected for wear, and detailed records maintained. Proper adjustment should be established for all operators when they start or return to each job, and when dies are changed. Once the proper adjustment is established for an individual operator's pullout device, the supervisor or setup person should lock the settings in place and record the information.

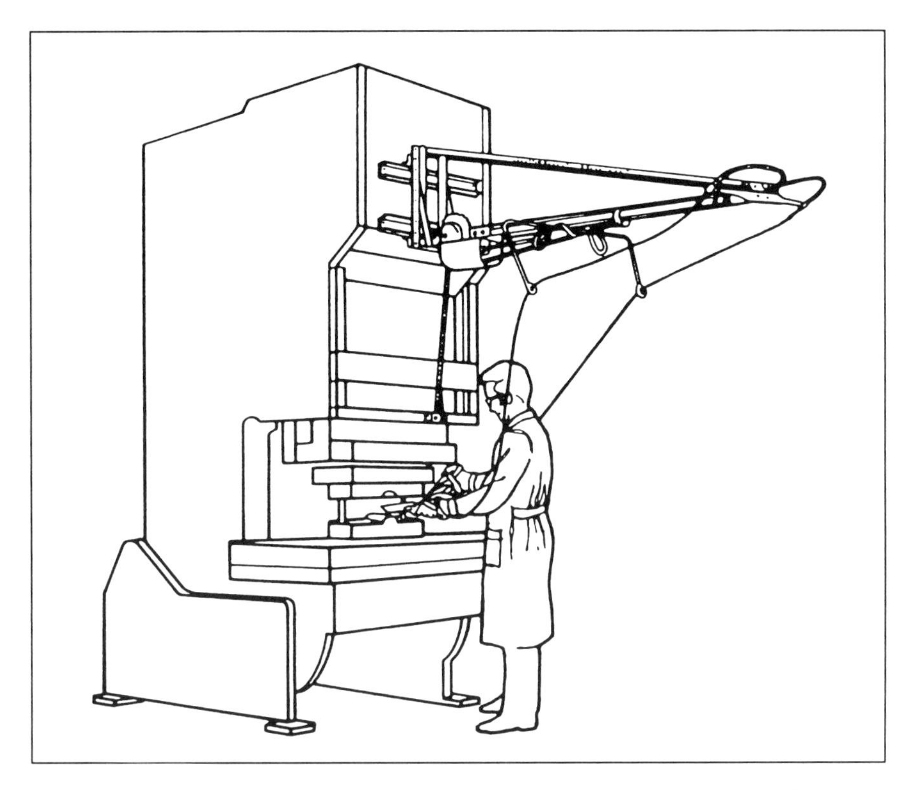

Figure 11-3. A ram-driven pullout device. If the operator's hands are in the point of operation as the dies close, they will be forcibly withdrawn. (Rockford Systems, Inc.)

The job should be inspected with the dies closed, and on the downstroke, to determine the point of operation hazard nearest to the operator. Such items as die cushion equalizer pins and diesetting bolts that are too long can constitute points of operation. They may also be dangerous projections that may catch the hand attachment of the pullout device upon descent of the slide. Should the operator's hands inadvertently be in the point of operation upon activation of the press, a clear path for their forceful withdrawal must be arranged.

Physical Barrier Guards

An example of a barrier guard that completely surrounds the point of operation of a strip-fed die is illustrated in Figure 11-4. Guards of this type are made of commercially available modular components, or fabricated from other materials. In the United

States, OSHA rules specify the size of opening permitted based on the distance to the point of operation hazard inside the guard. A physical barrier guard must make it impossible for anyone to place any part of the body in the point of operation hazard.

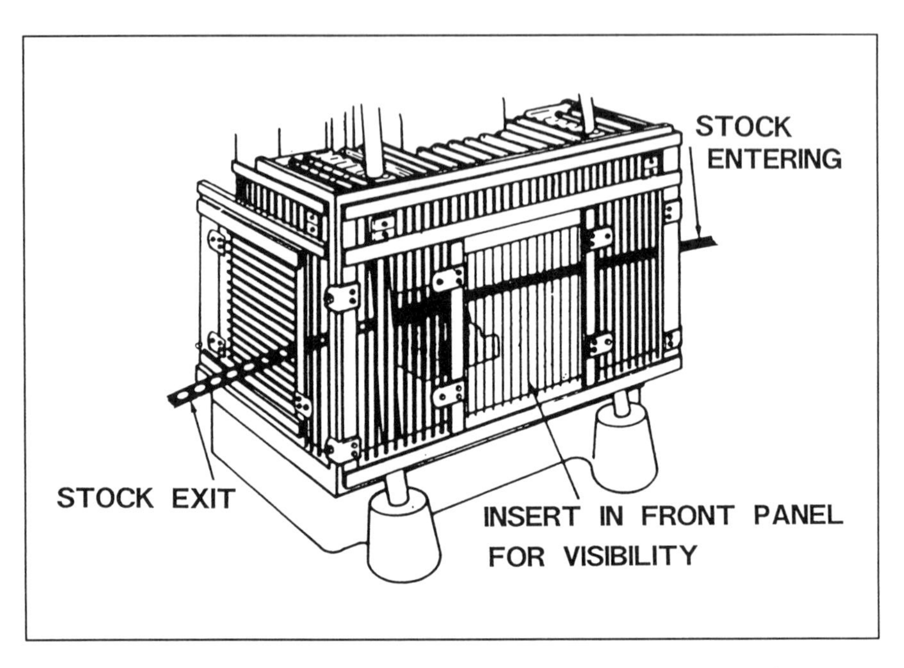

Figure 11-4. A physical barrier guard that completely surrounds the point of operation of a strip-fed die. (Rockford Systems, Inc.)

Security of Physical Barrier Guarding

It is important that the guard be designed, constructed, and installed in such a way as to completely safeguard the point of operation. This means that no one can place any part of the body into the point of operation by reaching through, over, under, or around the guard. Likewise, the guard itself must not form a pinch point between it and moving machine parts.

The employer must make sure that the guard is securely attached to the press, bolster, and/or die shoe. The fasteners must not be readily removed by the operator. For example, wing nuts are not recommended for guard attachment. Suppliers of press safety equipment can often provide tamper-resistant fasteners and special tools for their use.

The guard should provide maximum visibility of the point of operation consistent with its guarding function. A minimum guarding line of 0.5-inch (12.7-mm) spacing from the point of operation to the guard stock entry opening is an OSHA requirement for physical barrier guarding. Figure 11-5 illustrates point of operation guard opening sizes in inches based upon the distance to the nearest point of operation.

POINT OF OPERATION GUARD OPENING SIZES	
DISTANCE OF OPENING FROM POINT OF OPERATION HAZARD (INCHES)	MAXIMUM WIDTH OF OPENING (INCHES)
1/2 TO 1-1/2	1/4
1-1/2 TO 2-1/2	3/8
2-1/2 TO 3-1/2	1/2
3-1/2 TO 5-1/2	5/8
5-1/2 TO 6-1/2	3/4
6-1/2 TO 7-1/2	7/8
7-1/2 TO 12-1/2	1-1/4
12-1/2 TO 15-1/2	1-1/2
15-1/2 TO 17-1/2	1-7/8
17-1/2 TO 31-1/2	2-1/8
OVER 31	6

Figure 11-5. *An example of guarding requirements based on distance from the point of operation hazard.*

Figure 11-6 shows an especially convenient folding aluminum OSHA guard opening scale. The scale can be used to check the opening sizes in the guard for feeding material and operator protection to be certain that the OSHA guard opening spacing versus distance from the point of operation is correct.

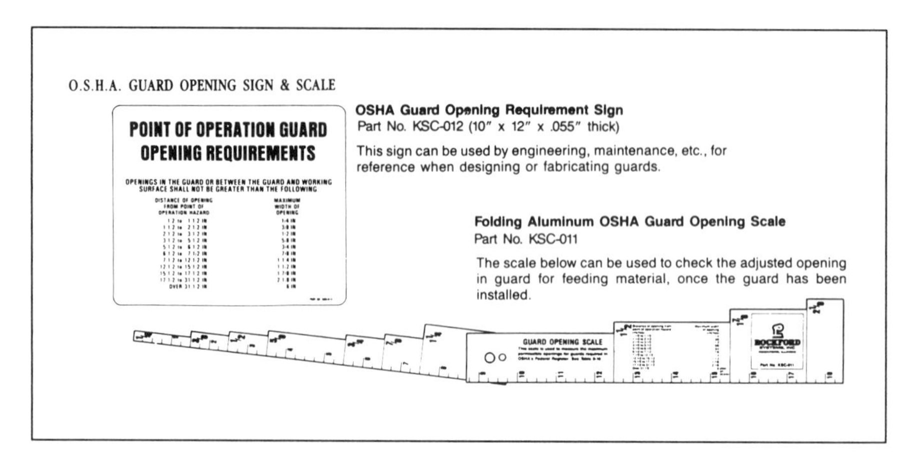

Figure 11-6. *A commercially available folding OSHA guard opening scale. (Rockford Systems, Inc.)*

Sliding Gate Physical Barrier Guard

A type of physical barrier guard having a sliding gate is illustrated in Figure 11-7. The gate is equipped with a limit switch wired to stop the press should the gate be raised. Physical barrier guards with movable gates having limit switches to prevent press operation unless the gate is closed, are among the most positive safeguards available.

It is absolutely necessary to make the safety switch as tamperproof as possible. However, virtually any safety switch can be defeated by a determined person. It is essential that all pressroom personnel be periodically instructed in safe practices.

Light Curtains

Safeguarding both automatic and manually loaded presses equipped with partial revolution clutches may be done with light curtains. Light curtains prevent the press from cycling should the plane of light be interrupted. Interfaced with the clutch/brake controls, they also stop the press if someone should inadvertently get into the plane of light. Breaking the light beam initiates stopping the press. This occurs only on the downstroke if the light curtain is muted on the upstroke to permit ejection of parts.

Light curtains can also be used to safeguard auxiliary equipment such as feeders, stock straighteners, and scrap choppers. Many light curtains employ infrared type light beams. To avoid false

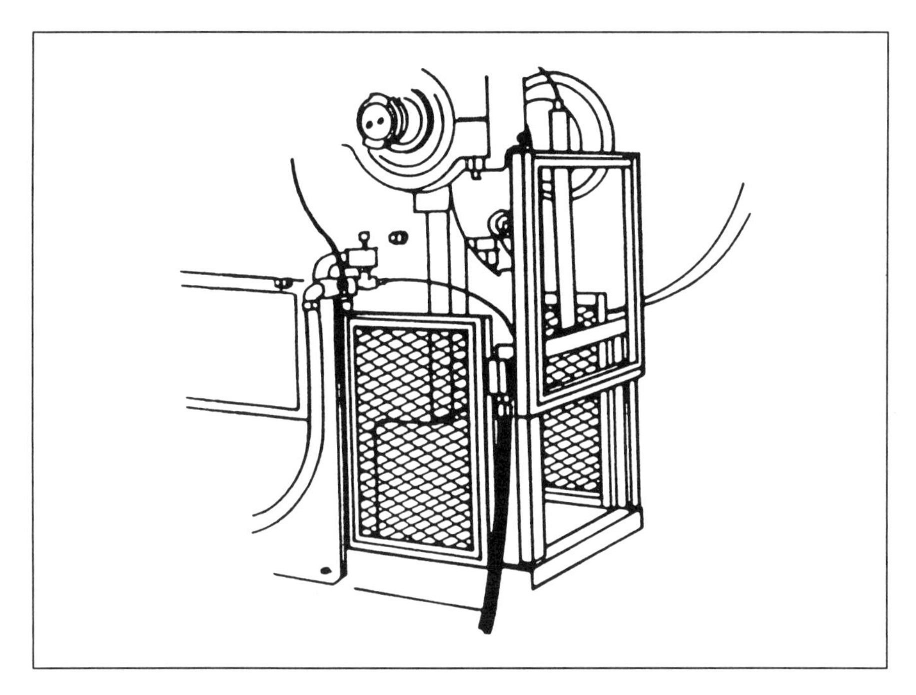

Figure 11-7. A physical barrier guard having a sliding gate. The press cannot operate unless the gate is in place; when the gate is fully closed, a limit switch is actuated, which permits press operation. (Rockford Systems, Inc.)

signals from stray light sources and ensure dependable operation, the invisible infrared beams may be modulated with a coded signal that must be properly received. To ensure light curtain control reliability, redundant and checking circuits are also employed.

Light curtains are a presence-sensing device. They must be strictly installed and maintained in accordance with the manufacturer's instructions. Properly installed, it is impossible to reach over, under, around, or through the safeguarding device without stopping the hazardous motion.

Radio Frequency Safeguarding Devices

Presence-sensing devices employing *radio frequency* (RF) energy are also used to safeguard pressworking operations. The point of operation hazard is surrounded by an antenna system made of metal tubes supported by insulated brackets.

The electronic presence-sensor control unit detects a change in the radio frequency energy field around the sensing antenna protecting the hazard. The machine must not cycle should anyone have any part of their body near the antenna surrounding the hazard. Also, the machine must stop should anyone inadvertently place any part of their body near the sensing device protecting the hazard.

Like any safeguarding device, RF presence safeguarding devices must be properly installed and maintained so that it is impossible to reach over, under, around, or through the safeguarding device without stopping the hazardous motion. The sensitivity of the control unit must be checked for each job and the control setting locked in place. The equipment must not be disturbed by other sources of RF energy such as RFI or EMI from contactors, induction heaters, and radio transmitters.

Two-hand Controls

The use of properly spaced two-hand controls is one way in which operators may actuate press operation. The method of actuation of two-hand controls is illustrated in Figure 11-8. The safety theory, simply stated, is that the operator cannot have a hand in the point of operation and initiate the press cycle with the two-hand control buttons at the same time.

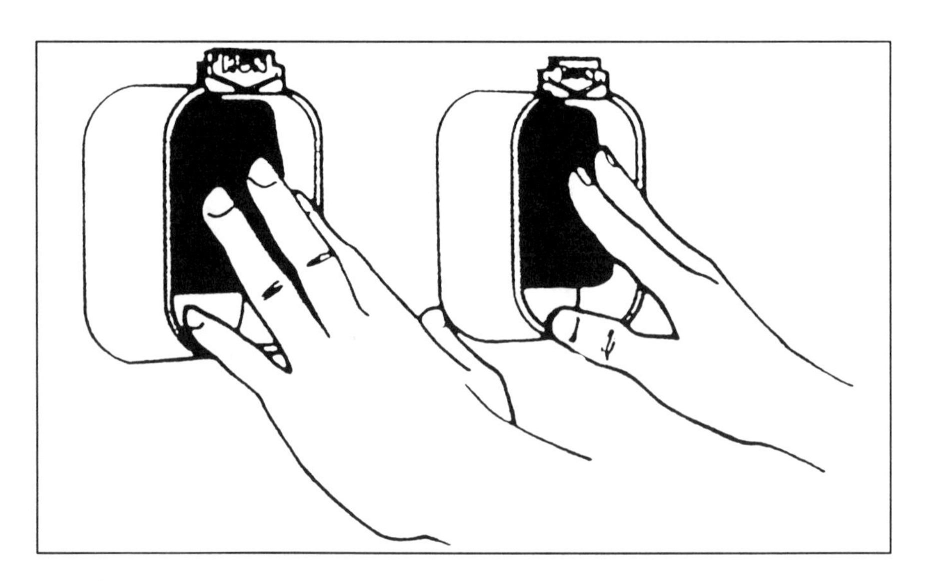

Figure 11-8. *Two-hand controls used to initiate and safeguard a press cycle. (Rockford Systems, Inc.)*

To ensure against injury, it is important to space the two-hand controls a sufficient distance from the point of operation. The spacing must ensure that the press will stop before the operator can remove a hand from a button and reach into the point of operation hazard.

The spacing between the two controls must be sufficient to prevent actuation by one hand and an elbow, for example. Also required is a guard to prevent accidental actuation by falling objects such as strips of stock. Mechanical switches are recommended to ensure RFI and EMI immunity.

The location of two-hand controls and trips from the point of operation hazard must be determined by the employer. The distance is determined by a combination of factors, including the type of clutch used on the press. For part revolution clutches, the measured stopping time of the press on the downstroke is a factor. For full revolution clutches, the number of engagement points and strokes per minute is needed.

Two good tutorials on compliance with the OSHA rules on spacing of two-hand controls and trips are listed in items 3 and 4 in the bibliography at the end of this chapter. The published data of Rockford Systems, Inc. of Rockford, Illinois is also recommended.

Multiple Operators on the Same Press

Each operator must have a set of approved two-hand controls. The danger of operating power presses without providing each operator his or her own two-hand control, is even more reckless than operating a ship without enough lifeboats to accommodate the entire crew. The certainty of a serious amputation occurring is so great that the persons responsible may be found guilty of criminal negligence by government authorities. There is no excuse whatsoever for allowing such a practice.

There is always a danger that one or more of the two-hand controls may be bypassed with a dummy plug or otherwise defeated. If more than one operator must be used on a press, it is essential that the operators systematically test the stopping function of their two-hand controls by removing one hand from each button on the down-stroke until all individual buttons have been tested and proper stopping assured. This procedure must be repeated whenever a new operator is assigned to the job or the operators return from relief or lunch. The use of parts transfer or automatic unloading devices is recommended to minimize the need for multiple operators.

Safety Distance

Many rules, including United States OSHA, are currently based on the assumption that the human hand travels 63 inches (1.6 m) per second. This hand speed constant is used to calculate safe distance from the point of operation hazard for placement of two-hand controls and safeguarding devices.

Actual press stopping time is normally measured and displayed by the brake stopping time monitor. It is also periodically tested with portable equipment that initiates an emergency stopping command at approximately 90 degrees on the downstroke.

The stopping time is measured in milliseconds. One millisecond is 0.001 second. The safety distance in inches, is determined by multiplying the stopping time in milliseconds times 63. Representative values of stopping times versus required safety distances are illustrated in Figure 11-9.

Safety Distance Theory

Generally, the assumption is that the operator may attempt to reach into the point of operation as the ram descends. A reason for doing this might be to reposition a part. The stopping time determines the required minimum spacing of press two-hand controls and safeguarding devices, such as light curtains and RF presence-sensing devices, from the point of operation.

It is highly recommended that a hand speed in excess of 63 inches per second be used in making these computations. Also, the operator's two-hand controls should not be located behind the operator when calculating safety distance.

Light curtains and RF safeguarding devices cannot be used on presses having full revolution clutches. Such machines will complete the entire stroke upon actuation.

Figure 11-5 illustrated guard opening size requirements based on distance from the point of operation hazard. It is important to remember that the point of operation is not necessarily where the metal is being worked in the die. Cam drivers, die cushion equalizing pins, and bolts that are too long can present a danger point for the operator. The point of operation then becomes the danger point closest to the guard. Stock feeders, scrap choppers, and other moving auxiliary equipment also are danger points and must be appropriately guarded.

Example of Safety Distance Requirements

Figure 11-9 is an example of representative safety distances required based on measured values of stopping times. It is recom-

SAFETY DISTANCE BASED ON 63 INCHES PER SECOND HAND MOVEMENT SPEED (REPRESENTATIVE VALUES)	
STOPPING TIME SECONDS	SAFETY DISTANCE INCHES
0.050	3.150
0.100	6.300
0.150	9.450
0.200	12.600
0.300	18.900
0.500	31.500
0.750	47.250
1.000	63.000

Figure 11-9. *Representative values of stopping times versus required safety distances.*

mended that a generous safety factor be included when setting up the job.

In the author's experience, some shops have ignored this requirement. In cases where the press stopping time was found to be very long, the light curtains were spaced so far away from the press that the operator could stand inside of the light beams and run the machine. Practices of this type must never be permitted.

If needed, extra photoelectric sensing devices or safety mats which are actuated by the operator's presence inside the main light curtain can be used to safeguard the operation.

Safety Training and Shop Rules

Training in power press safety is beyond the intent and scope of this work. Here we have presented a brief overview.

Video safety training materials that are highly recommended for pressroom personnel are available from Rockford Systems, Inc., Rockford, Illinois and The Precision Metalforming Association, Richmond Heights, Ohio.

Both Rockford Systems and Link Systems of Nashville, Tennessee conduct regular training classes in power press safety. In addition safety consultants trained in occupational safety such as Gordon Wall of Waterford (Detroit), Michigan provide in-plant training and safety evaluation services.

The OSHA power press safety laws in force in the United States are based on common sense good practices. The law is fairly brief and not difficult to understand. Generally, safety catalogs and the publications of manufacturers of safeguarding equipment have useful information on power press safety. Trade publications are another source of up-to-date safety information.

Power Press Law, Training, and Shop Rules

Training in how to operate power presses in conformity with the law is an absolute necessity for all pressroom employees. It is highly recommended that formal rule booklets be developed. Both applicable law such as United States OSHA and any special state, company, or other requirements should be included. The rule book should explain the procedures required to perform all pressroom tasks in a safe manner.

In addition to applicable law, supervision, lead persons, and the engineering department should have input and review the booklet before the rules are finalized. If the workers are represented by a labor union, the officers should have an opportunity to participate in the process.

Once the rule booklet is finalized, based on applicable law, all employees should be periodically instructed in each item. Willful or careless disregard of the rules should result in prompt corrective action by management. Safe work practices in pressworking operations are not an option.

Every employee should be familiar with shop rules. All rules and warning instructions should be made available in English as well as any other language in use in the shop. The same multilingual requirement should be followed when posting machine warning signs. All such warnings should be conspicuously posted on the machine.

PRESS AUXILIARY EQUIPMENT

Stock decoilers, straighteners, feeders, part handling, and scrap removal systems are known as press auxiliary equipment. Other examples of auxiliary equipment may include robots and dedicated die change carts. Appropriate safeguarding measures are required to prevent injury.

There are a variety of commercially available coil handling, decoiling, straightening, and feeding equipment used in coil-fed die operations. The equipment, available from many manufacturers, can be used interchangeably in a variety of configurations.

In some cases, the entire system is delivered as a turnkey package by the press builder or equipment supplier. However, it is very common to find a mixture of used equipment working as an integrated system. Cost conscious stampers often retrofit older equipment with modern drive systems and controls at a fraction of the cost of new machinery.

Example of Coil Feeding Equipment

Figure 11-10 illustrates the essential elements of one system. Safety guarding and necessary electrical controls are not shown.

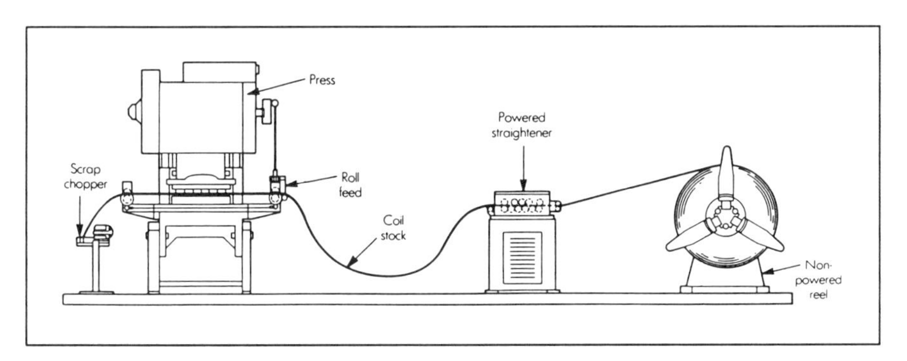

Figure 11-10. *Coil line containing a nonpowered reel, powered stock straightener, roll feeder, press, and scrap chopper. (Cooper-Weymouth-Peterson)*

Decoiling Systems

The stock must be decoiled (unwound) from the stock reel in a smooth manner. Uneven stop-go operation may cause kinks in the stock that can result in variations in the parts being produced. Both power and nonpower driven systems are used. Powered systems incorporate controls to ensure smooth actuation of starting and stopping functions.

In the example illustrated in Figure 11-10, a nonpowered decoiler is used. For light-duty applications, nonpowered decoilers have the advantage of low cost and simplicity. A mechanical drag brake may be used to prevent excessive stock being fed out. Should the use of a nonpowered decoiler result in kinked stock, or an overload of the pulling capacity of the stock straightener or feeder, a powered decoiler is needed.

Quick Coil Change

There should be a rapid means to band and remove a partial coil of stock left over from the job being removed. Time is saved if the new coil is pre-staged at the decoiler.

Cradle type decoilers may be movable on a track to center different widths of stock on the press center line. Here, markings of the correct settings should be provided to avoid trial-and-error adjustment.

Decoilers having expanding arbors may require shoes or inserts to accommodate widely differing coil inner diameters. All changeover parts and needed tools should be ready as part of the pre-staging or external dieset function.

Decoilers having double arbors permit a new coil to be loaded while production continues — good way to improve up-time. The decoiler base rotates 180 degrees. This permits a new coil to be loaded or an old coil removed while production runs.

Stock Straightener

When the stock is unwound from the coil, a normal curvature or coil-set often remains. For smooth feeding, and to reduce product variation, coil-set and minor material flatness problems are usually removed by a stock straightener. This is done by subjecting the stock to a series of up and down bends as it passes through a series of rollers. The bending action must be severe enough to exceed the yield point of the stock as the outer fibers of the metal are alternately stretched and compressed.

Figure 11-11 illustrates the principle of operating of a powered stock straightener. Depending upon the application, a greater number of straightening rollers may be used: 9, 11, or 17. For normal operation, the straightening rollers on the entry end of the machine are set to bend the stock more severely than those on the exit end. When correctly adjusted, the stock will exit the machine with an equal amount of residual stress on both sides of the neutral axis and be very straight.

Stock straighteners incorporating simple leveling rolls can do little to correct problems such as stock camber material and

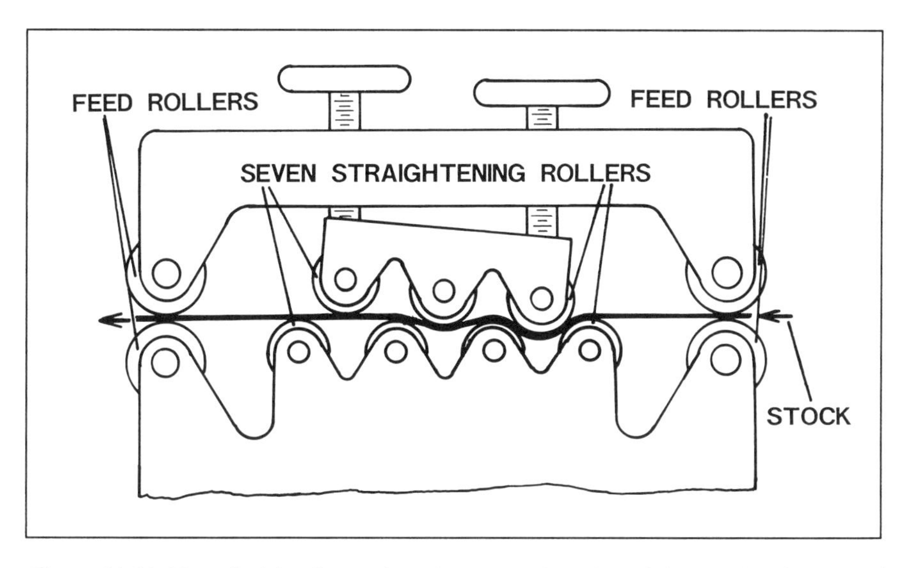

Figure 11-11. *The principle of operation of a powered stock straightener. The first pair of powered rolls feeds the stock into a series of seven straightening rollers. A second set of powered rollers operating in synchronism with the first set pull the stock through the straightener. (Smith & Associates)*

crowning. Specialized leveling equipment incorporating adjustable back-up rolls is required for such applications.

Quick Setup Considerations

A rapid means to set the stock straightener roll positions to correct values rapidly is an important way to reduce setup time. The adjusting mechanisms should have built-in position scales, turn counters, or transducers to permit presetting the straightener to values that are known to be correct based on the history of previous runs. The settings can be made automatically from a computerized data file kept at the press or in the pressroom in a file cabinet for ready reference. An increasingly popular way to automatically accomplish many pressworking setup parameters is through the use of computer integrated manufacturing.

Roll Feeds

Roll feeds are frequently used to advance the stock into progressive dies. Often, the stock is momentarily released upon pilot entry to permit the pilots to correctly align the stock. A slight overfeed is often helpful when using pilot release feeders as it is usually easier for the pilot to shove the stock back than to pull it forward.

Some applications employ two pairs of synchronized rolls (Figure 11-10) with one pair pushing and the other pair pulling the stock across the die. Figure 11-12 illustrates a pair of feed rolls driven by the press crankshaft.

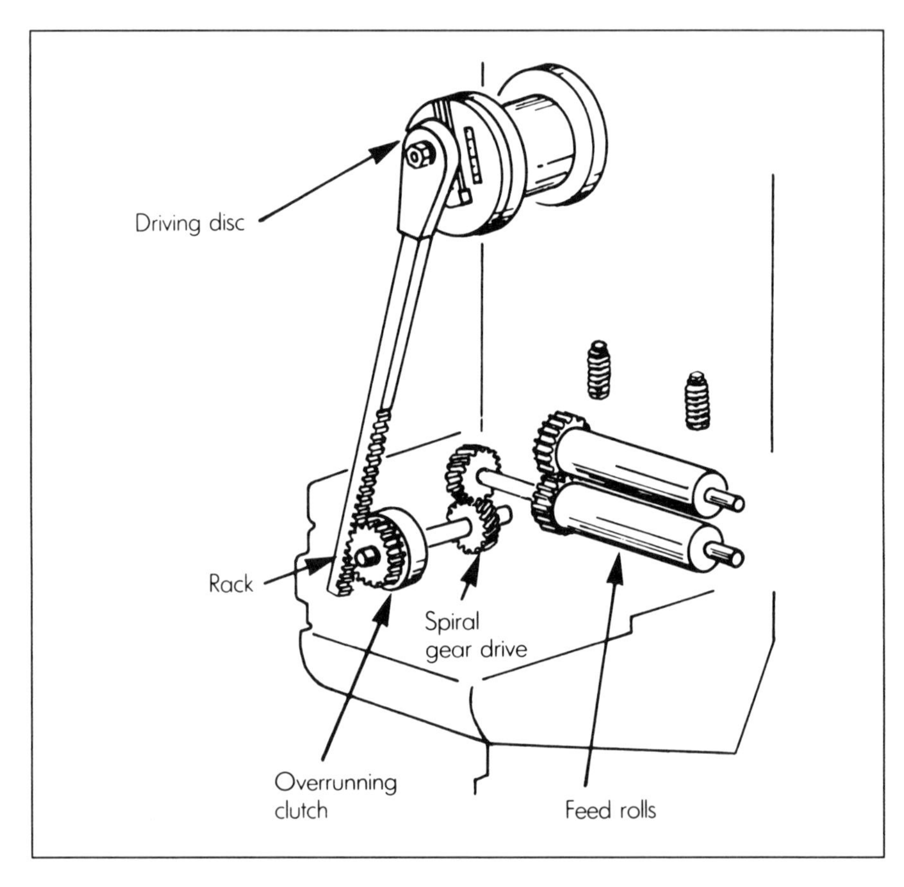

Figure 11-12. *A rack-and-pinion driven roll feed driven by the press crankshaft. (F.J. Littell Machine Co.)*

Crankshaft-driven feed rolls usually require some trial-and-error adjustment of the feed mechanism to achieve the correct pitch setting. To avoid die damage, the feed advance per stroke can be measured before the die is placed in the press. Precise settings can be accomplished by scribing a mark on the stock and measuring the total advance with a tape measure after cycling the press 10 to 20 times. The pitch is determined by dividing the total measured advance by the number of times the press is cycled.

Setup and Accuracy Considerations

Modern roll feeders are powered by electrical servo motors that provide very precise control. The feed pitch can be preprogrammed and stored as a job number in a computer that can also preset press functions such as counterbalance pressure, shut height, and tonnage limits.

It is important to mark the correct pitch setting on the die itself as a check in case the figure from the quick die change database is in error.

The correct information for automatic setup of the straightener and many other press functions can be accomplished automatically by inputting bar-coded information from the die or dieset work order.

Roll feeder accuracy can be improved by using feed measuring rolls that are separate from the main feeding rolls. This is especially useful when feeding heavy stock which may slip slightly on the feed rolls. The measuring rolls turn a digital pulse generator.

As the preselected length is reached, the digital control initiates the slowdown and stopping action of the main feed rolls. Close feed accuracies and excellent repeatability can be obtained.

Example of Servo Roll Feed Retrofit Package

It is practical to retrofit some existing crankshaft driven roll feeders with servo motor drives and electronic control packages. This provides much easier setup of accurate feed pitch lengths. Replacing an old feeder or adding a new unit as a complete package is a straightforward modification.

Figure 11-13 shows a P/A Industries microprocessor-based AC servo roll feed retrofitted as a turnkey package to a Minster 60-ton OBI press. The illustration is shown without guarding in place for clarity. One feature that contributes to the cost effectiveness of this unit is the mechanical pilot release actuated by an outboard driver attached to the press ram. It is simple, easy to adjust, and easily understood. The control package is simple to program and has job setup memory capability.

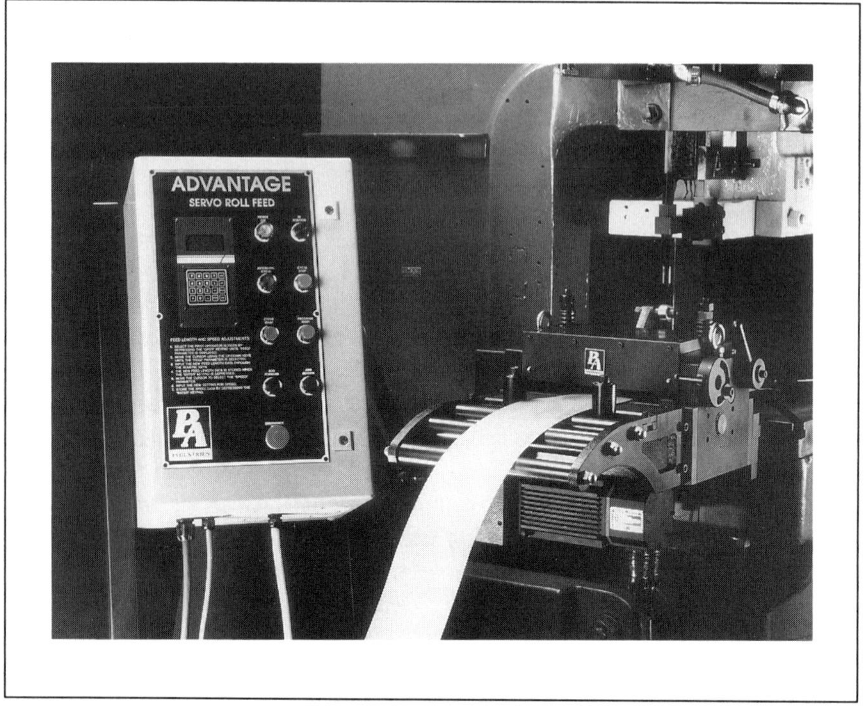

Figure 11-13. *A microprocessor-based AC servo roll feed retrofitted to older OBI press. (P/A Industries)*

Limitations of Combining Roll Feeding and Straightening

While economy and space limitations may dictate otherwise, the best feeding practice is to separate the roll straightener from the feeder. Feeding is naturally an intermittent operation. Better results of removing coil curvature in a roll straightener are achieved if the machine runs at a uniform velocity. This is accomplished by providing space for a material storage loop (Figure 11-10) that maintains a relatively constant supply of material between the feeding and the straightening operations.

Progressive-die operations that require a pilot release work best if the feeding and straightening operations are separate. Upper feed rolls are easily and automatically lifted to allow the pilots to shift the material. The material may not have the desired flatness if the upper straightening rolls are lifted after each progression of the material.

Hitch Feeds

Figure 11-14 illustrates one type of press-powered hitch feed. This type of hitch feed has a reciprocating head with a gripper unit. There is a similar gripper on the stationary unit. On the downstroke of the press, a cam attached to the press slide contacts a cam roller on the reciprocating head. Continued downward motion of the press slide pushes the reciprocating head outward, compressing a spring.

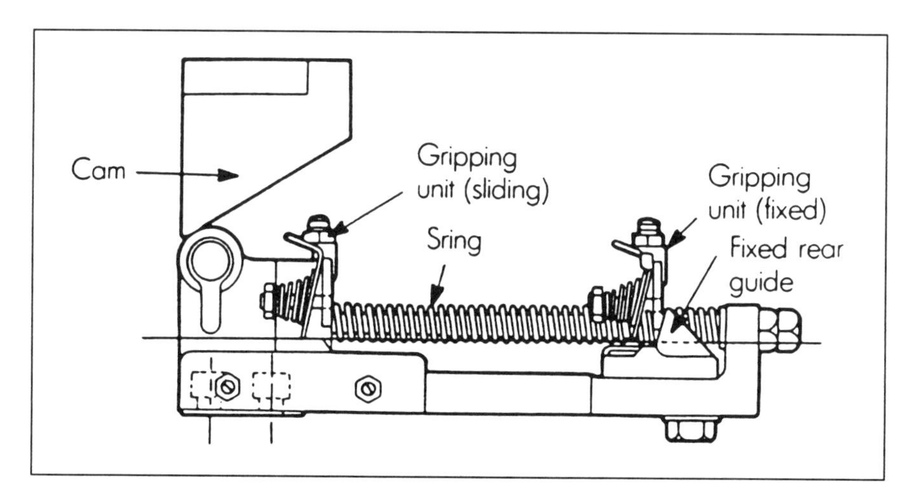

***Figure 11-14.** A press-powered hitch feed is actuated by a cam attached to the press ram or upper die shoe. (Dickerman Div., Reed National Corp.)*

During the downward press stroke, the gripper plate on the stationary head prevents the stock from moving backward. On the upward stroke of the press, the stock is held by the movable gripper plate as the head moves inward propelled by the compressed spring. The amount of feed advance is set by a feed-length adjustment nut.

Air-powered hitch feeds feature an air cylinder that powers the back-and-forth feed motion. Short-stroke cylinders actuate both the stationary and moving grippers. The gripper may have a provision for pilot release by releasing both grippers upon pilot entry.

Quick Setup of Hitch Feeds

A proven technique for quickly adjusting the correct pitch setting of air-driven hitch feeders is to use a measuring bar equal to the pitch length to adjust the device. Power to the device (usually compressed air) is first locked out and residual air drained. The reciprocating member is then moved by hand between the adjustable stops and the correct adjustment of the stops made. An excellent way to store the measuring bar is to bolt it to the die shoe.

Aligning a Coil Feeding System

The state-of-the-art method of aligning press coil handling machinery uses laser sighting equipment. If laser alignment equipment is not available, a taut length of music wire stretched from the press through the feeder and decoiler provides an excellent straight-line reference. Conventional hand tools such as squares and scales are then used to verify alignment.

Limitations of Hitch-feed Application

Both press-driven and pneumatic hitch feeds provide accurate and dependable feeding of stock. Both types work best in light- to medium-duty applications.

Pneumatic units which grip only one side of a wide strip are subject to binding and erratic operation. Adding an unpowered roll straightener adds to the unbalanced pull on the pneumatic grippers. Inaccurate feeding, scrap, and die damage may result.

Starting Strips in Progressive Dies

A major portion of the mishit damage to progressive dies occurs during the starting of the strip or coil of stock. A correctly designed progressive die has a sight stop mark or mechanical starting stop to permit the operator or diesetter to correctly position the incoming end of the stock for the first hit.

The stock must be carefully advanced until the strip is correctly started for automatic operation. Success factors in correct strip starting include:

- Easy to use positive starting stops;
- Avoiding making loose pieces of scrap when starting;
- Providing a pitch notch stop where feasible;
- Providing for easy strip removal; and
- Planning a good starting sequence to avoid lateral forces.

Joining Coils by Welding

Welding equipment specifically designed for joining coil ends at the press is commercially available. This procedure is especially useful for jobs that are difficult to start. A means of cutting the coil ends to provide square edges to weld is required. The joining process must produce a weld that will not damage the die. Usually, any parts containing the weld are discarded.

Chutes and Conveyors

In order to set jobs quickly, all required portable chutes and conveyors should be at the press in advance of changing dies. A written plan should be followed. No randomly conceived jury-rigged setup techniques should be permitted. Any chutes that do not remain with the die should attach without the use of tools.

Proper sheet metal fabrication equipment such as a shear and box forming brake are needed for fast economical chute fabrication. Proper equipment and neat organized storage of unused chutes is an important factor in good housekeeping.

Often, the limiting factor in producing large stampings is the capacity to convey scrap away from the die. Trimming dies for large irregularly shaped parts such as appliance cabinets and automotive body components produce scrap that is difficult to handle manually or with floor level conveyors. Good scrap handling equipment is especially important in the case of tandem lines and large transfer presses. The best solution often is a basement under the line equipped with large rugged conveyors. Easy access and good lighting are necessary to permit preventive maintenance.

Conveyors and powered chutes used in metal stamping are available in several basic types which include:

- Cog-driven construction employing slats or hinged flights;
- Metal mesh friction driven by steel rolls;
- Rubber or elastomer belt conveyors including flexible magnetic types;
- Motor-driven nonmagnetic stainless steel fully enclosed conveyors having moving internal permanent magnets;
- Vibrating chutes powered by compressed air or electricity; and
- Reciprocating chutes driven by special air powered devices.

Figure 11-15 illustrates a die having three chutes driven by air powered shakers. This type of conveyor is very versatile, combining locally fabricated chutes with commercially available air powered shakers. Many clever adaptations of this system are possible.

Dual level chutes can convey both parts and scrap from the same die opening. Chutes having slots or perforations can discharge small slugs, while permitting the conveyance of scrap-free parts to the finished work hopper.

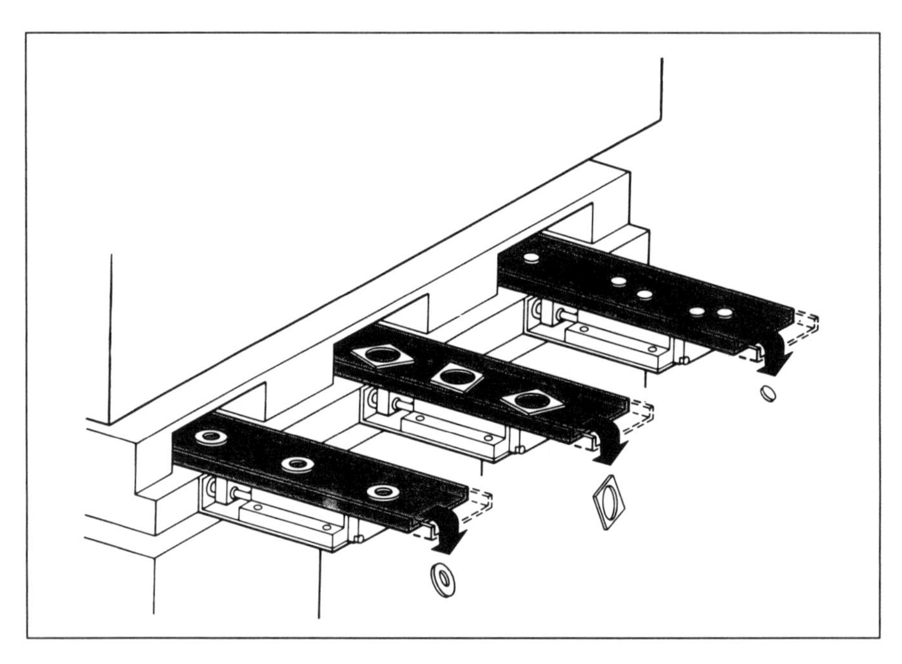

Figure 11-15. *A die equipped with pneumatically powered chute shakers. The reciprocating shaker has a slowly moving forward stroke out of the die to convey the parts and scrap. The return stroke is very rapid. The chute slides under the material, which remains in place due to inertia. (P/A Industries)*

Part and Scrap Containers

The exact type of container as well as its exact placement should be part of the written instructions for each job. The manner in which the parts are to be packaged in boxes and other shipping containers is very important in many cases to prevent shipping damage. Some parts can be damaged by simply permitting them to fall several feet (1 m) into a tub.

Many shops train the operators to inspect the parts being run. This is a reasonable expectation, especially if the press and die has sensors for mishit protection. Properly trained and motivated employees take pride in their work and are willing to certify both the quality and amount of parts in every container. Making sure that the correct bar code information is on the container is also often an operator responsibility.

Air Blow Off

If used, the air blow off should be connected and positioned properly. The noise generated by simple air blow off devices may contribute to hearing loss. In addition to approved hearing protection for pressroom personnel, several other measures can reduce employee exposure. These include:

- Using a timed air blast rather than a continuous flow;
- Equipping the end with a commercially available nozzle that increases the air volume while muffling noise; and
- Enclosing the point of operation with a soundproof physical barrier guard having a clear plastic window.

Generally a timed blast is more effective than a continuous one. The savings in compressed air can repay the cost of installing a good electronic cam limit switch in a few months.

BIBLIOGRAPHY

1. E.V. Crane, *Plastic Working of Metals and Non-Metallic Materials in Presses*, Third Edition, John Wiley and Sons, Inc., New York, 1944. (Out of print.)
2. D.B.M. Ehlke, *Metal Forming*, "You and the Law," Precision Metalforming Association, Richmond Heights, Ohio, October, 1990.
3. R. Harrison, *Safe Mounting Distance For Two-Hand Press Actuation, Metal Forming*, Precision Metalforming Association, Richmond Heights, Ohio, presented in two parts, December, 1992 and January, 1993.
4. R. Harrison, "Point-of-operation Safeguarding Mechanical Power Presses," *The Fabricator*, The Fabricators and Manufacturers Association International, Rockford, Illinois, May, 1993.
5. D. Smith, *Die Design Handbook*, Section 23, Society of Manufacturing Engineers, Dearborn, Michigan, 1990.

6. C. Wick, J. Benedict, R. Veilleux, *Tool and Manufacturing Engineers Handbook*, Volume 2, Fourth Edition — Forming, Figure 10-24. The appearance of a failed crankshaft due to a catastrophic overload. Entire fracture surface displays the rough torn metal typical of a tensile failure. (Smith & Associates) Society of Manufacturing Engineers, Dearborn, Michigan, 1984.
7. D. Smith, *Die Design Handbook*, Section 16, Society of Manufacturing Engineers, Dearborn, Michigan, 1990.
8. Dr. B. Bahr and Z. Ali, *Microcomputers and Flexible Manufacturing Systems*, SME Technical Paper MS91-338, Society of Manufacturing Engineers, Dearborn, Michigan, 1991.
9. A. Miedema, Reference Guide for the videotape training course, *Progressive Dies*, Society of Manufacturing Engineers, Dearborn, Michigan, 1988.

CHAPTER 12

DIESETTING PRINCIPLES AND TECHNIQUES

The ability to set a die safely, clamp it in the press securely, and stamp good parts with little or no trial-and-error adjustment, is the major pressworking goal. There is much talk about "Quick die change." However, how quickly dies can be exchanged is not the main quantifier of stamping efficiency. It is more important to be able to consistently change over an operation in a reasonable amount of time and produce good parts, than to strive for speed alone when the results are dangerous shortcuts.

LOT SIZE

There are many benefits to making only the stampings needed to assemble products in a short time. Large-lot production requires expensive storage space and inventory control procedures. Also, there is always the danger of part obsolescence and deterioration due to corrosion.

In general, the lot size is based on the number of parts required for production in hourly, daily, weekly, biweekly, or monthly quantities. For example, an integrated automobile manufacturing facility may stamp all large body panels at the assembly plant in one- or two-day lot sizes. The small stampings typically are produced by contract stamping shops who deliver the required parts in similar small-lot sizes. Only the required number of parts are produced and delivered on schedule as needed. A common term for this type of low-inventory system is Just-In-Time, or JIT, manufacturing.

The JIT system requires that the stamping process be strictly controlled. The stock must be on hand in the correct quantity and it must satisfy material specifications. The dies and presses must be capable of high-quality production, which requires good maintenance procedures, and the time required for diesetting should be low to maximize stamping flexibility and press uptime. Every activity's goal must be to deliver high-quality stampings on time.

Economic Order Quantities

The correct *economic order quantity* (EOQ) is defined as the amount of parts to produce for the setup and storage costs to be equal. This production amount reflects the lowest total part cost. Figure 12-1 is a graphic representation of these three factors.

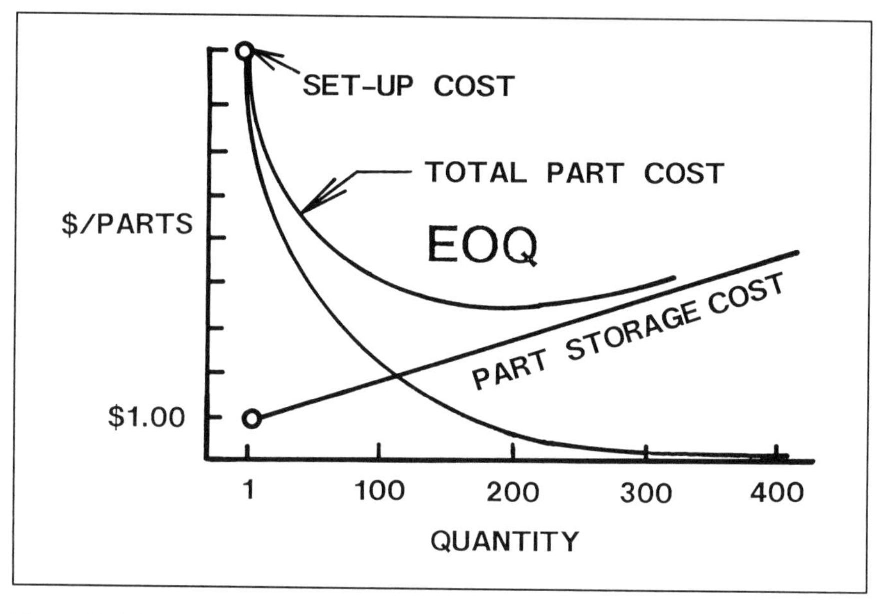

Figure 12-1. *The U-shaped curve describing how the total cost of setup and storage varies as a function of the number of pieces produced.*

The part cost alone has a linear upward slope, reflecting the cost of storing the parts increasing with time. The setup cost has a curved downward slope, reflecting amortization of this expense over an increasing lot size. The EOQ curve minimum value is the lowest per part cost.

Storage costs are very difficult to reduce because they are the sum of all costs associated with keeping inventory. These include:

- Warehouse space;
- Heat and light;
- Taxes; and
- Possible part obsolescence.

Setup time is usually a candidate for reduction. There are many case studies where die changeover times were reduced from several hours to a few minutes. Figure 12-2 illustrates how a reduction in setup time and cost reduces the size of the EOQ.

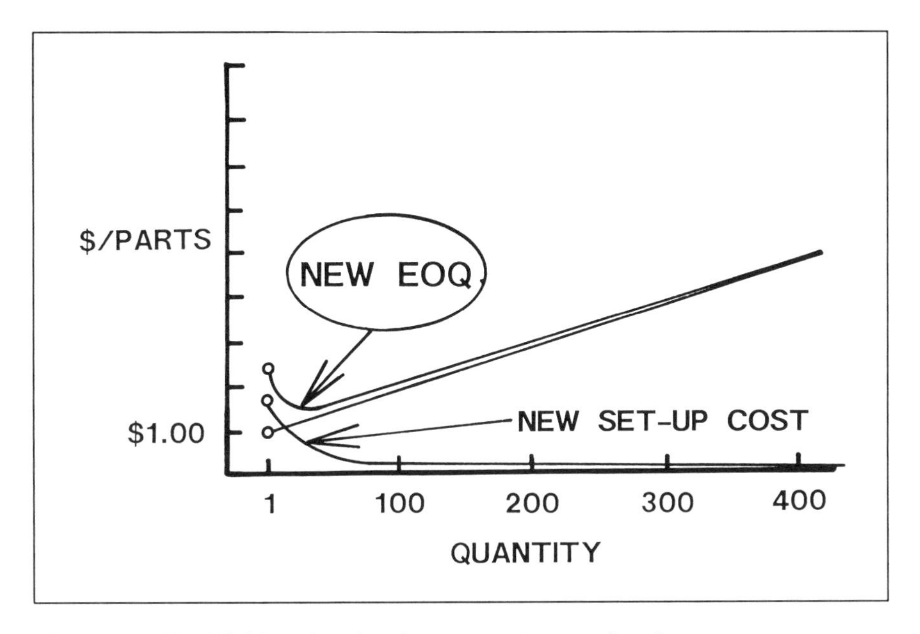

Figure 12-2. *The EOQ is reduced as the setup costs are reduced.*

The actual time elapsed between producing the last old part and producing good new parts is the true setup time. It includes delay factors such as:

- Finding the correct fasteners;
- Adjusting by trial-and-error;
- Performing any needed die repair; and
- Obtaining quality control approval.

Reducing Setup Cost

Setup costs are candidates for reduction in many cases. It is common for the setup of a coil-fed progressive die to take two hours or more. Tandem press line changeovers often involve a full shift.

There are several reasons why diesetting can take hours. The usual cause is lack of planning, haphazard preparation and, finally, poor execution of the changeover. Industrial and manufacturing engineers divide the setup time into two categories: external versus internal setup and the conversion of internal to external diesetting.

External versus Internal Setup. The external or preparatory phase of diesetting is the work to be done before the changeover begins. External setup is accomplished while production is running. The internal portions of the diesetting activity are the tasks to be done while production is stopped. Figure 12-3 illustrates the difference between internal and external diesetting tasks.

Converting Internal to External Dieset. Tasks required to prepare for and carry out diesetting activities should be identified as separate elements and the required times noted. Low-cost home videotaping and playback equipment is an effective tool. It is best

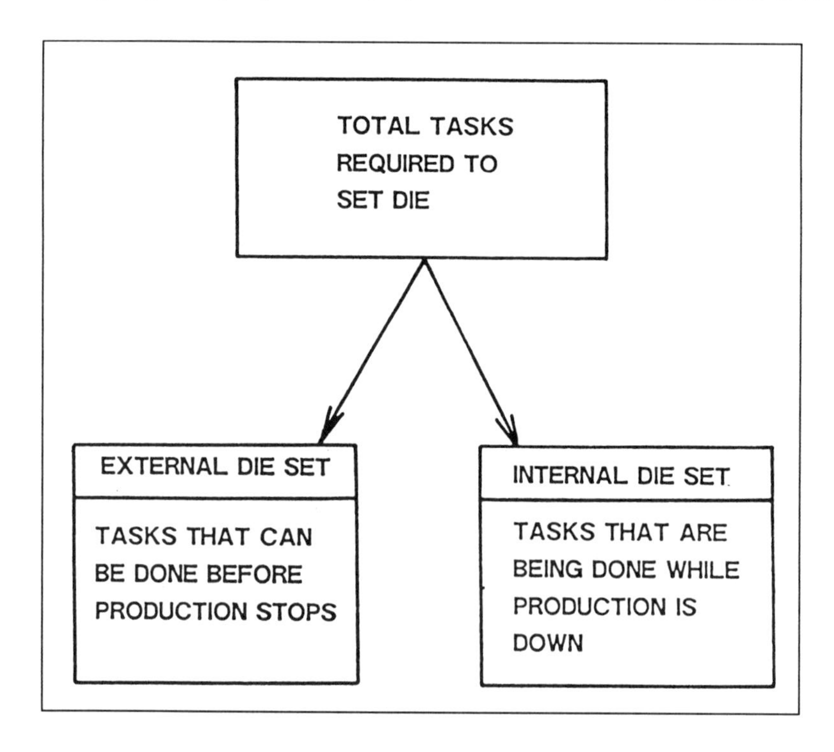

Figure 12-3. *To improve diesetting efficiency, as many internal dieset items as possible must be converted to external dieset.*

to have the persons doing the diesetting review the tape and work with engineering management to improve diesetting efficiency.

Enhancing Pressroom Efficiency

Improving pressroom efficiency was once mainly the duty of industrial and manufacturing engineers. Increasingly, this responsibility is being shared with the hourly work force.

Terms such as "employee involvement" and "employee empowerment" are applied to manufacturing management systems in which employees are encouraged and empowered to improve the functioning of manufacturing activities. Of course, top management still retains responsibility for workplace safety and efficiency. However, it is a system where employees are given time to meet and jointly plan improvements with management.

It is key to developing mutual trust. For this system to work, everyone must benefit. The rewards for such cooperation must be tangible, such as increased job satisfaction, employment security, and often a share of the increased profit.

DIE HANDLING, TRANSPORTING, AND STORAGE

For pressworking to be carried out efficiently, all aspects of die handling, transportation, and storage must be accomplished with ease and safety. Serious injuries and damage to tooling can result if proper equipment, procedures, and training are not provided.

For example, when opening and closing dies with a crane during bench repair operations, there is a risk the die may be damaged. This is especially a problem with progressive and other dies containing cutting stations.

Do not allow the die to cock and bind, especially when closing the die. Should the die shoes not remain parallel during closure, cutting details may be damaged should interference occur.

Dies having ball-bearing equipped guide pins and bushings have several advantages over plain bushings. They provide better alignment, have longer wear life, operate with less friction, and make the die easier to open and close without cocking. The latter factor favors specifying them in place of plain bushings in applications not otherwise justifying extra expense.

Dies especially difficult to open may be opened in a press. Often the die repair area is equipped with a tryout press. Once opened, the lower die can be removed with a die cart or fork truck. If the upper die is needed, it often must be removed with a fork truck.

The latter procedure requires great caution to avoid die damage and injury.

Resetting the upper die in the press is even more difficult than removing it. Great care must be exercised to ensure that it is properly located with respect to the lower die.

Die Openers

To avoid damaging dies and to help ensure safe die handling, die opening devices are practical to open and close dies in the die room. Figure 12-4 illustrates a hydraulically actuated die opener.

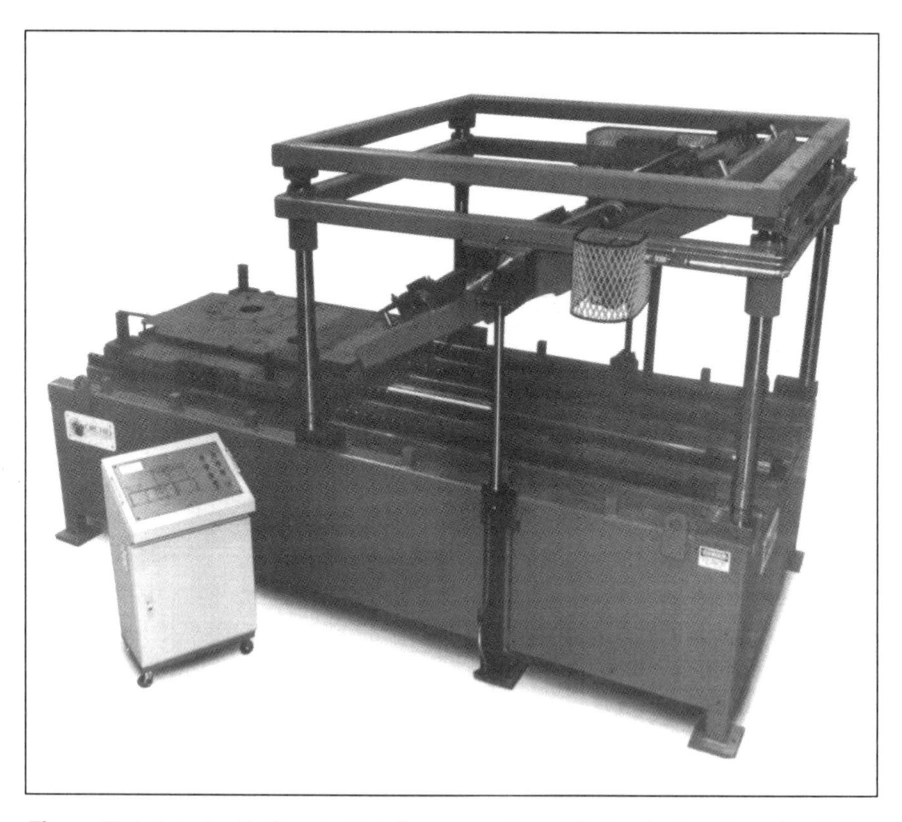

Figure 12-4. *A hydraulically actuated die opener shown flipping the top shoe of a die just opened. The sequence is reversed to assemble the die. (Orchid Automation Group, Inc.)*

After the die is loaded onto the opener it is conveyed into the machine, and the upper shoe is attached to the opening mechanism. Parallelism of the mechanism is maintained by four guide posts. The die is then opened and the upper half flipped. Finally, the lower half is conveyed out of the way and the top half lowered into position for repair work or removal. To assemble the die, the sequence is reversed.

Another type of die opener is illustrated in Figure 12-5. Four synchronized hydraulic jacks powered by a cart-mounted hydraulic power source and control unit open the die. Parallelism should be maintained within ±0.040 inch (1 mm) during the opening and closing cycle. This is accomplished by a linear position transducer built into each jack cylinder. A microprocessor-based controller operates a servo control valve for each cylinder to maintain parallelism.

***Figure 12-5.** A hydraulically actuated die opener featuring four synchronized cylinders, used to open and close large dies on the shop floor without cocking or binding. (Enerpac Division of Applied Power, Inc.)*

Turning Dies and Die Shoes

In addition to dedicated die openers, some of which are designed to flip the upper die shoe, dies and die shoes may be turned over with an overhead crane. This is easily done with die shoes typically weighing no more than 300 lbs (136 kg), provided proper swivel hooks or hoist rings are used.

Turning large die shoes safely poses a problem. In some cases, a fork truck supports the lower end of the shoe, requiring a degree of coordination between the fork truck and the crane. The coordination is difficult to achieve safely. A far better approach is to equip a bridge crane with an auxiliary hoist to handle the free end of the die shoe. This permits turning the entire shoe in mid-air.

The capacity of the auxiliary hoist is typically one third to one fourth the main hoist. The auxiliary hoist is normally required to lift one half of a shoe, while the main hoist is usually designed to handle the total weight of the largest die, including any attached buildup components.

The die opener shown in Figure 12-4 is also capable of flipping die shoes. Several other equipment suppliers also provide die openers with shoe flipping capability. In addition, specialized equipment is available for turning machine parts, including die shoes.

Handling Dies with Cranes

Figure 12-6 illustrates one system for handling large dies used by the Ford Motor Company. A full set of corner lift pins is kept with the die to speed die handling.

Whenever any object is to be lifted with a crane, it's essential the hook be centered over the load before the lift is begun. Otherwise, the load will swing when lifted, endangering personnel and equipment.

Another way in which equipment damage or personal injury can occur, is through careless lifting or crane movement after removing the cable or chain slings from the die. The danger is that the sling may unexpectedly catch on a die shoe. This may be avoided by placing the slings on top of the die before the hoist is lifted clear. Another procedure, especially useful for very heavy chains, is to place them all on the same side of the die after removing them. The hoist is then moved in that direction so the chains are free before the hoist is raised. Constant caution is required by everyone in the area. Everyone should stand clear, to avoid injury in case of an unexpected crane move.

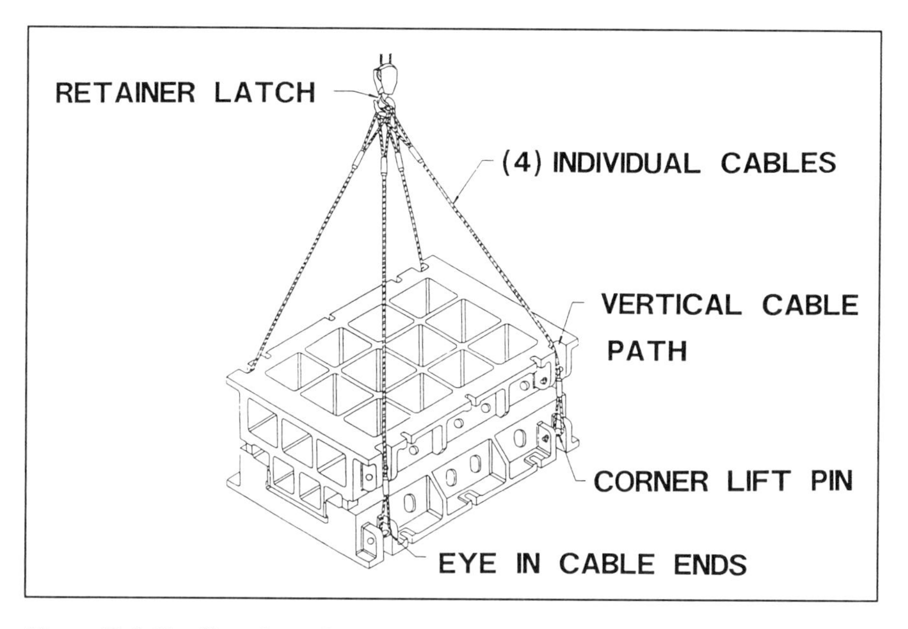

Figure 12-6. *Handling a large die with cable slings. (Ford Motor Company)*

Eyebolts, Hoist Rings, and Die Hooks

At one time, eyebolts such as illustrated in Figure 12-7 were the choice for most die handling operations. They are designed to be used at full-rated capacity only when making a vertical lift. The threaded portion of many commercially available eyebolts is several times longer than the thread diameter. This permits their use with soft tooling materials such as zinc alloys. Washers should not be used under the shoulder. The threads must be fully engaged for safe use. The manufacturer's recommendations and applicable safety rules should be followed as to permissible loads, especially those involving moves other than a straight lift.

Figure 12-8 illustrates a Jergens center-pull hoist-ring, which has many advantages over eyebolts. The screw length can be selected for proper thread engagement in the material being lifted. Jergens also makes a special side-pull hoist ring (not illustrated) designed to withstand large side loads. Special die hooks are commercially available in large tonnage capacities from suppliers specializing in rigger's equipment.

Any lifting device can be damaged if abused. Figure 12-9 illustrates how a center-pull hoist ring may be damaged if misused.

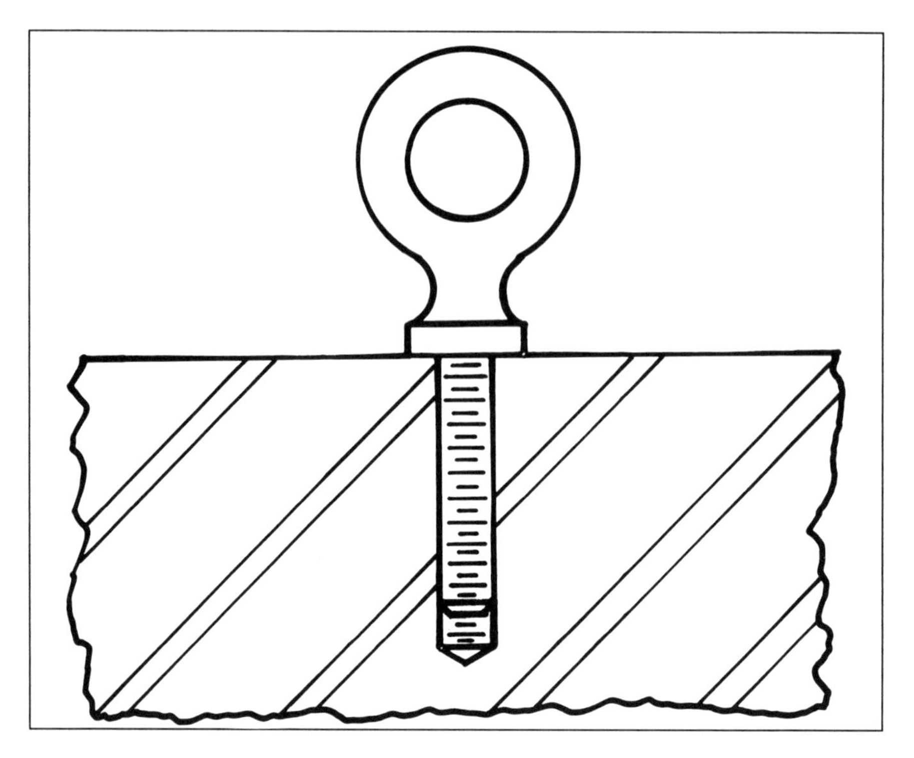

Figure 12-7. *Commercial eyebolt threaded into a heavy die section. Note that the threads are fully engaged.*

Using a hook that is too large for a hoist ring or eyebolt opening is another common form of abuse.

Lifting slings also must be derated for the angle of the chain or cable. Figure 12-10 illustrates how a sling at a 30-degree angle from the horizontal is subject to twice the load as the vertical pull. Suppliers of slings and lifting devices provide derating tables for their products subjected to lateral loads.

DIE STORAGE AND RETRIEVAL

Appropriate die storage methods vary. Dies, set frequently for small-lot production, should be stored as near to the point of use as possible to minimize prestaging or external setup costs.

Each die should have a designated storage location. Dies may be identified according to the following:

- Job number;

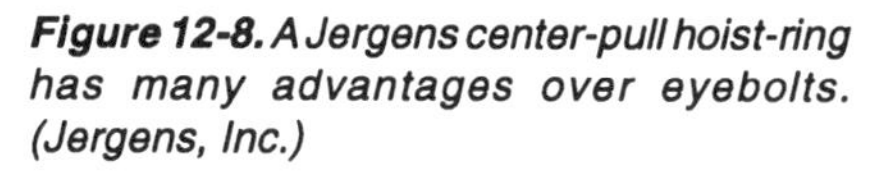

Figure 12-8. *A Jergens center-pull hoist-ring has many advantages over eyebolts. (Jergens, Inc.)*

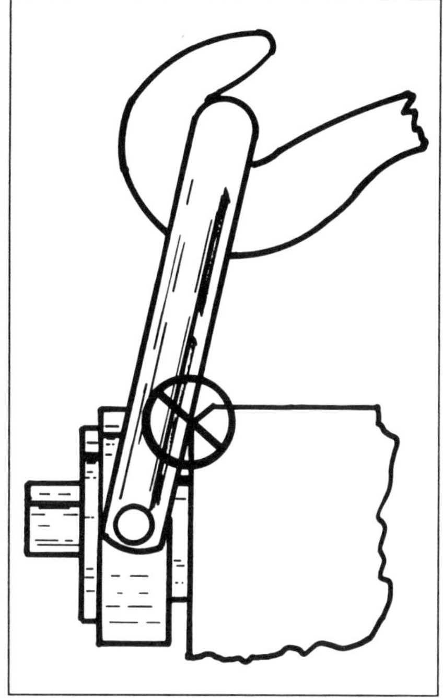

Figure 12-9. *A center-pull hoist-ring may be damaged if misused. The use of hooks too large for hoist-ring or eyebolt openings must be avoided. (Jergens, Inc.)*

- Customer;
- Product line die number;
- Description of the part(s) the die produces; and
- Storage location.

Shops with many dies usually assign a letter and numerical designator to each die and corresponding storage location. Both the storage place and the die are conspicuously identified. Color coding dies to indicate the product produced is also helpful.

A computerized database is a simple way to keep track of each die's location. This data can be made available at data terminals through the plant or printed out as it is periodically updated. This data can also tie in with a computerized die maintenance management system.

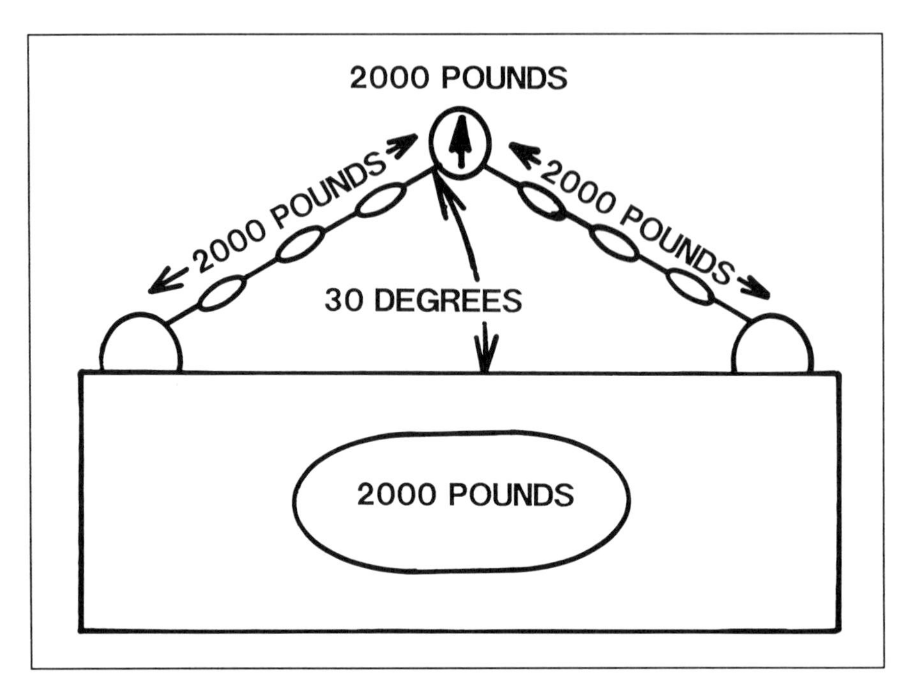

Figure 12-10. *Lifting slings must also be derated for the angle of the chain or cable. A sling at a 30-degree angle from the horizontal is subject to twice the load of the vertical pull.*

Racks for Die Storage

Die storage racks must be of sturdy construction. Generally, they are constructed of structural steel and must be designed with a load-bearing capacity including a generous safety factor. Often, the die is put on a pallet to permit easy placement and retrieval from storage with a fork truck.

It is important not to exceed the lifting capacity of any lifting or storage device. For this reason, the weight of both the upper and lower die half as well as the total die weight should be stamped on the die.

Dies used for high-volume current production are seldom stored outdoors. Dies needed to run service parts are an exception. Such dies must be carefully protected by coating surfaces with heavy rust-preventive grease. A die in outdoor storage for several years will require extensive cleaning, and often a day or more of bench refurbishing prior to running service parts. Only when the future production of service parts is unlikely should dies be stored in unprotected outdoor areas.

Effective Use of Fork Trucks

Fork-lift trucks, or *fork trucks* as they are usually referred to, remain popular for both handling and setting dies. They are especially effective for setting dies having parallels that are permanently attached to the lower die shoe. The die is retained on the forks with the same security as a pallet or basket of parts designed for fork-truck handling. In addition to setting dies, the fork truck and driver can accomplish many other material handling activities when not setting dies.

Many shops routinely set coil-fed progressive dies with fork trucks and produce new parts in under 20 minutes. This includes the use of manual bolts rather than hydraulic clamps. To accomplish this, procedures must be planned in advance.

For example, the die can be prestaged in the area on a wooden pallet to ensure the parallels' bottom remains clean. A spare pallet can be placed on the floor to receive the old die. While production runs, the fork-truck driver, diesetter, and operator work together to get the new coil, parts baskets, and other needed items prestaged for rapid changeover. When the old job is complete, everyone assigned to the changeover proceeds to perform the assigned tasks. Often the operator bands the old coil and replaces it with the new one. Other operator tasks might be to assist in changing parts and scrap containers as well as assisting the diesetter.

The fork-truck driver promptly removes the old die after it is unclamped and the chutes removed. Both the diesetter and operator may work together to clean any slugs and debris from the bolster before the new die is set.

Quick Changeover Success Factors

A cooperative effort to ensure correct prestage while production is running helps avoid delays. Written checklists are an aid in accomplishing this.

Once the changeover is started, training and teamwork are the keys to quick completion. Increasingly, old restrictive work rules are being changed by mutual agreement, permitting cross training and multifunctional work teams with the incentive of increased job security and employability.

It really doesn't matter who sets the shut height, feeds the strip, checks the tonnage meter, sets the new counterbalance pressure, and does the other tasks required to complete the dieset. It is key that each task is done correctly, safely, and results in good parts.

Simple press and die modifications to ensure die location repeatability, such as bolster and die keyways, are low in cost and can save trial-and-error adjustments. A constant clamping height to permit the use of the same bolts or clamps on every job also will increase diesetting speed and safety.

Unsafe Fork-truck Applications

The truck must have enough capacity to lift the die in one piece, handle it securely, and set it safely. Using more than one fork truck to handle the same die, or setting the die one half at a time, are dangerous practices. One problem is that the load may shift, possibly damaging the die, press, and other equipment, as well as injuring personnel.

Additional counterweight material must not be added to the rear of the fork truck to increase lifting capacity without the manufacturer's or competent engineering approval. Good engineering planning includes ensuring that the dies scheduled for a given press can be handled safely with existing equipment.

Battery-powered Die Carts

Figure 12-11 illustrates a self-propelled battery-powered die cart, which features support rollers and a push-pull module. The operator walks behind the cart and guides it by means of the operator control handle. Battery-powered diesetting carts are available to set dies ranging in sizes from small OBI dies to very large dies set in straightside presses. For setting and transporting very large dies, they are much safer to use than fork trucks of marginal capacity.

A safety feature shown in Figure 12-11 is the attachment hook that securely engages the press bolster before the die is moved into or out of the press.

Figure 12-12 illustrates a simple die cart that was locally fabricated, made of commonly available structural and hardware components. The wheels have a groove in the center to permit the cart to be guided by a track made of strap iron fastened to the floor. The table rotates on eight small wheels attached to the framework.

Adapting a Commercial Material Handling System

Figure 12-13 displays a commercial storage rack and material handling system adapted for diesetting by an employee involvement group. The team consisted of experienced diesetters, supervision, and a manufacturing engineer.

Each steel pallet location and die is conspicuously identified. The storage and retrieval device shown is attached to a suspension

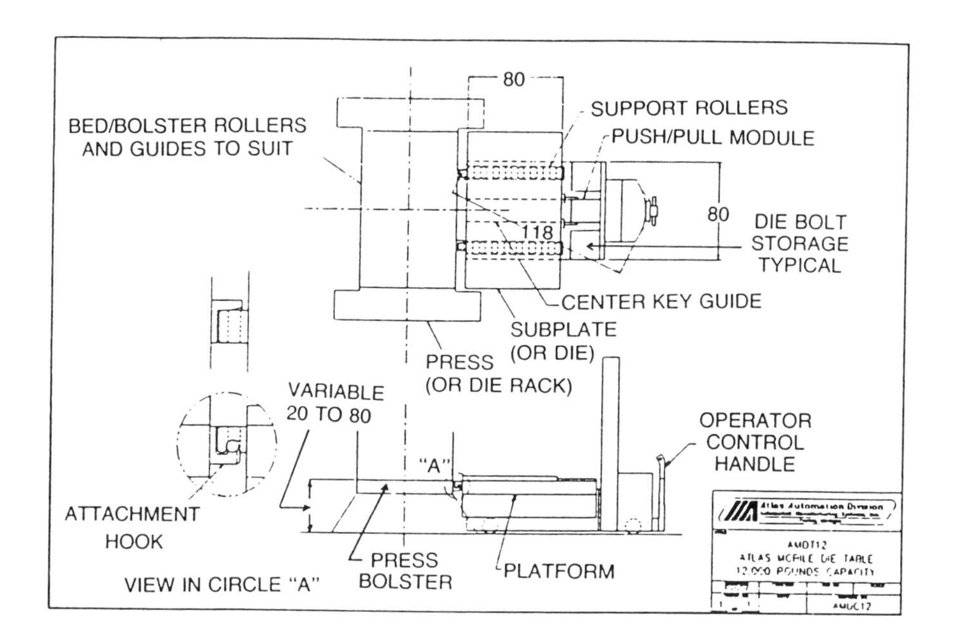

Figure 12-11. A mobile self-propelled walk-behind battery-powered die setting cart. (Atlas Automation)

system similar to a conventional overhead crane bridge and trolley.

The employees adapted a commercial material handling system designed for double-sided palletized storage. The bridge and trolley permit in-and-out as well as left-to-right movement between the storage racks and the OBI presses serviced by this system. The racks on one side of the system were replaced with structural steel to permit ease of access to the line of presses.

AUTOMATED STAMPING WORK CELLS

Figure 12-14 illustrates a cellular stamping system featuring automated die storage and retrieval. A die can enter the patented vertical die storage system from either the front or rear side. A central die elevator serves a total of 16 storage locations. The die is elevated to the proper level and discharged to either the left or right side according to programmed instructions from a computer.

While eight dies can be stored on either side of the system, the bottom level is reserved for prestaging the new die and temporary

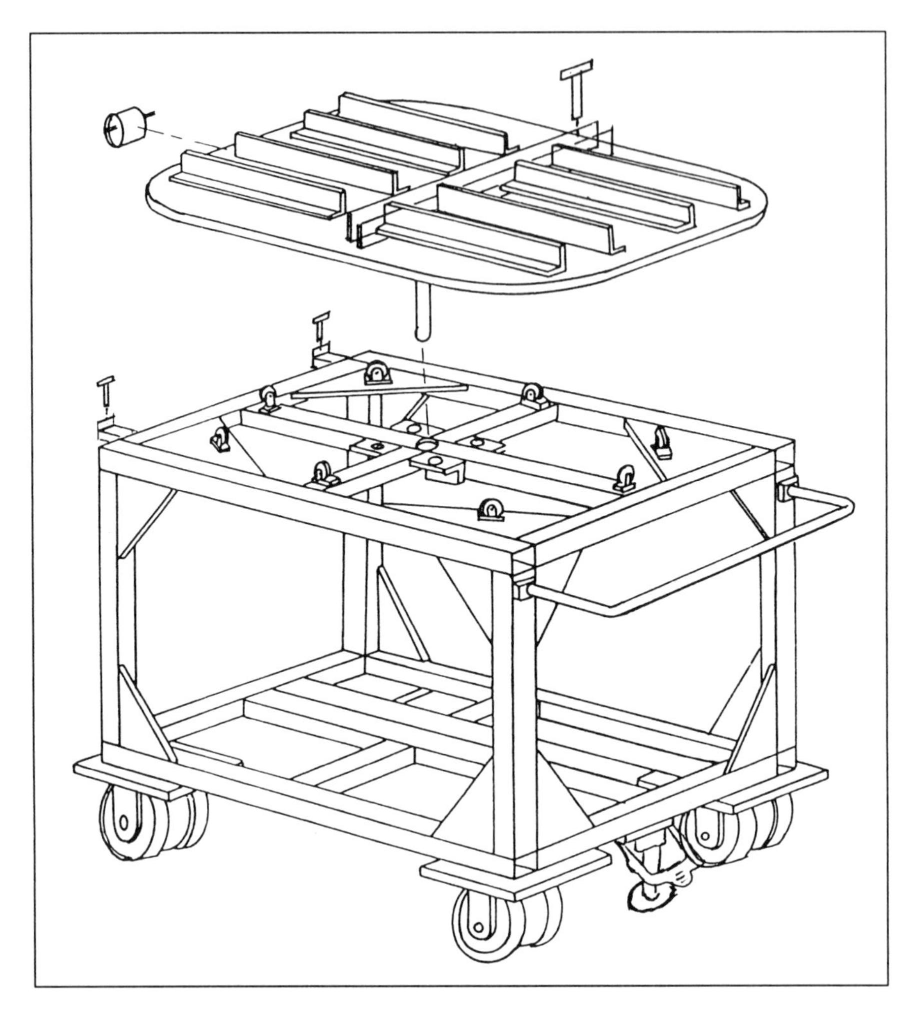

Figure 12-12. An exploded view of a simple light-duty locally fabricated die cart. (Calsonic Yorozu Corp.)

storage of the old die during changeover. The new die is automatically retrieved by entering work order instructions, and is stored in the prestaging location while production continues.

During changeover, after the die is unclamped, an automatic grasping device on the transfer table at the rear of the press removes the die and stores it temporarily in the die storage rack. The prestaged die is set in reverse order. Finally, the old die is either stored in its designated location, or removed for inspection and maintenance.

Figure 12-13. *A commercially available palletized storage system adapted to store and set small OBI dies. (Rowe International)*

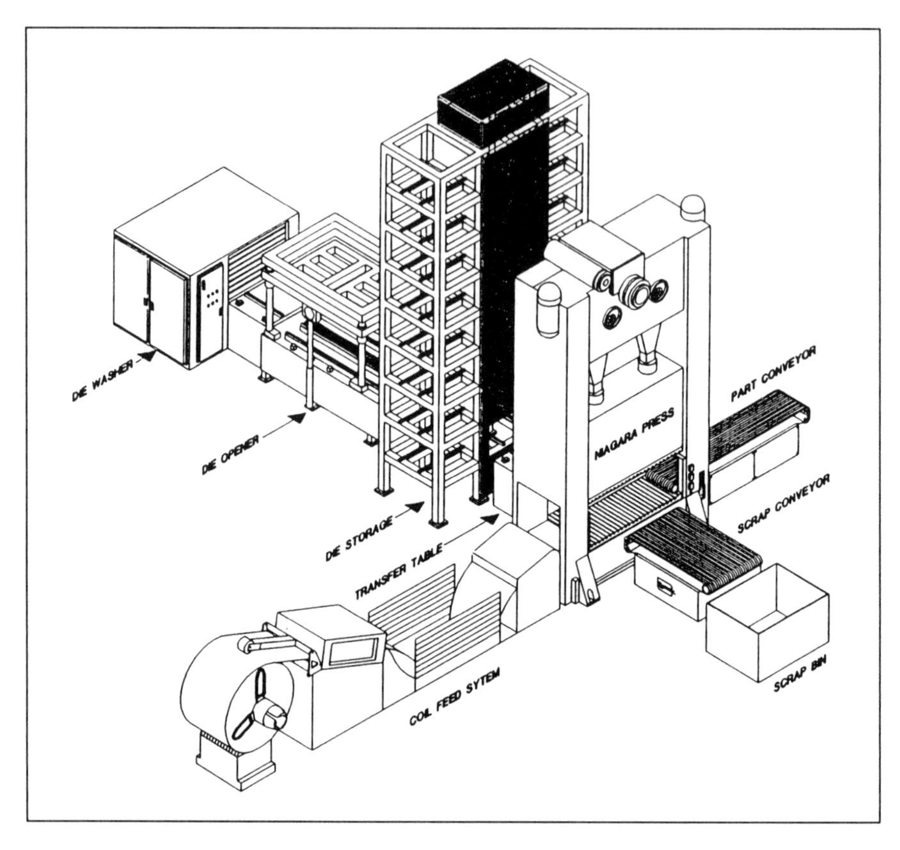

Figure 12-14. *A stamping work cell featuring automatic die storage and retrieval. Some additional features are a die opener and die washing system. (Orchid Automation Group, Inc.)*

Should the die require inspection and maintenance, it is opened with a die opener like the one shown Figure 12-4. Next, each half is placed into an automatic die washer. Here, heated high-pressure water jets and detergent combine to remove dirt, excessive lubricant buildup, slivers, and slugs.

Once the die is cleaned, it is ready for inspection. If no repair work is needed, it is lubricated and returned to storage.

POSITIVE DIE LOCATING METHODS

A precise, repeatable die locating method is essential if rapid die changing and setup repeatability are to be achieved. Accurate die locating should not depend upon complicated measuring or trial-

and-error methods. Adjustments with fork trucks, bumpers, or pry bars do not guarantee precision. Even though someone sights down through the feed line of a blanking or progressive die operation and gives instructions, the stock often buckles due to misalignment, requiring repeated trial-and-error adjustments.

Figure 12-15 illustrates a very common method of locating dies and die subplates on the bolster. A V-locator and a machined flat pocket engage bolster pins spaced at standard locations. A close-up of the method is illustrated in Figure 12-16.

To accommodate differing die widths in the same press, several pairs of pin holes may be bored in the bolster (Figure 12-17) and the pins moved to the appropriate pair of holes for the die being set. For frequently set dies, hardened inserts (Figure 12-18) are used to avoid wear problems.

The machining of the subplate or die shoe can be simplified if only one pocket is provided, as illustrated in Figure 12-19. Exact front-to-back and side-to-side location is achieved. If the plate is flame-cut, both contact areas should be machined to give a precise distance and offset from the center line. The disadvantage is that the pin holes must be offset from the center line a fixed amount.

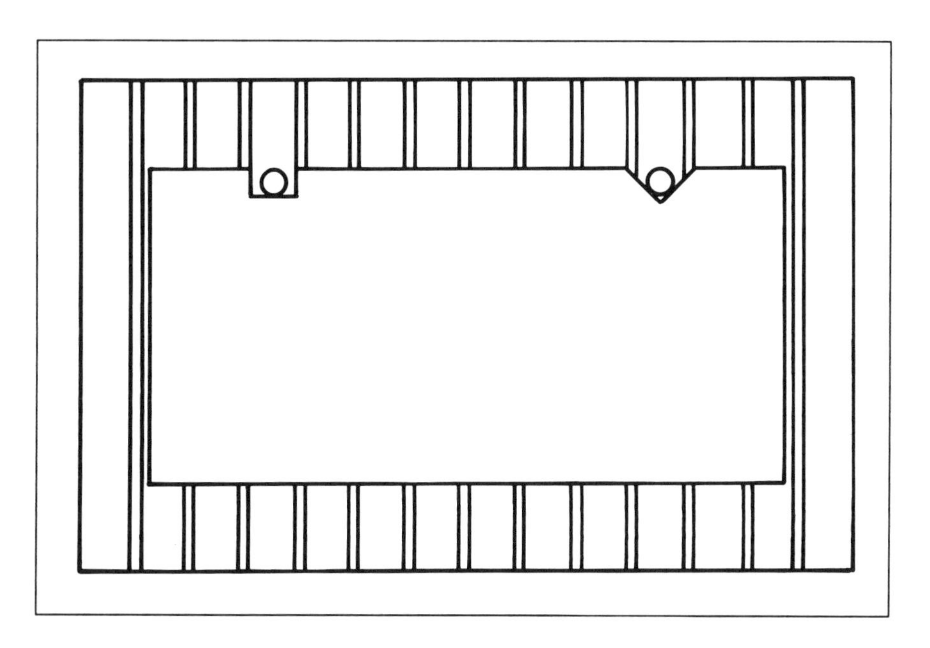

Figure 12-15. A popular method of locating die subplates on the bolster. A V-locator and machined flat pocket engage pins in the bolster.

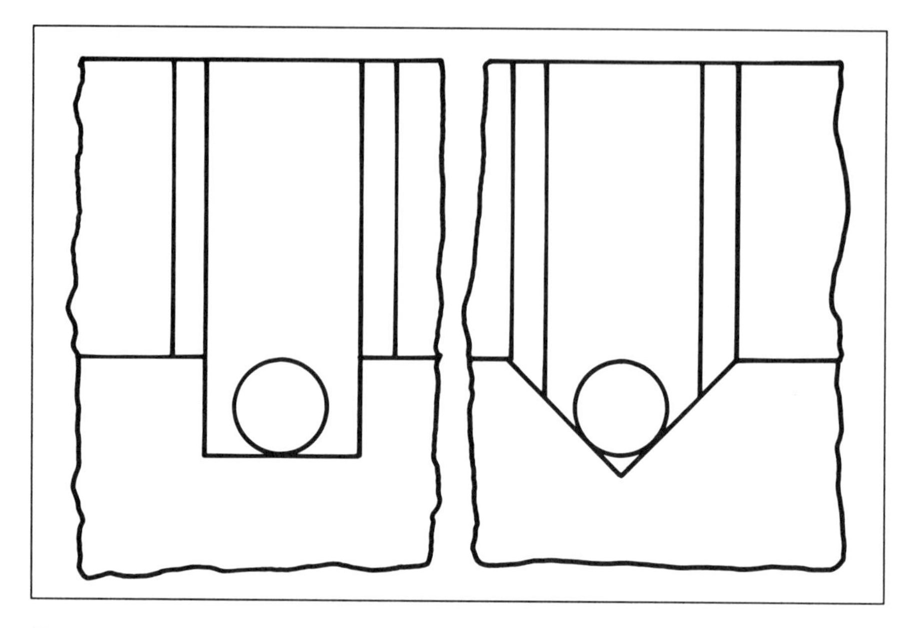

Figure 12-16. *Close-up view of die locators in Figure 12-15.*

This is not a disadvantage provided a standardized system is adhered to throughout the shop.

Locating pins fitted into bored holes on the center line of the press bolster is illustrated in Figure 12-20. In this very common method of locating dies on moving bolsters, the die bottom shoe has one round and one slotted hole machined on standard locations. Several pairs of bored holes may be provided on the center line of the bolster to accommodate dies of various lengths.

For large dies, a typical pin diameter is 3 inches (76.2 mm). A conical point permits easy engagement as the die is lowered with a crane.

Generally, 0.060 inch (1.524 mm) clearance is provided to prevent binding and pin breakage as the die is lifted off the bolster. Figure 12-21 illustrates a section through a pin, die shoe, or subplate and bolster. The necessary clearance around the pin is illustrated in Figure 12-22.

This locating method is frequently employed for moving bolster applications in automated tandem and transfer lines. For press automation to function reliably, every effort should be made to minimize locational errors from all sources. For this reason, the V-

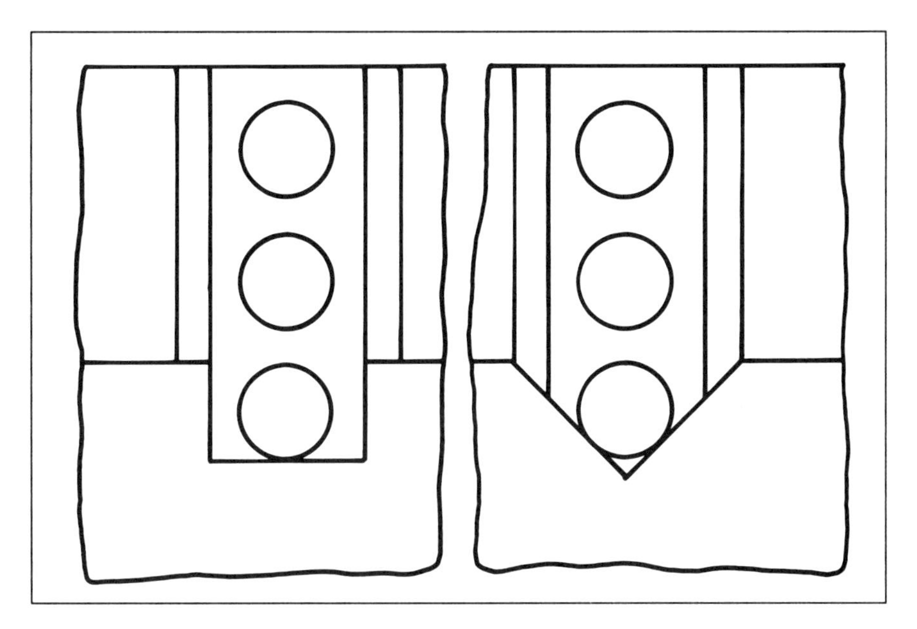

Figure 12-17. *Extra bumper pin holes accommodate differing shoe or subplate widths.*

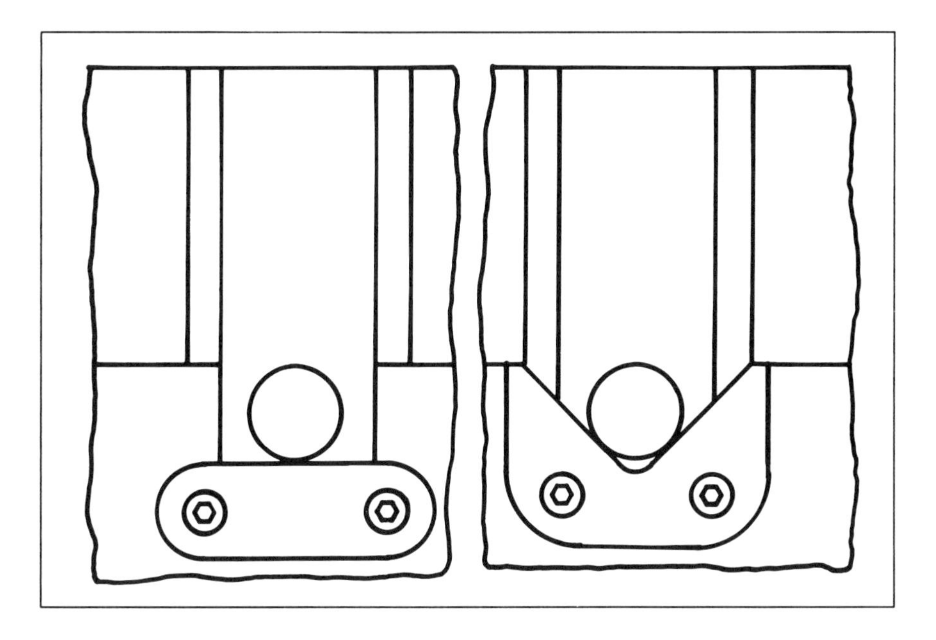

Figure 12-18. *Hardened inserts in the subplate or lower die shoe to lessen wear.*

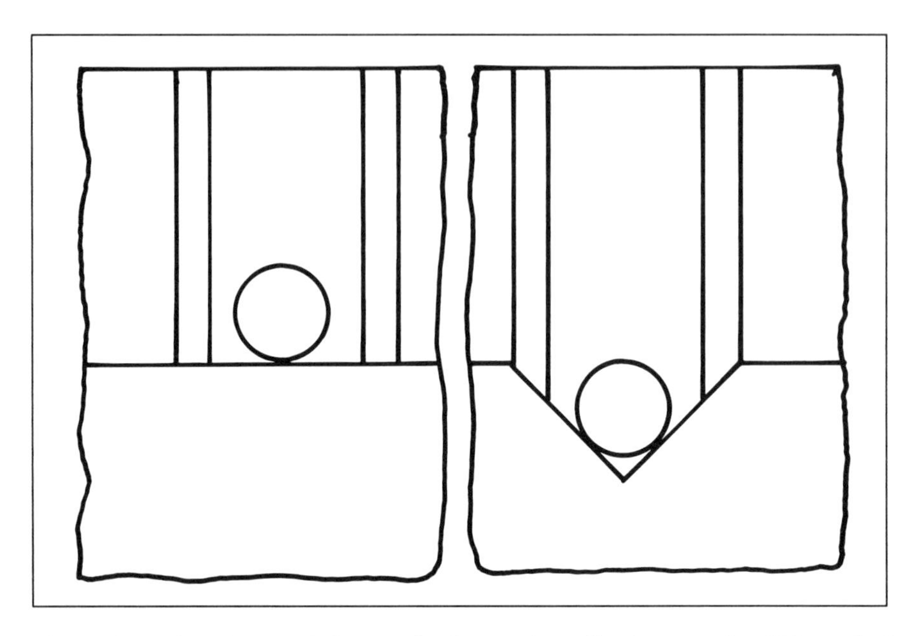

Figure 12-19. *Locating a subplate or die shoe against offset bumper pins. This method reduces machining requirements.*

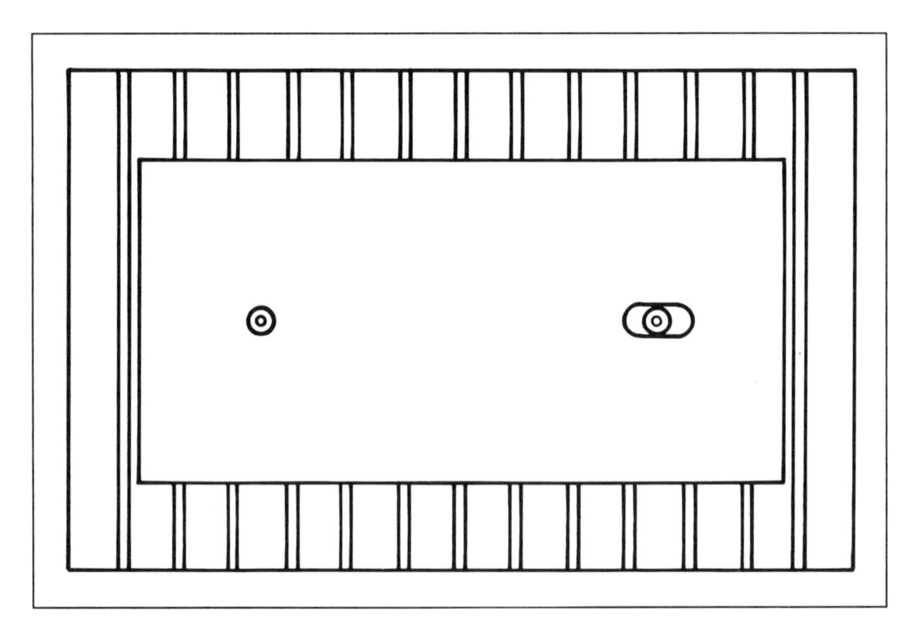

Figure 12-20. *Locating a subplate or die shoe with pins fitted into bored holes located on the center line of the bolster.*

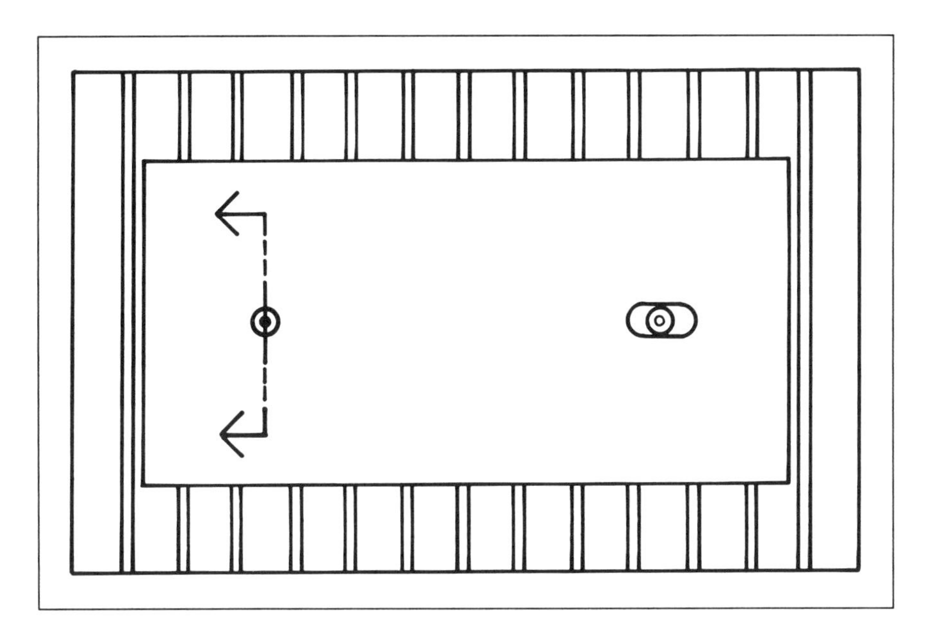

Figure 12-21. Section through a round locating pin, subplate, and bolster.

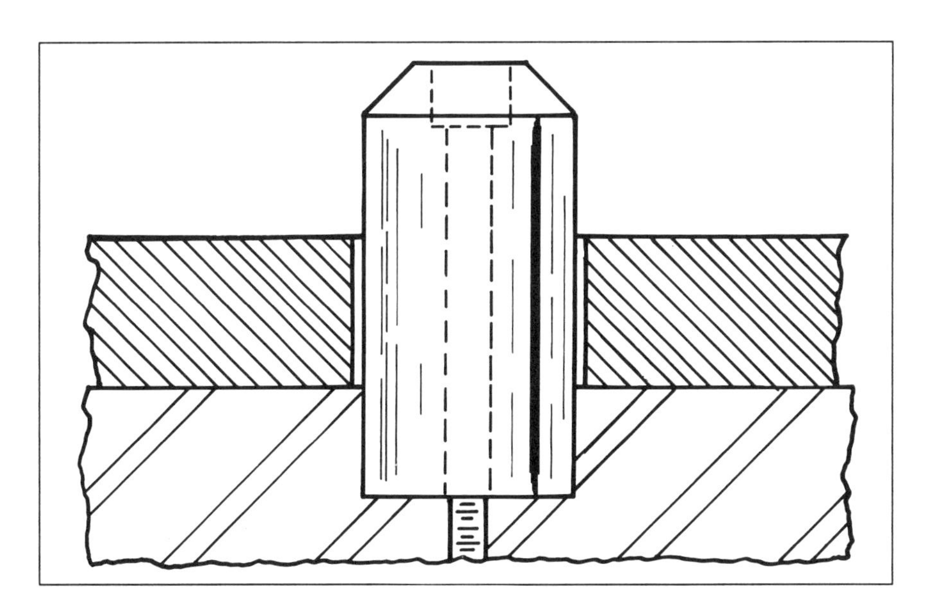

Figure 12-22. Sectional view through the large conical locating pin illustrated in Figure 12-21.

locator system can provide more precise location, and is often favored for moving-bolster as well as fixed-bolster applications.

Removable Pin Locators

Figure 12-23 illustrates the use of removable pin locators to obtain positive location for large moving bolster applications. The die is positioned over the bolster, and very carefully lowered to just above the bolster surface. The two locating pins are then inserted. This method provides very accurate location. Only a small pin-to-hole clearance is required, generally no more than 0.010 inch (0.25 mm). Figure 12-24 illustrates a sectional view of a subplate locating pin inserted into the press bolster.

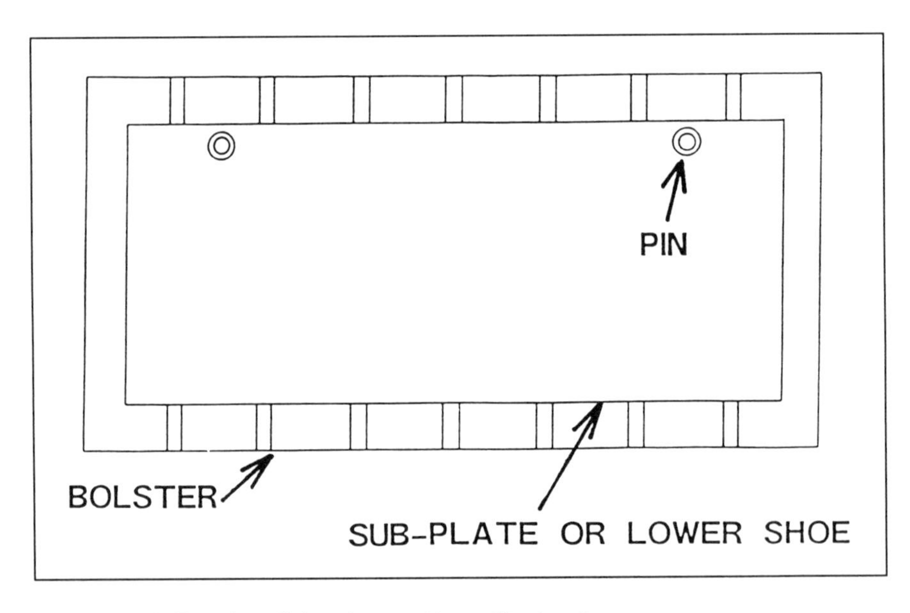

Figure 12-23. *Two close-fitting pins provide positive location.*

Wide spacing between pins helps assure precise location. Some location systems maximize the spacing between locating pins by placing them across diagonal corners of the bolster. A safety concern when setting large dies is that the diesetters cannot see each other.

Small pin locators are used to assure positive location of dies that are light enough to be manually positioned once placed on the bolster. Typically, this is limited to die weights of no more than 200

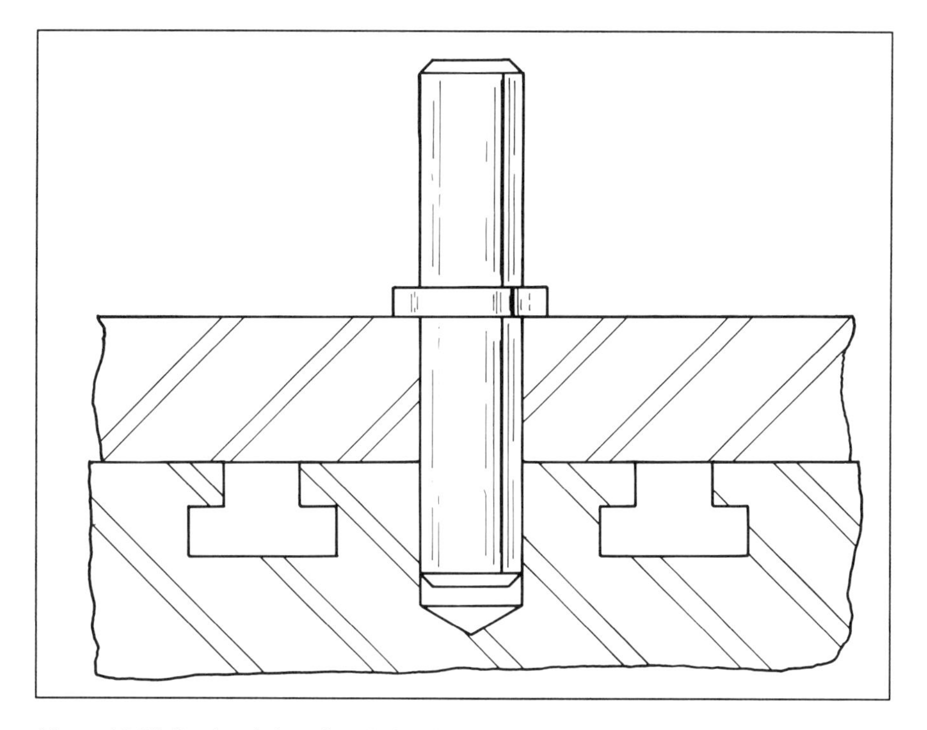

Figure 12-24. Sectional view of a subplate locating pin inserted through the subplate and into the press bolster.

lbs (90.7 kg), based on a typical coefficient of friction of steel on steel of approximately 0.20.

If bolster lift rollers are used, the coefficient of rolling friction is reduced by a factor of approximately 10 from the coefficient of sliding friction. Thus, a die weighing up to 2000 lbs (907 kg) can be positioned by hand and positively located with pins.

Pins can be used to locate to a milled keyway in a press bolster. This application is illustrated in Figure 12-25. The pin seen in Figure 12-26 locates to a drilled and reamed hole in the bolster. Welding a T-handle on the pin makes the pin easy to insert and provides a means to pry it out if it gets stuck in the bolster.

Corner Positive Locating Systems

The subplate positive locator system featured in Figure 12-27 is unusual, but effective as used here. Although the two locators attached to the bolster position the die accurately, the disadvantage is that all subplates must be the same size. The closeup of the

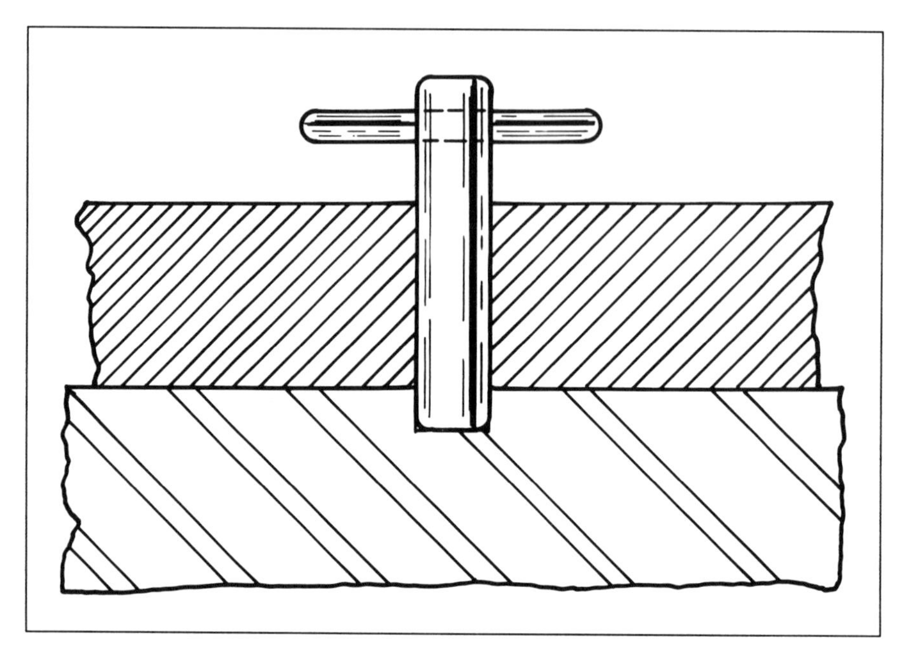

Figure 12-25. *Locating pin for setting a small die in line with a milled bolster keyway.*

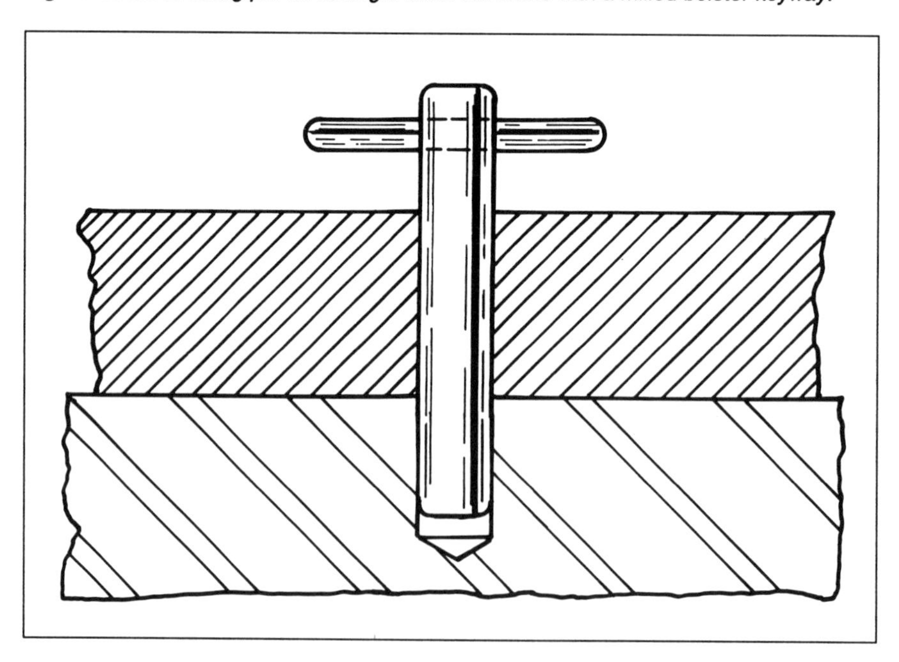

Figure 12-26. *Small pin locator inserted through a die shoe or subplate, and into a hole in the bolster. Hardened bushings should be used if wear is a problem.*

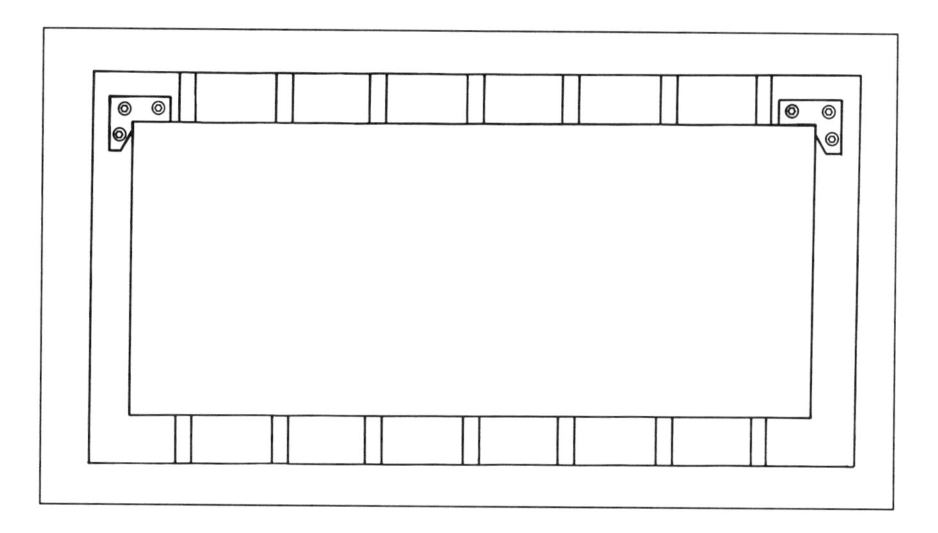

Figure 12-27. *Positive die subplate corner locating system.*

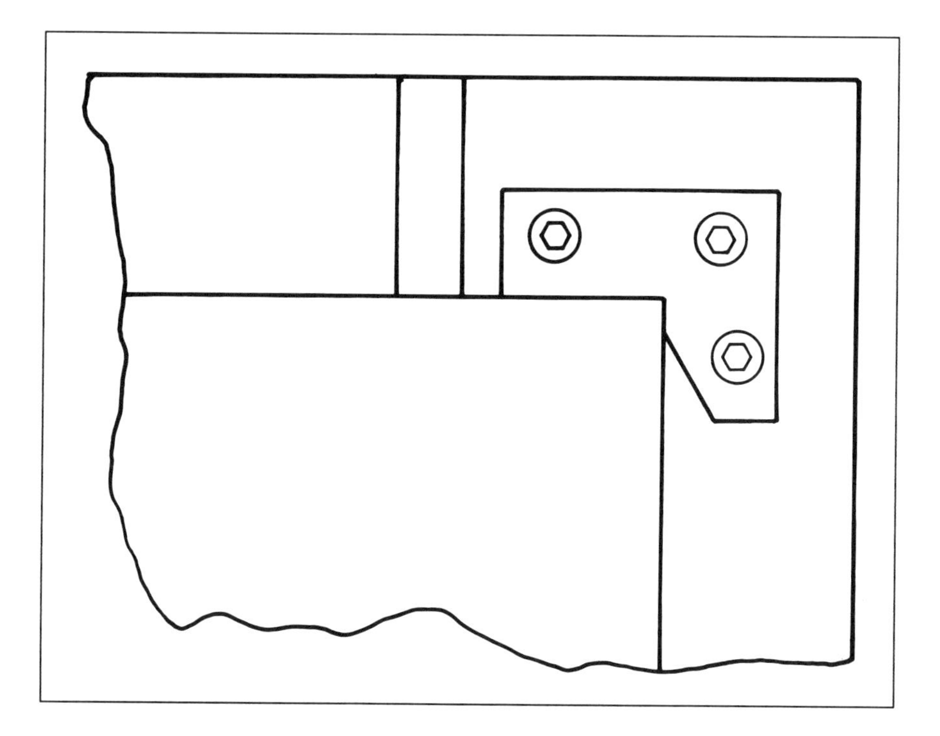

Figure 12-28. *Close-up showing detail of positive subplate corner locator.*

locator illustrated in Figure 12-28 shows the simplicity of the locator. It can easily be installed with a portable drill.

Figure 12-29 illustrates how one of the locators can be redesigned to accommodate different shoe lengths. However, this type of locating system is less positive than the type shown in Figure 12-27. A fork truck may be needed to nudge the die into correct side-to-side, as well as front-to-back, position.

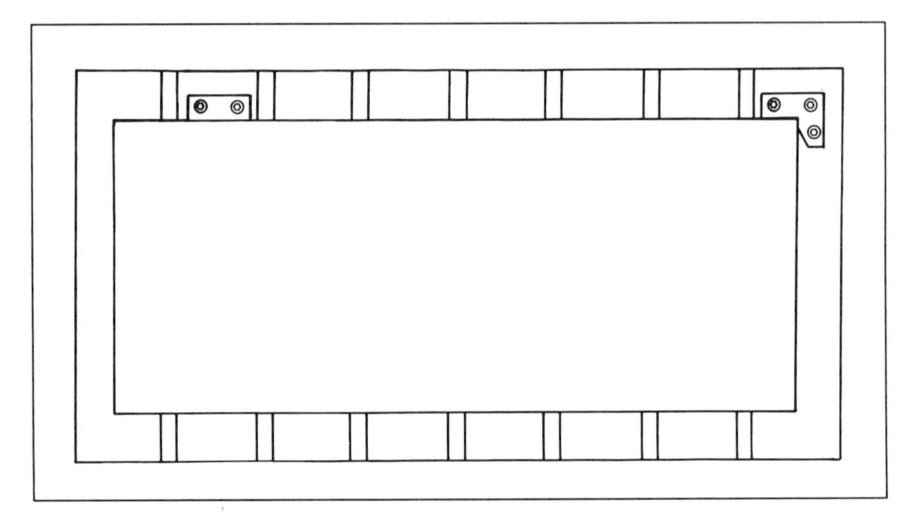

Figure 12-29. By replacing one corner locator with a flat stop, the locating system illustrated in Figure 12-27, can accommodate different subplate lengths.

CENTER LINE KEYWAY SYSTEMS

Many examples shown so far use standardized sizes of subplates attached to the lower die as a part of the die handling and locating system. Subplated tooling has been popular for many years for the large dies used in the appliance and automotive industries.

However, subplates add both cost and weight to the tool. While they may be necessary for large dies changed with dedicated die handling systems, they may not be appropriate or cost effective for smaller tooling.

Contract or Job Shop Practice

The contract stampers supplying large stampings to the automotive industry may find subplated tooling advantageous. However, the majority of contract work involves smaller dies. A typical contract shop may run from 50 to over 500 dies on a monthly basis. Often, there are many other jobs set less frequently to meet service part or very small lot requirements. The total number of dies having active production requirements typically ranges from several hundred to over 2000 tools. Increasingly, the dies are of coil fed progressive design. Secondary pressworking operations are usually manually fed, and performed in OBI or OBS machines.

In most cases, dies belong to the stamping customer. The contract to produce the parts often can be canceled on short notice. The contract stamper almost universally has two conflicting prerequisites: to produce high-quality parts on short notice and to provide them as economically as possible. Any trial-and-error adjustment, tool damage, or scrap directly reduces the profit margin.

Two Quick Changeover Requirements

The two prime requirements for quick die changeover are positive location and fast secure clamping. Coil-fed tooling must be located accurately on a common center line of the feed path of the tool and press.

While the feed path requires very close front-to-back alignment, the left-to-right die location need only be close enough to equalize the load on the press ram. Failure to keep the load equal may result in the ram tipping as the operation is performed. Ram tipping problems, in turn, accelerate press and die wear, and often result in product variation.

If the die has tie-down slots in the lower shoe or parallels, the correct T-slots to use should be identified in the diesetting instructions. Typically, this will assure correct left-to-right location, within 0.125 inch (3.176 mm) or closer.

Locating to a Center Line Keyway

Figure 12-30 illustrates a key placed in a 1 inch (25.4 mm) wide keyway milled on the center line of the entire length the bolster. Locating dies to this type of keyway can be accomplished in several ways.

Figure 12-31 illustrates a simple method of positively locating a die front-to-back by means of a keyway milled in the bolster. A wider keyway is milled in the lower die parallel to permit easy engagement of the key when the die is placed in the press with a fork truck.

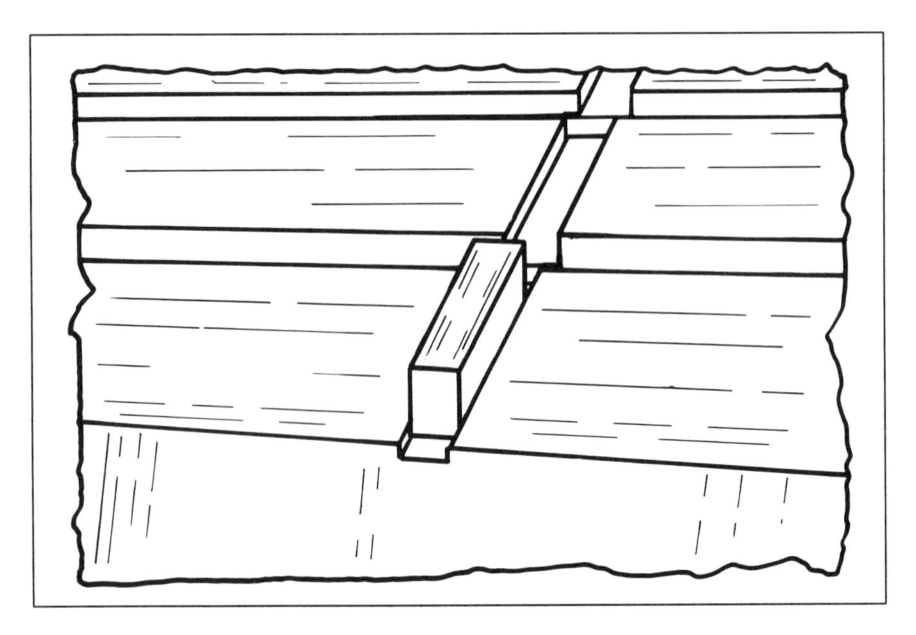

Figure 12-30. An end-view of a modified straightside press bolster illustrating a 1 inch (25.4 mm) wide keyway milled on the center line. One of two keys are shown in place to locate a progressive die accurately. (W.C. McCurdy Company)

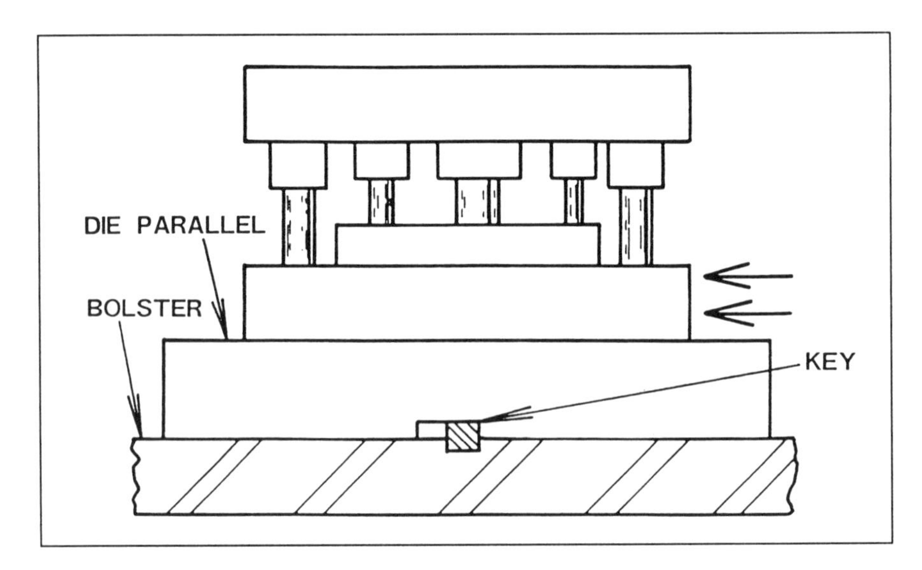

Figure 12-31. Locating a die front-to-back by means of a keyway milled in the bolster. A wider keyway is milled in the lower die parallel to permit easy engagement of the key when the die is placed in the press with a fork truck. (Smith & Associates)

Once the die is lowered into approximate location, the standard practice is to shove the die toward the rear of the press against the key. If the keyways are accurately machined, this ensures the center line of the feed path of the die is exactly on the press center line.

An advantage of the method shown in Figure 12-31 is that the key may be in place before the die is put in the press. The extra clearance makes die placement less critical, yet achieves accurate location. A skilled diesetter and fork-truck driver can place the die over the key and nudge it into exact front-to-back location on the first try. Any minor side-to-side location correction to line up the tie-down slots in the parallel feed can be easily done with a pry bar.

Placing the Die with a Pry Bar

Except for moving bolster, V-locator, and die carrier systems designed to achieve positive die location, some positioning of the die is often required once it is in the press. This can be done manually for small dies.

Final positioning of dies weighing several tons (3000 kg) or more may be accomplished by means of a pry bar. The locating keys may be inserted after visually aligning the die. This is an especially good procedure when the diesetter must set the die in the press without another worker present to give directions as to exact placement.

Care must be used to avoid damaging bolster surfaces such as T-slots when prying dies into location. A snug fitting steel plate can often be inserted into a T-slot or a special pin placed in cushion pin hole to quickly provide a safe fulcrum for prying.

Some other presses use a more conventional center line bolster keyway locating system. Figure 12-32 illustrates a sliding captive key in a lower end parallel. Many variations of this basic design have evolved. A close-up view is seen in Figure 12-33.

USING JIGS AND TEMPLATES FOR ADAPTING DIES

If only a few dies are to be adapted to positive die locating and rapid clamping methods, laying out parallel locations, center line keyways, and ram tie-down holes may be done in the production press. This requires scheduling coordination between the tool room and production departments to avoid delays or missing an opportunity to make an accurate layout.

After the adaptation is completed, it should be checked during the next dieset to ensure the work's accuracy. This is especially

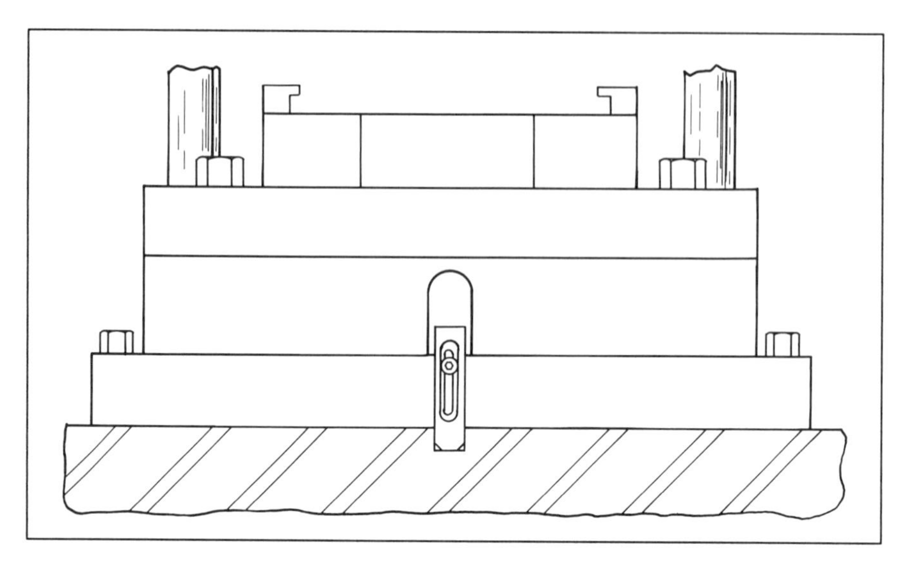

Figure 12-32. *A sliding captive bolster-center line key installed in an end parallel. (Olson Metal Products Company)*

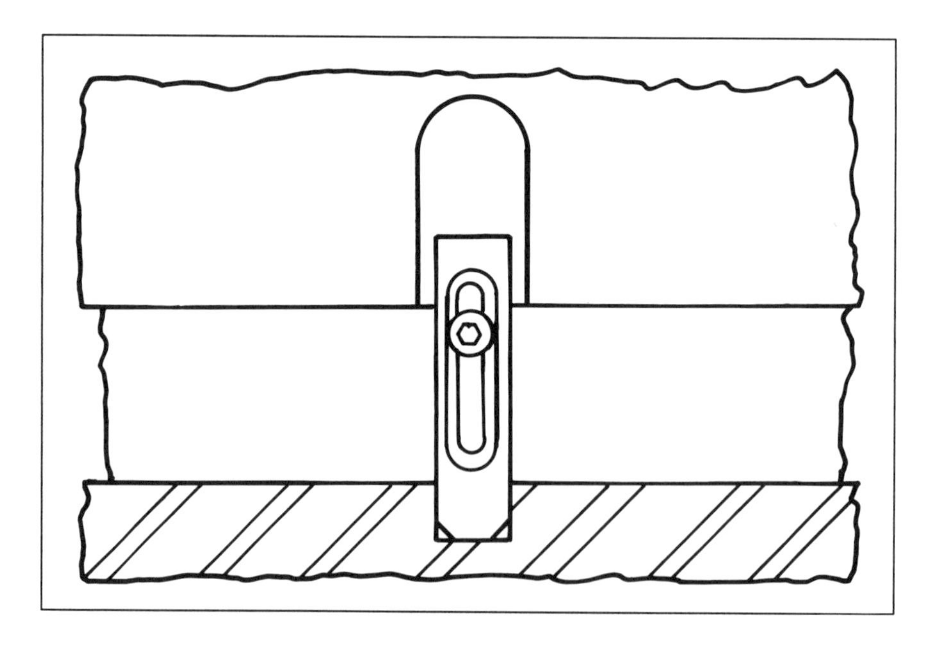

Figure 12-33. *A close-up view of the sliding key locator shown in Figure 12-32. (Olson Metal Products Company)*

important on coil-fed tooling where the feed path of the die must accurately match the center line of the press bolster, feeder, stock straightener, and decoiler. Any problems should be resolved by a cooperative effort of the maintenance department production supervision, and the tool room.

Figure 12-34 depicts a method to correctly place permanently attached parallels having center line locating keys, as well as top shoe adapter plates on small precision dies. This method, developed by a large diversified contract stamper, enables them to adapt large numbers of electronic component dies sourced to their facilities on short notice.

Figure 12-34. *A used four-post die set made into a jig. It is used to place permanently attached parallels having center line locating keys, as well as top shoe adapter plates on small precision dies. (Olson Metal Products Company)*

The jig seen in Figures 12-34 and 12-35 has a center line keyway and bolster cutout identical to the OBI or OBS presses that run the dies. It is fabricated from a surplus die shoe the same size as the press bolsters.

T-slot locations are marked on the jig by milling shallow grooves the same width on the standard Joint Industry Conference (JIC)

Figure 12-35. A close-up view of the quick die change adaptation jig shown in Figure 12-32, illustrating the center locating keyway and T-slot layout on 6-inch (152.4-mm) centers. (Olson Metal Products Company)

locations as that of the presses. This permits accurate location of parallels for permanent attachment to the lower die shoe. A progression strip with a scribed center line is needed. It ensures the feed path of the tool lines up with the center key in both the jig and the press.

This diesetting improvement work is an excellent assignment for new apprentices. The student will have an opportunity to become proficient at locating, drilling, and tapping holes. By following up on the job in the pressroom, the function of the tool in the press will be learned also.

Using the Jig to Adapt the Upper Die

Current United States OSHA rules do not permit dies having shanks to be retained by the press ram solely by that means. An additional fastening method such as toe-clamps is required. Unless the shank is needed for part knock-out purposes, it is milled off and a steel plate attached to the upper die shoe. The upper half of the jig has the standard press ram tie-down holes drilled on location.

It is assembled, and the holes are transferred to the steel plate attached to the upper die shoe.

With this system, small-to-medium size progressive dies are virtually always ready to produce good parts when first set. The goal of good diesetting is to avoid trial-and-error adjustments. Using this method provides quick die change adaptations efficiently and at low cost.

Figure 12-36 shows keys used to provide front-to-back location in a knee, or OBI press. If the optional stem is used, only one key is needed, and precise left-to-right location is also ensured.

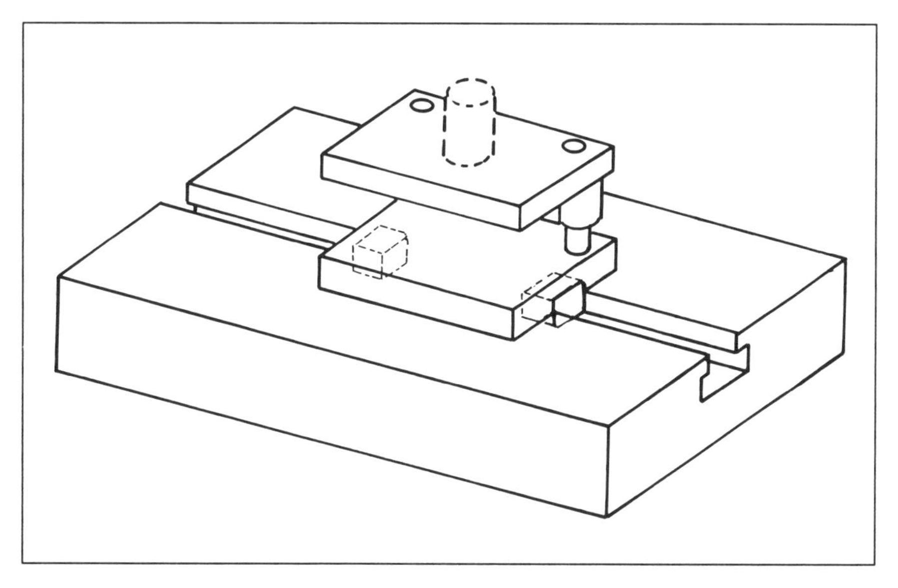

Figure 12-36. Bolster T-slot keys used to provide front-to-back location in a knee or OBI press. If the optional stem is used, only one key is needed, and precise left-to-right location is also ensured. (American Aerostar)

OBI and OBS JIC T-slot Pattern

Figure 12-37 illustrates a typical JIC T-slot pattern for an Open Back Inclinable (OBI) or Open Back Stationary (OBS) press. It provides locations to tie down large dies covering most of the bolster. Locating keys can be used with the T-slots provided die shoes or subplates extend over them.

Not every press builder routinely supplies the standard JIC T-slotted bolster. T-slots require a thicker bolster for an equal amount of deflection, all other factors being equal. Thicker bolsters also reduce the available shut height of the press. Many presses presently in use were purchased with blank bolsters (Figure 12-38). Providing die fastening methods then becomes the responsibility of the user. To tie down dies, the usual practice is to tap holes in the bolster plate as illustrated in Figures 12-39 and 12-40.

TAPPED BOLSTER HOLES

When the press user buys a press with a blank bolster, the initial cost savings may seem attractive. However, the bolster soon begins to resemble Swiss cheese, and many hours will be wasted tapping additional holes for new jobs.

There are several major problems associated with tapped bolster die tie-down holes. First, the use of SAE grade eight bolts or equivalent is highly recommended for diesetting applications. These bolts are made of heat-treated alloy steel. The minimum yield strength is 130 ksi (896 MPa) or greater. For metric applications, an approximately equivalent fastener is an ISO property class 10.9 or 12.9. The strength of these bolts is far superior to the bolster material.

Generally, bolsters are made of cast iron, iron alloy, or low to medium carbon steel. The bolster material has less than 50% the strength of the recommended fastener materials. The holes will tend to strip out with repeated tightening of the diesetting bolts.

Slug Problems

The tapped holes in bolsters seem to have a magic attraction for slugs (Figure 12-41). The proper way to keep slugs out of unused holes is to plug them, which can be done in a number of ways. Set screws, especially those having screwdriver slots, work quite well (Figure 12-42). Those having hex key holes tend to have the key recess fill with small slugs and debris. Threaded nylon rod is an easily worked material from which to make plugs. The screwdriver slot can be cut with a hacksaw. If the nylon plug should become stuck, it is easily drilled out. Other materials for plugging holes include:

- Wooden dowel stock;
- Corks; and
- Rubber stoppers.

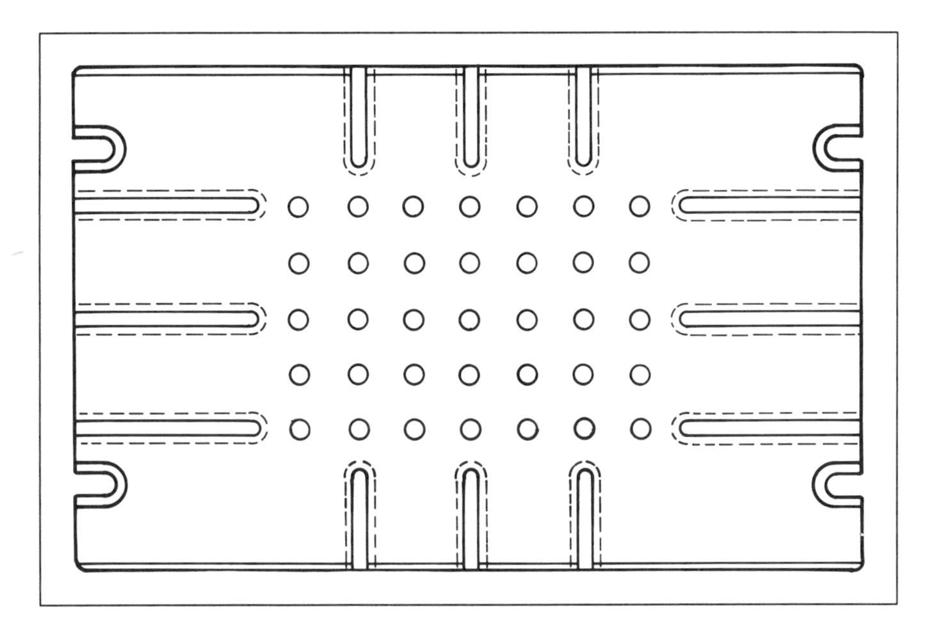

Figure 12-37. *A typical Joint Industry Conference (JIC) T-slot pattern for an open-back inclinable (OBI) or open-back stationary (OBS) press.*

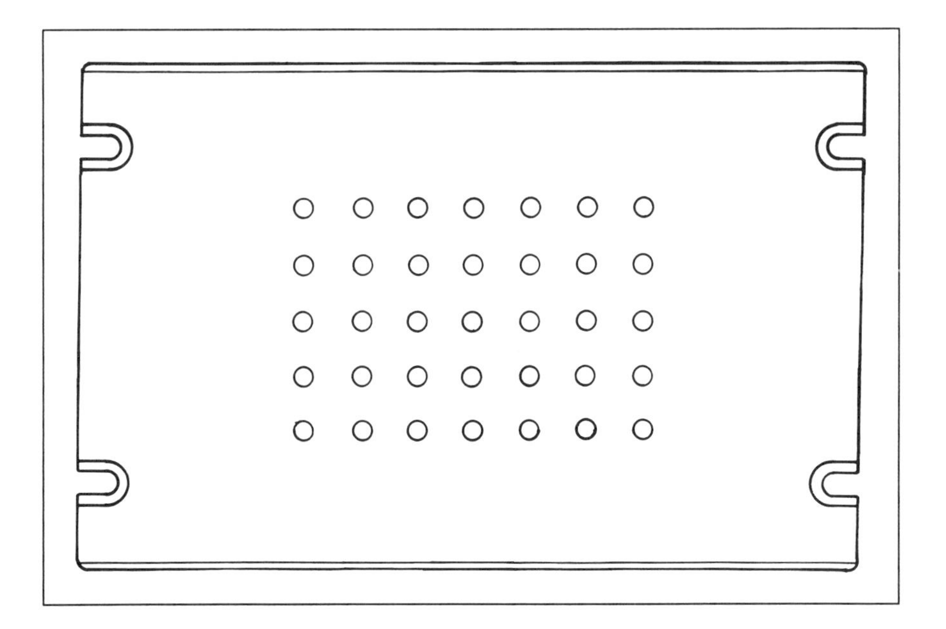

Figure 12-38. *A blank bolster plate having die cushion pin holes, but no provision to tie down dies.*

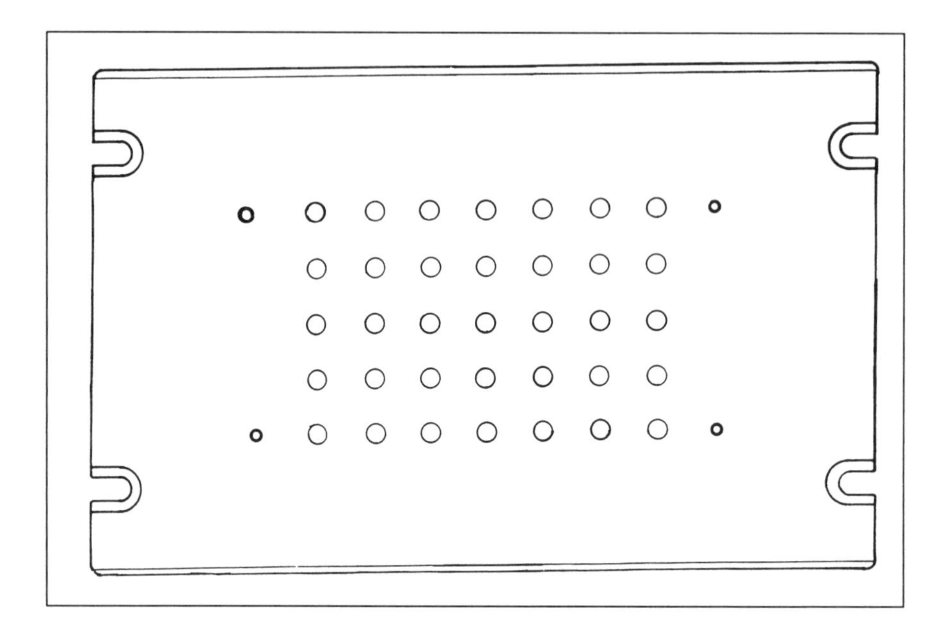

Figure 12-39. *Example of user-provided tapped holes in a blank bolster plate.*

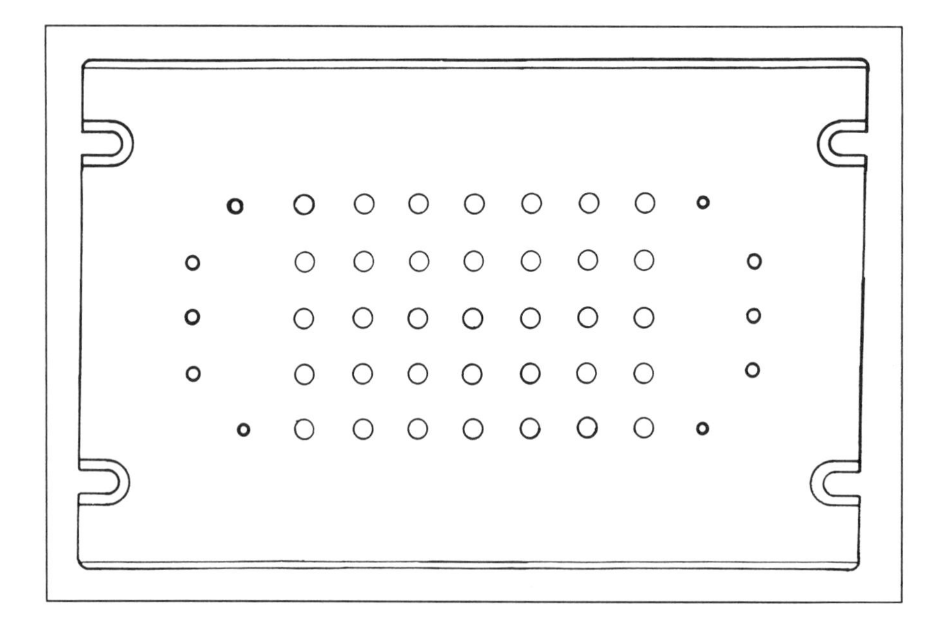

Figure 12-40. *Additional tapped holes added to accommodate new jobs.*

Figure 12-41. *Threaded bolster holes have a "magic" attraction for slugs.*

Figure 12-42. *A set screw or plug used to keep slugs out of tapped bolster holes.*

Slugs and debris get into die tie-down and die cushion pin holes by careless bolster brushing and cleaning during diesetting. One way to help avoid the problem is to minimize the slugs remaining on the bolster through the use of effective scrap removal methods during production. Remaining slugs should be carefully brushed or wiped away — before the bolts that fasten the die are removed, if possible.

Steel slugs can be removed from tapped bolster holes with a small magnet. However, the work is slow, delaying production. Should the slugs and debris be compacted in the bottom of the hole, a portable drill may aid their removal.

Nonferrous materials such as aluminum and brass, as well as nonmagnetic stainless steels, are best removed by picking them out with a hooked scriber (Figure 12-43). Another method is to blow them out with compressed air (Figure 12-44).

Using compressed air as a cleaning tool is very common in the stamping industry, but the practice can be dangerous. The slugs can be propelled at high velocities resulting in potential eye injuries and cuts. In addition, normal shop air pressures of 80 psi (552 KPa) exceed OSHA regulations for air nozzle pressures. Should a nonpressure-limited nozzle be used, it has the capability of injecting air into the bloodstream. If accidentally actuated when in contact with the skin the result can be fatal.

Figure 12-45 illustrates the insertion of a bolt into a hole filled with slugs. When the bolt is tightened, the first few threads strip. Even after the hole is cleaned out, the remaining threads may not provide enough holding ability to properly retain the lower die in the press under normal loads.

Figure 12-46 depicts a bolster having so many holes drilled to replace failed threads that it resembles Swiss cheese. The use of thread repair bushings or special helical inserts is a frequently used repair procedure. However, it is not a long-term solution. While the thread repair inserts may provide superior mechanical strength compared to the original tapped hole, the problem of accumulation of slugs and debris will continue to delay diesetting activities.

T-SLOT LAYOUT

Figure 12-47 illustrates a bolster with only four T-slots. This is false cost savings. A minimum requirement for flexibility of clamping is the standard JIC pattern for an OBI or OBS press bolster shown earlier in Figure 12-37. To locate with keys in a way shown

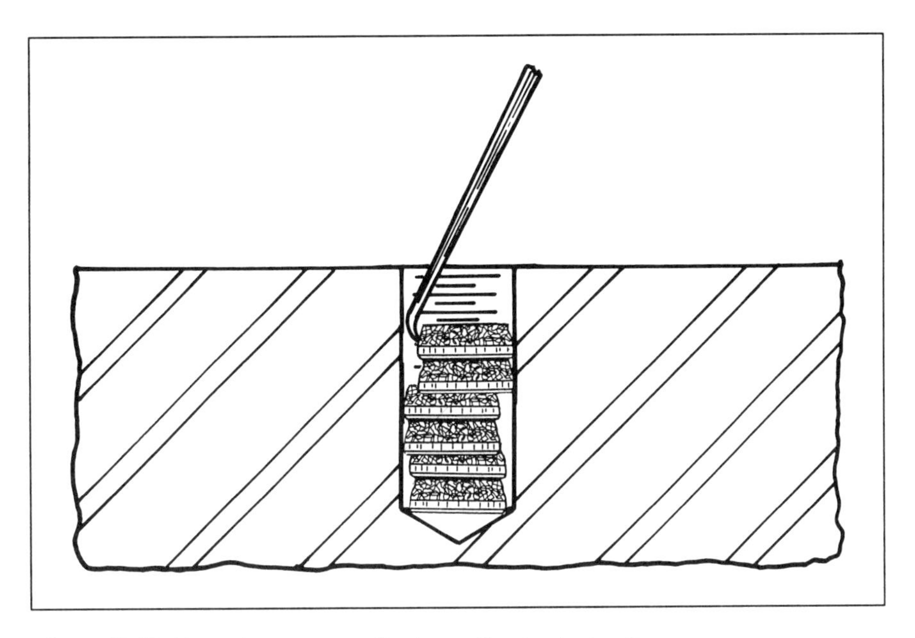

Figure 12-43. *Removing nonmagnetic slugs with a hooked scriber.*

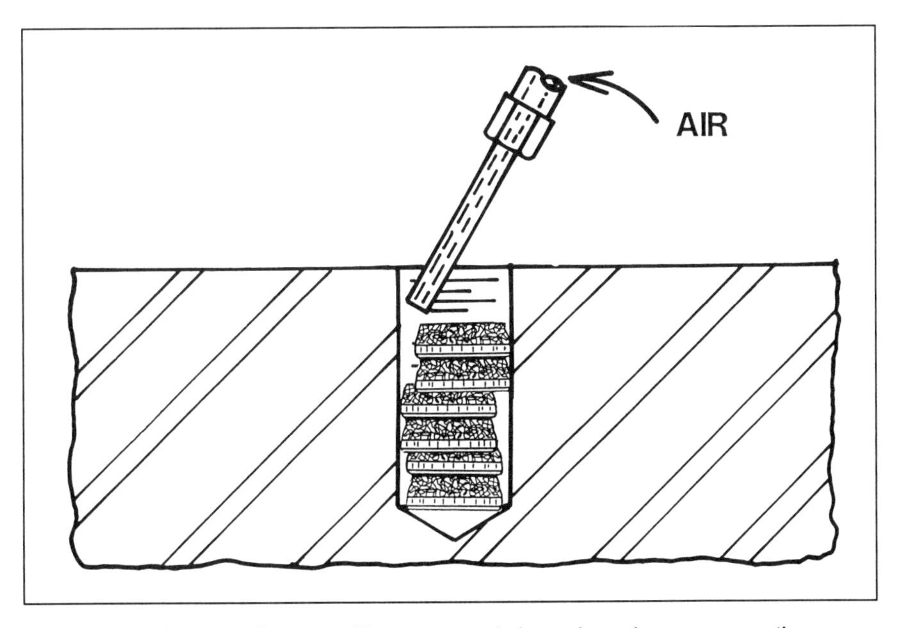

Figure 12-44. *Blowing slugs out with compressed air can be a dangerous practice.*

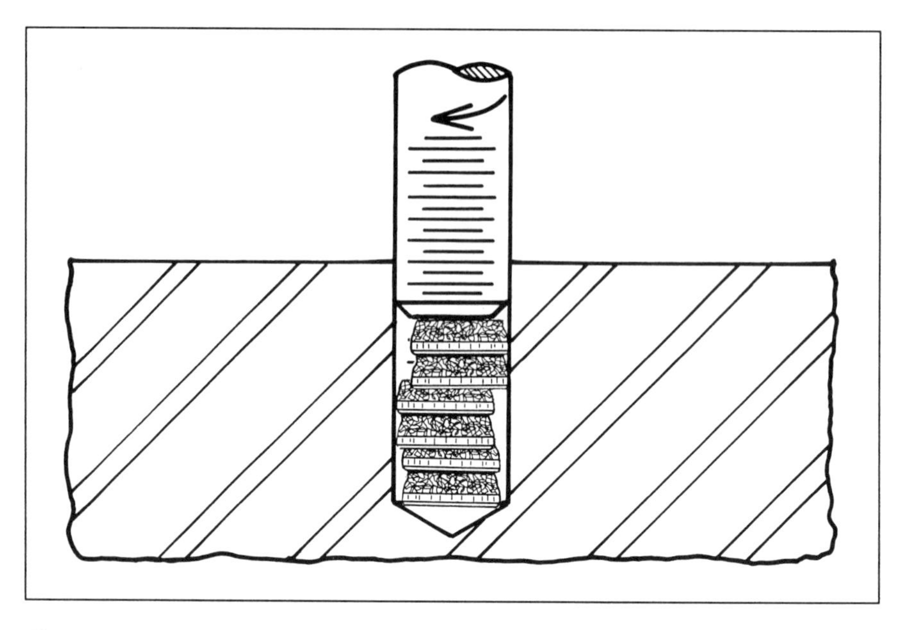

Figure 12-45. *Inserting a bolt into a hole filled with slugs. When the bolt is tightened, the first few threads strip.*

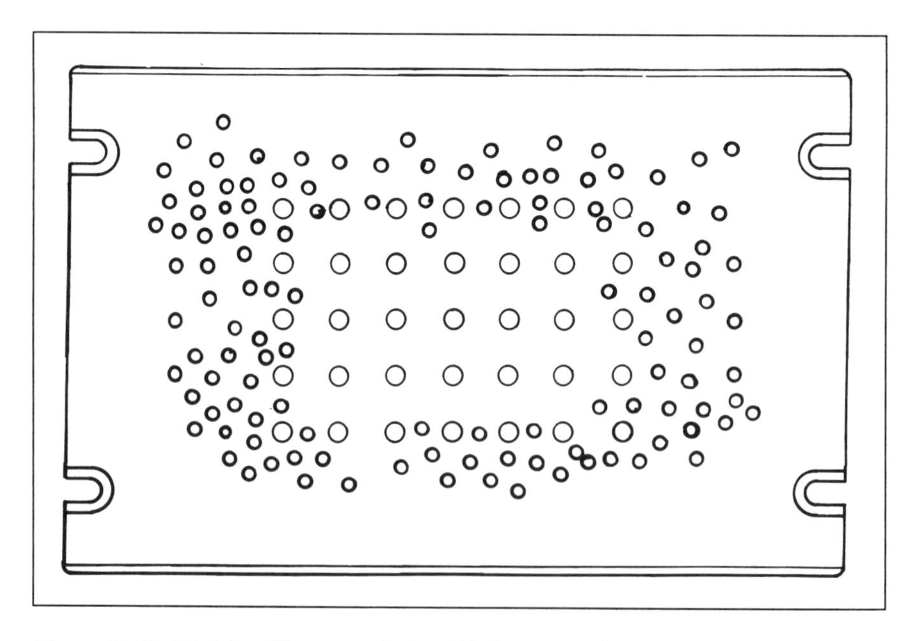

Figure 12-46. *A bolster with so many holes drilled to replace failed threads that it resembles Swiss cheese.*

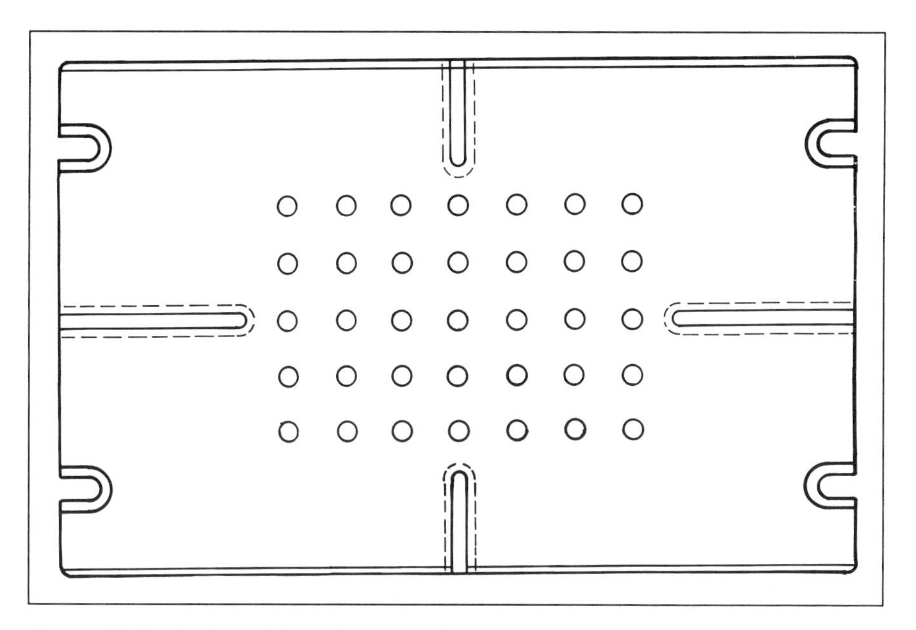

Figure 12-47. *A bolster with only four T-slots. This is false cost savings; there is little flexibility in how dies of differing sizes can be tied down.*

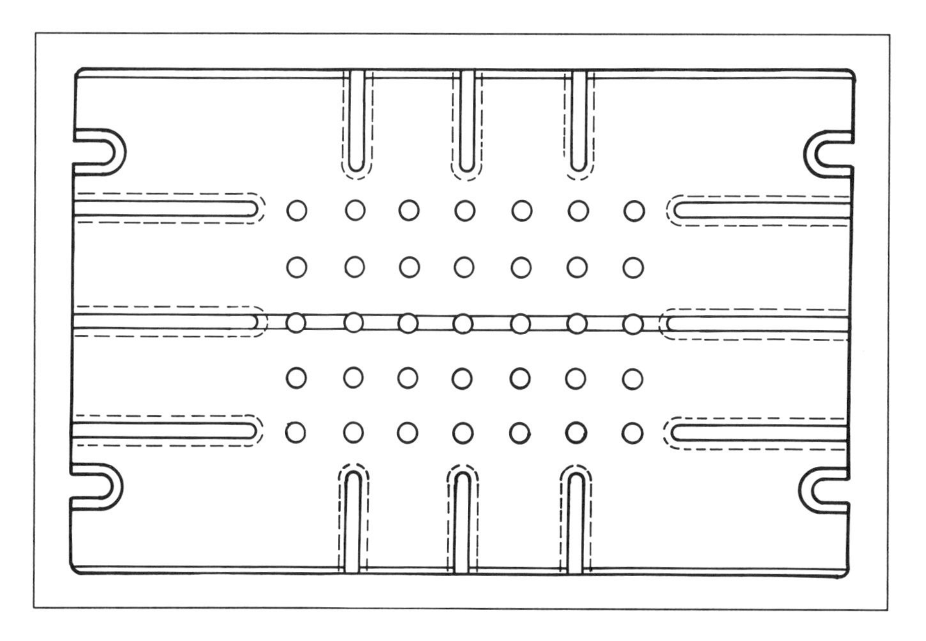

Figure 12-48. *A shallow keyway cut across the bolster center line provides a means to locate with keys. This method results in less bolster deflection than cutting a full-width T-slot.*

in Figures 12-30 through 12-35, a keyway can be cut across the bolster center line (Figure 12-48).

Many OBI and OBS press bolsters and rams are slotted from front to back only. Limited access to the rear of the press causes the diesetter great difficulty when tying up the back side of the die. Providing the diesetter a right-angle-drive air impact wrench or a ratchet wrench designed for use in tight situations can reduce the time required to tie up the die from 15 or more minutes or more to 2 or 3 minutes. A key factor in reducing the difficulty in such cases is to provide a uniform clamping height that permits the same fasteners to be used on each job.

Providing a T-slot or slots completely across the bolster may result in excessive deflection. If this is to be done, there must be sufficient thickness. The minimum shut-height required may limit the bolster thickness.

Figure 12-49 illustrates how to locate the lower half of a die by means of a bolster keyway of the type shown in Figure 12-48. U-shaped cut-outs are provided in the end parallels to permit tying the die down with four T-bolts.

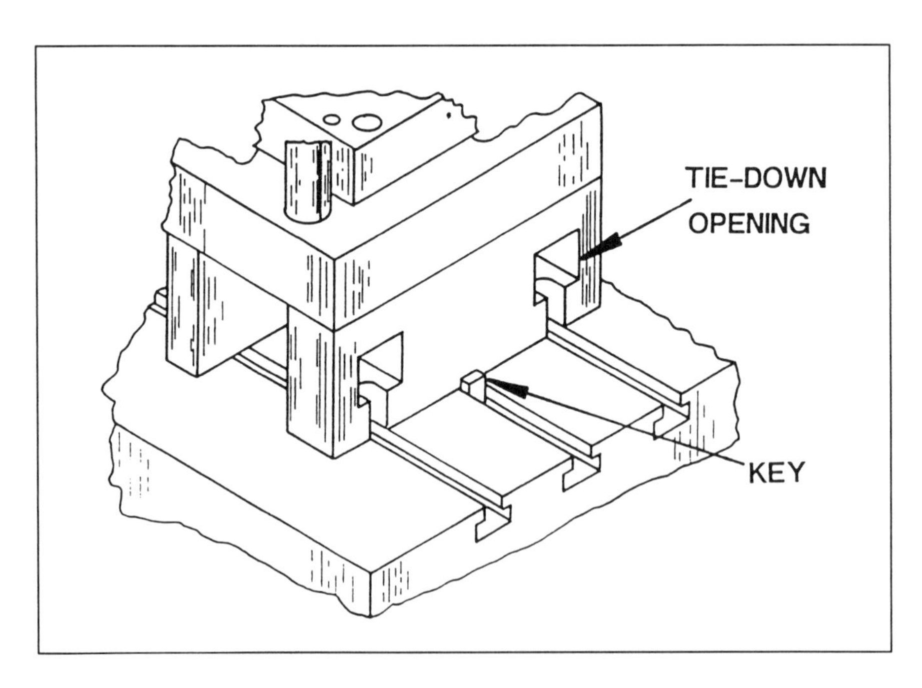

Figure 12-49. *Locating the lower half of a die by means of a bolster keyway of the type shown in Figure 12-48.*

Locating Dies to the Press Ram

The press ram may be used as a means to locate coil-fed dies on the press feed-path center line. Two methods are in common use.

The first method, illustrated in Figure 12-50, uses locating stops attached to the front edge of the upper die shoe. These stops are shoved against the front edge of the ram to provide positive front to back location. The following procedure is used to set the die on location:

- Adjust the press shut height to the die shut height plus a small amount, usually 0.25 inch (6.35 mm);
- Visually locate the die in correct left-to-right position;
- Move the die into the press until the stops contact the ram;
- Lower the slide until contact with the die is established; and
- Clamp the die in place, and complete the dieset.

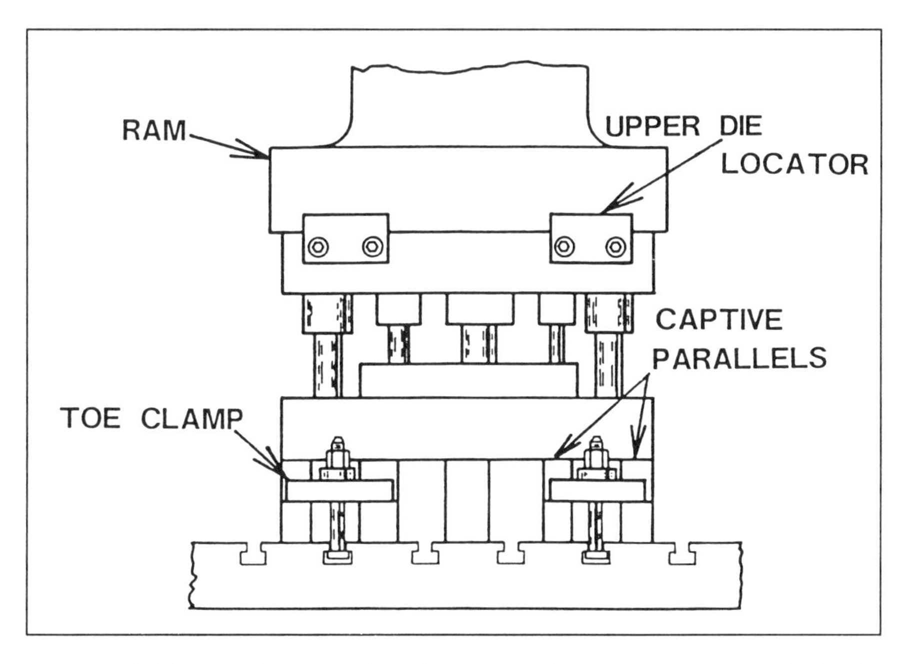

Figure 12-50. Locating the upper die to the press ram by means of stops attached to the upper die shoe. Note the manner in which each of the toe-clamps engages two parallel ledges, eliminating the need for setup blocks.

This method works well with small dies set in OBI and OBS presses. The front edge of the ram must have a parallel machined surface and offset a known distance from the press center line. The same center line-to-edge relationship must be established when installing the stops on the die.

If all dies are of similar width, the two attachment areas on the upper shoes may be cleaned up by milling, and the stops installed. Slight differences in edge-to-center line or feed-path to edge-of-die distances may be compensated for by milling offset stops or shimming.

If the ram edge is a cast surface or not parallel to the center line, it should be machined to ensure an accurate locating surface. Repeatable left-to-right location can be provided in a number of ways. One of the stops can locate in a pocket on the ram face. A V-locator can be used in one of the stop positions. Also, a hole may be drilled on location in one of the die stops to engage a pin installed in the ram locating surface.

The second method uses the T-slot some presses have in the ram located on the press center line. This T-slot can be used to locate the die by means of keys or locating pins installed in the upper die shoe. To avoid press and die damage if the press is closed with the die out of location, spring-loaded pins are preferable to fixed keys or pins.

Figure 12-51 illustrates a spring-loaded pin installed in an upper die shoe and used to engage a ram T-slot.

The spring should be stiff enough and provided with sufficient preload to prevent the locating pin from floating when the die is used in high-speed service. It is permissible to have the spring compress more than the amount allowable in continuous service, provided it does not go solid before the end of pin travel is reached. A common pin height above the die shoe is 0.5 inch (12.7 mm). The diameter is governed by the width of the T-slot. If frequent die changes are to be made, T-slot wear may be a factor. In such cases, the T-slot may be fitted with a hardened tool-steel insert to avoid wear on the T-slot. When setting the die, the usual procedure is to:

1. Adjust the press shut height to the die shut height plus a small amount, usually 0.25 inch (6.35 mm).
2. Place the die into approximate location.
3. Carefully inch the press onto bottom.
4. Move the die until the locators snap into place.
5. Lower the slide until contact with the die is established.
6. Clamp the die in place, and complete the dieset.

If the press is maintained at a common shut height, the die may be placed in its approximate location, the press inched nearly closed, and the die moved until the locators snap into place.

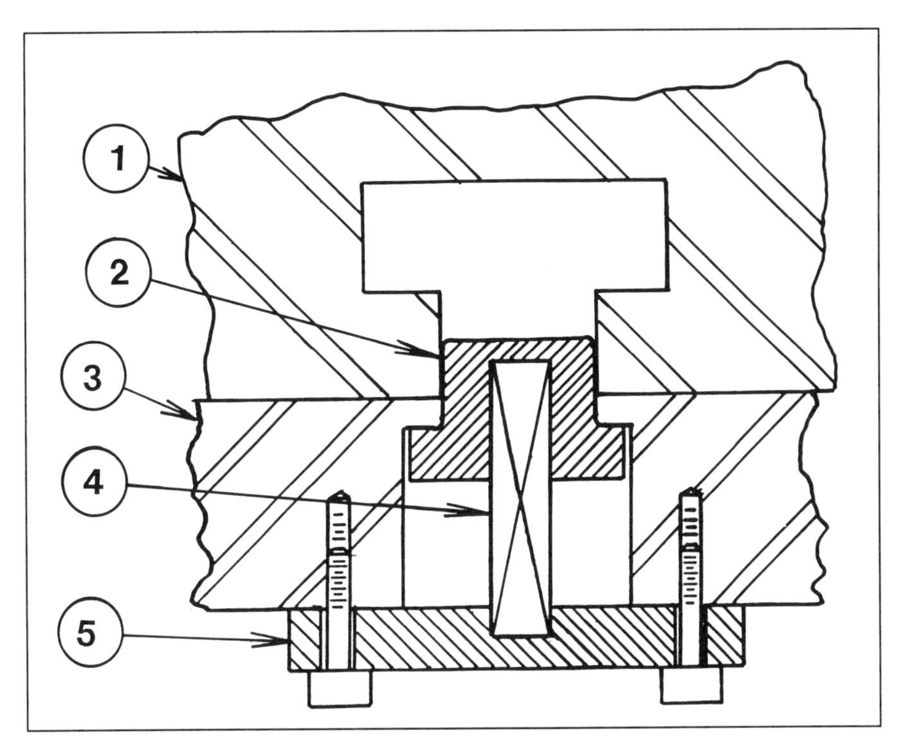

Figure 12-51. *A spring-loaded pin installed in an upper die shoe to engage a ram T-slot for accurate front-to-back location. (1) is the press ram. (2), the locating pin, is typically made of ASTM-SAE 1020 steel and carburized for wear resistance. A hole to provide a sliding fit on the pin body is bored through the upper die shoe (3) in two places on the center line of the feed-path through the die. It is important the feed center line not coincide with the die center line. The top die shoe (3) is counterbored to provide a clearance for the locating pin. A spring (4) preloads the pin and permits it to be forced flush with the die shoe if the press is closed with the die out of location. The spring is retained by a plate (5) made of cold-rolled steel. The retaining plate may be made of aluminum to save weight in high-speed applications.*

This method is limited by the diesetter's ability to move the die manually into location. Assuming the coefficient of friction for steel-on-dry-steel to be 0.19, a 200 pound (91 kg) die would require a lateral force of 38 lbs (17.3 kg) to move it into place. Dies weighing much more than this can be moved into place by a single diesetter, provided a pry-bar is used. Prying the die into place will add very little to setup time if a proper fulcrum is in place and the pry bar is close at hand when the die is set.

Bolster rollers reduce the coefficient of friction to a much lower value than that of steel against steel. Dies weighing 3000 lbs (1,364 kg) or more can often be moved into place by a single individual.

BIBLIOGRAPHY

1. A paper by Phillip A. Gibson, Regional Manager, Atlas Technologies, Atlanta, Georgia. The paper was presented at the SME Die and Pressroom Tooling Clinic held in Chicago, August 29-30, 1989.
2. William J. Stevenson, *Production Operations Management*, Richard D. Irwin, Inc., Homewood Illinois, 1990.
3. David A. Smith, *Quick Die Change*, Chapter 6, "Relationship of Quick Die Change to EOQ and JIT," Society of Manufacturing Engineers, Dearborn, Michigan, 1991.
4. Jerry Claunch and Philip Stang, *Setup Reduction; Saving Dollars With Common Sense*, PT Publications, Inc., Palm Beach Gardens, Florida, 1990.
5. Ford Motor Company, *Die Design Standards for North American Operations*, Body and Assembly Division, Dearborn, Michigan.
6. David A. Smith, *Die Design Handbook*, Section 13, "Dies for Large and Irregular Shapes," Society of Manufacturing Engineers, Dearborn, Michigan, 1990.
7. David A. Smith, *Die Design Handbook*, Section 27, "Press Data," Society of Manufacturing Engineers, Dearborn, Michigan, 1990.

CHAPTER 13

ESTABLISHING GOOD CLAMPING PRACTICES

The primary purpose of any die clamping system is to retain the die in the press in a safe and secure manner. A good clamp must be strong and quickly applied. It must resist breaking or loosening under repeated shock loading. High-quality fasteners provide the best overall safety and economy.

Many die clamps use threaded fasteners. The clamping system may be as simple as a T-bolt, nut, and washer, if U-shaped cutouts are provided in the die. Hydraulically powered clamps may be cost effective when many die changes are performed daily.

SAFE FASTENER STANDARDS

A basic responsibility of pressroom management is to provide a safe workplace. The number, size, and location of fasteners should be specified in the design of the die. Every company having pressworking operations should have die design standards covering each class of work performed.

Safe pressworking depends on the use of a sufficient quantity of clamps to securely hold the die in the press. The attachment of the die buildup, including parallels and subplates, also requires an adequate size and number of high-grade fasteners to ensure against failure.

Some classes of work may require fastener load safety factors several hundred times the static weight of the die and buildup alone. High-speed work involving rapid cyclical loading is an example. The shock and impact loads occurring in slow operations can also be quite severe. Typical causes are snap-through energy release in heavy punching operations as well as cam and pad return impact.

Recommended Fasteners

It is highly recommended only SAE grade eight or equivalent fasteners be used for diesetting applications. These bolts are made of heat-treated alloy steel having a minimum yield-strength of 130 ksi (896 MPa). An ASTM equivalent fastener is designation number A354, grade BD. Both of these fasteners are identified by six markings on the head illustrated in Figure 13-1 (A).

All specification-grade fasteners made in the United States have a manufacturer's mark or logo in addition to the grade marking. If there is no manufacturer's identification, the bolt may be imported. Some imported bolts fail to meet specifications. The safety issue is too critical to leave fastener safety to chance. Require the seller to properly identify all fasteners and supply certified test results.

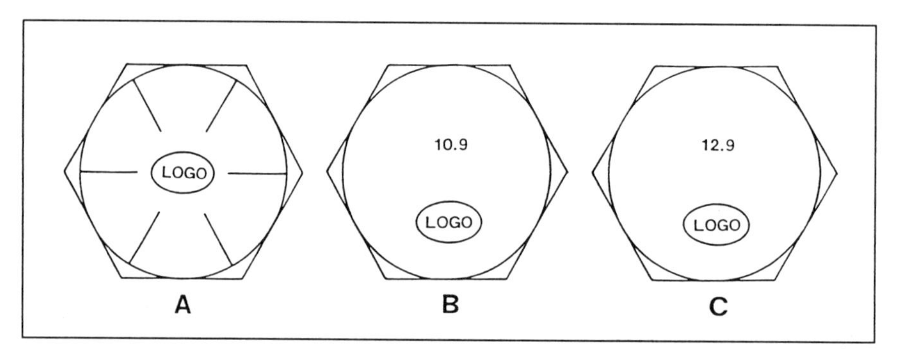

Figure 13-1. *The bolts often specified for diesetting have distinctive head markings: SAE-ASTM grade eight bolts (A) are identified by six markings; the metric fasteners are ISO 10.9 (B) or ISO 12.9 (C) property class bolts; all SAE-ASTM and ISO fasteners meeting these specifications are identified with the manufacturer's logo or identification.*

Many metric die standards specify ISO 10.9 or 12.9 property class bolts for diesetting. These have yield and tensile strengths similar to the recommended SAE and ASTM fasteners. Metric fasteners have the property class numerical designator and maker's identification stamped on the head as illustrated in Figure 13-1 (B) and Figure 13-1 (C).

Applicable safety rules, including those of some government regulatory agencies, may only require the die to be securely mounted to the bolster and slide. An important part of management's duty when providing a safe workplace is to train the employees in setting each die securely.

When threaded fasteners are used, specifying a sufficient size and quantity of heat-treated alloy-steel bolts is highly recommended. Avoiding the use of plain carbon-steel or soft bolts actually saves money. High-grade bolts last much longer, because the heads do not round off, nor do the threads stretch and wear out rapidly, which can be a frequent problem with inexpensive fasteners. Commercial all-thread rod should not be used. There is no simple way to know if it has the required mechanical strength to securely set the die.

Fabricating special threaded fasteners for diesetting and die-handling by welding should never be done. The properties of the weld itself, and of the heat-affected zone, are difficult to determine. There is an unacceptable likelihood of failure in service.

High-strength fasteners for bolting structural members are typically tightened to a one-time value equal to 70% of yield strength, to develop the specified holding force. Some fasteners tightened to a large percentage of their yield strength will fail with repeated reuse. The correct value of torque for a given diesetting service should be based on the application and the engineering data for the fastener.

Diesetting fasteners are subjected to cyclical loading from both the dynamic action of the stamping process and retorquing with reuse. The die clamping system should incorporate redundant fasteners. Should any one fastener fail, the cause should be analyzed and corrected. Many factors affect the endurance limit and fatigue strength of mechanical systems. Standard engineering reference sources should be consulted to establish safe fastening standards for each class of work, and to analyze any failures occurring.

Bolt Styles

There are two systems in widespread use for bolting dies in presses having T-slots. The preferred method is the use of a T-slot bolt (Figure 13-2). Another popular method, illustrated in Figure 13-3, is to use a bolt and a T-slot nut.

An advantage of the T-slot bolt is that a larger fastener can be used than is the case with the hex-head bolt and a T-slot nut system. A 1.000 inch (25.4 mm) diameter T-slot bolt has approximately twice the strength of a comparable 0.75 inch (19.05 mm) hex-head bolt and T-slot nut.

Nut Safety

To avoid stress concentration, approximately one and one half threads should extend beyond the end of the nut. This condition

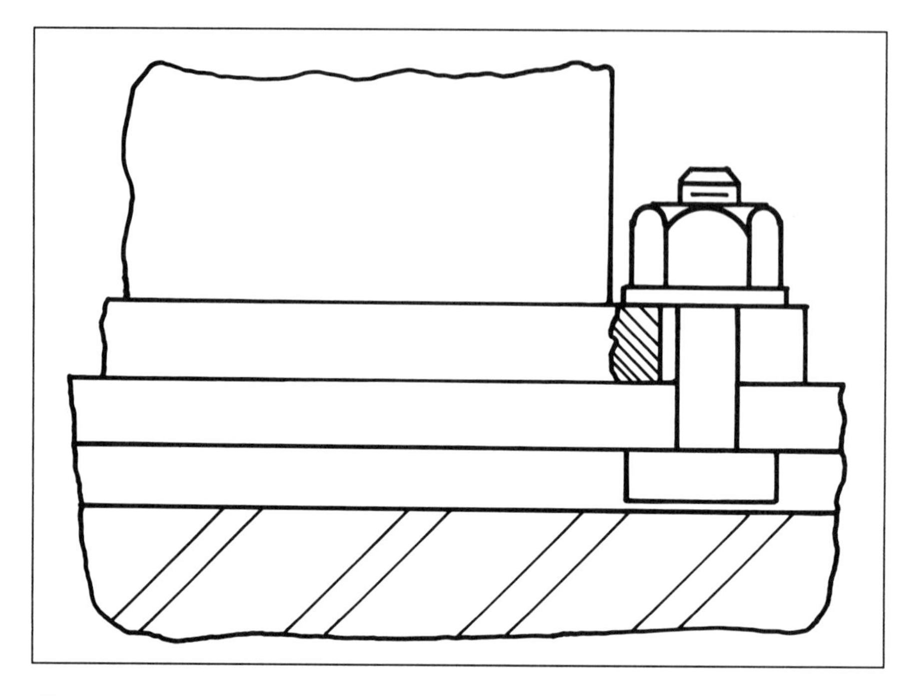

Figure 13-2. A high-strength alloy-steel T-slot bolt and nut permits the use of a large strong fastener, and visual inspection of proper thread engagement.

is easily determined by visual inspection with the T-slot bolt system. Once in place, there is no easy way to determine the thread engagement in a T-slot nut. Also, there is a danger that the screw may interfere with the bottom of the T-slot before the screw is completely tightened.

Nuts used for diesetting applications should be made of the same high-grade heat-treated material as the bolts. Nut thickness should be great enough to permit a length of thread engagement of at least one and one-half times the thread diameter.

Clamping Methods to Avoid

To institute good clamping practices in the plant, it is necessary to:

- Provide enough proper clamping equipment;
- Provide proper storage for the equipment when not in use; and
- Train the diesetters in the proper use of the equipment.

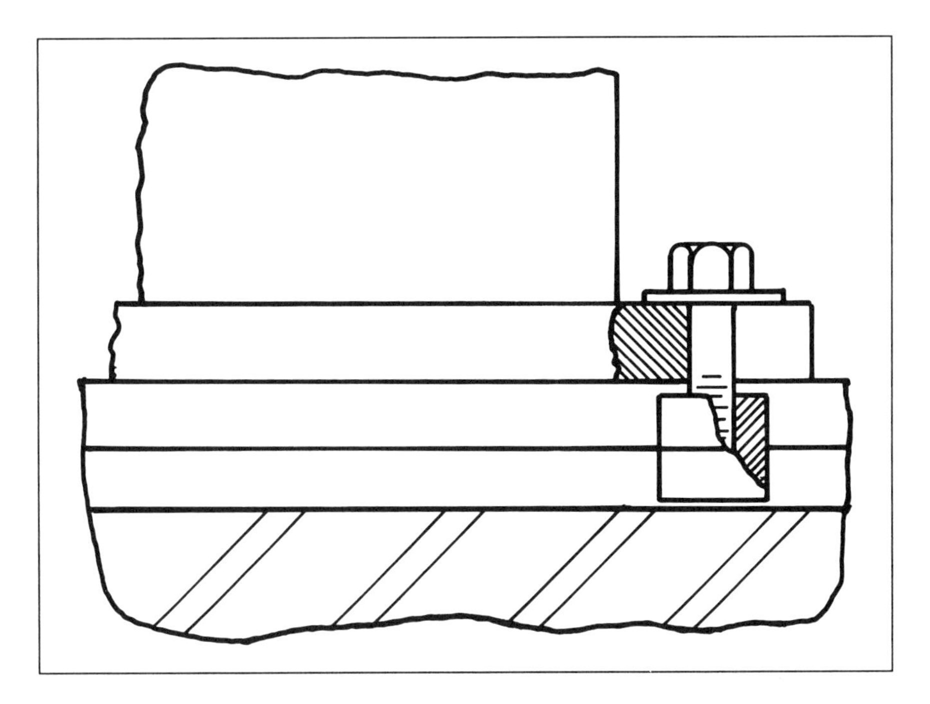

Figure 13-3. *Using a T-slot nut requires a smaller hex-head bolt, and does not permit visual inspection of thread engagement.*

Relying solely on good ratings by insurance inspectors and government enforcement officers can create a false sense of security. Bad methods and individual poor habits quickly grow into everyday unsafe practices. Visiting inspectors often lack the time and knowledge to spot these problems.

Safe pressworking involves many disciplines. The die designer and process engineer must apply accurate pressworking formulas when determining press requirements. A mechanical engineer's knowledge of materials strength is required to avoid overload and metal fatigue problems in manufacturing systems. The industrial or safety engineer's skill is needed to determine proper operator safeguards.

Of course, it is necessary to follow government and insurance regulations. Safe pressworking requires the application of sound engineering principles to analyze and avoid potential problems. Pressroom personnel must deal with many factors to ensure safe operations, and that may require written safety procedures. Inspectors from regulatory agencies might be unaware of these safety procedures. The die design and engineering department always should consider the requirements for safe diesetting, as well as safe part loading and unloading, when the process for a new stamping is approved.

Figures 13-4 through 13-10 illustrate some poor die securing practices observed in various pressrooms over the years. Today, most shops strictly forbid these practices. Adopting safe methods requires more than conducting safety training for pressroom personnel. Management must be willing to:

- Supply the correct fasteners and clamping devices;
- Provide proper storage for all needed equipment;
- Scrap inferior diesetting fasteners;
- Provide hands-on training in proper methods; and
- Plan and achieve the goal of clamping standardization.

Standardizing Clamping Height

Standardized clamping heights and positive location methods are keys to quick repeatable die changes. Considerable savings in changeover time and die fastener inventory can be achieved by adopting a standardized clamping height for all dies. This is true whether the die is fastened with straps and setup blocks or hydraulic clamps.

In cases where more than one tie-down height is required, a system of uniform increments of clamping heights should be used. For example, the distance from the clamping surface to the ram and bolster can be standardized in one-inch or one-centimeter steps.

Figure 13-11 depicts three simple methods of correcting differing die shoe or subplate thicknesses to a constant clamp height. In the case of a shoe or subplate that is too thin (A), small pockets can be milled to provide a common clamping height. If it is too thick (B), the correct thickness spacers are attached to the edge of the die shoe or subplate to provide the correct dimension. In some cases, tack welds are used, but screws are preferable to avoid warping the shoe. In cases where T-slot bolts and washers are used (C), differing heights can be corrected by milling, or attaching, a horseshoe-shaped spacer as shown.

Constant Height Clamping Ledges

It is basic that a constant clamp height be employed for any rapid clamping system when changing dies. This can be as simple

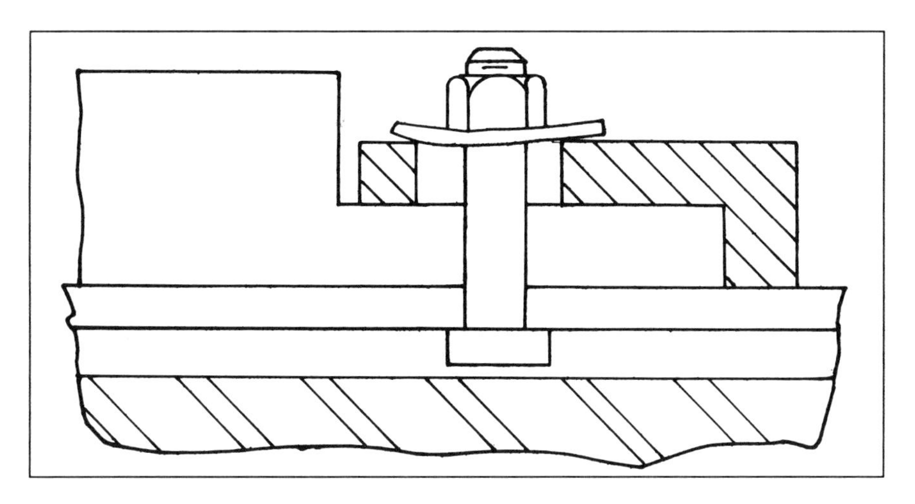

Figure 13-4. *The hole in a good toe-clamp should only be slightly larger than the bolt. A washer will deform and allow the clamp to loosen if the hole is too large.*

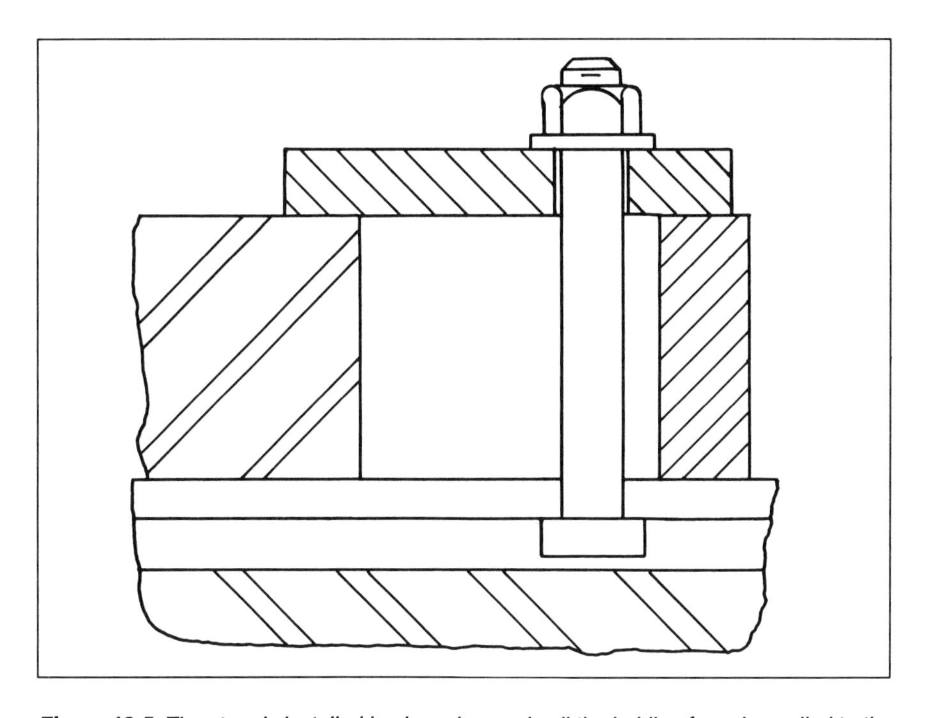

Figure 13-5. *The strap is installed backwards; nearly all the holding force is applied to the setup block rather than the die shoe.*

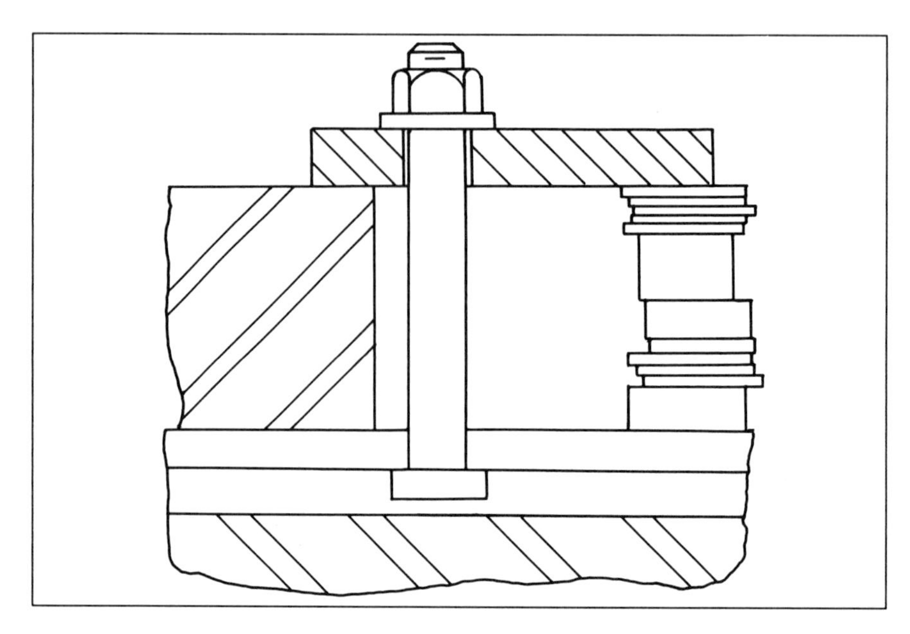

Figure 13-6. *The use of slugs and washers to make up the required spacer block height is unsafe. They are seldom flat and often have burrs that will compress and cause the clamp to loosen.*

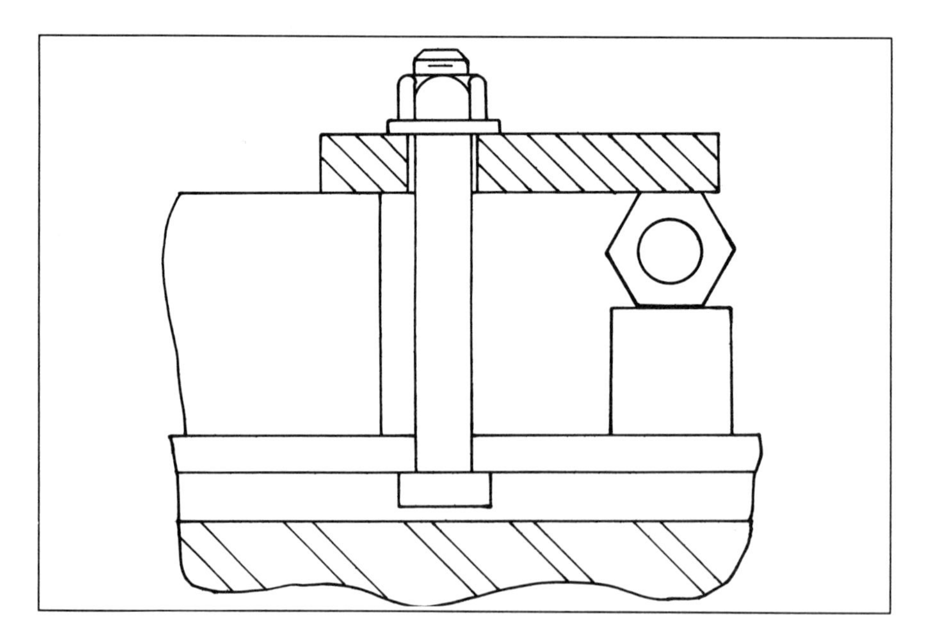

Figure 13-7. *A good alloy steel nut intended for diesetting applications is expensive. They are not suited for setup block applications, and the practice must never be used.*

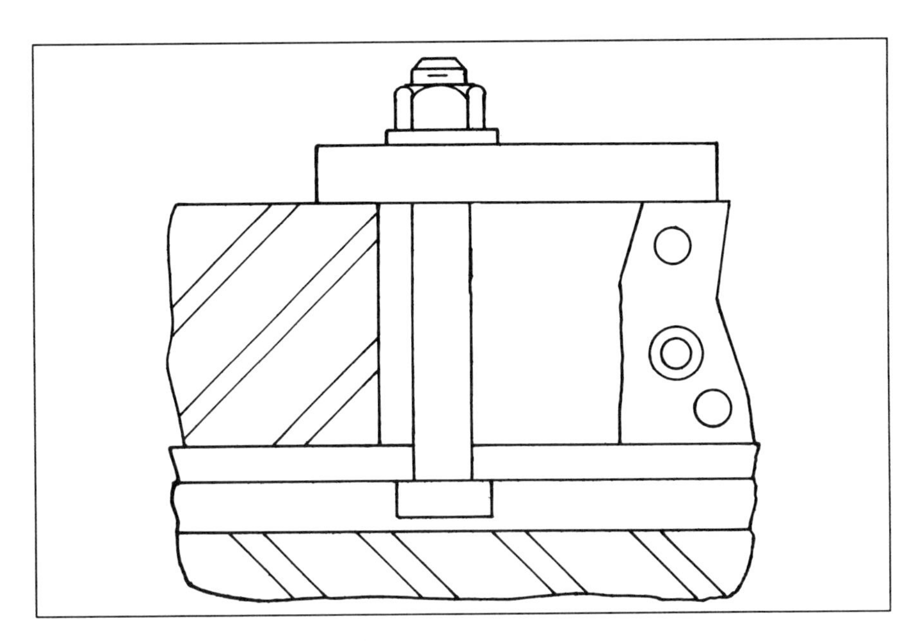

Figure 13-8. *Using hardened steel die sections for setup blocks leads to poor housekeeping practices and the use of expensive interchangeable die details as diesetting aids.*

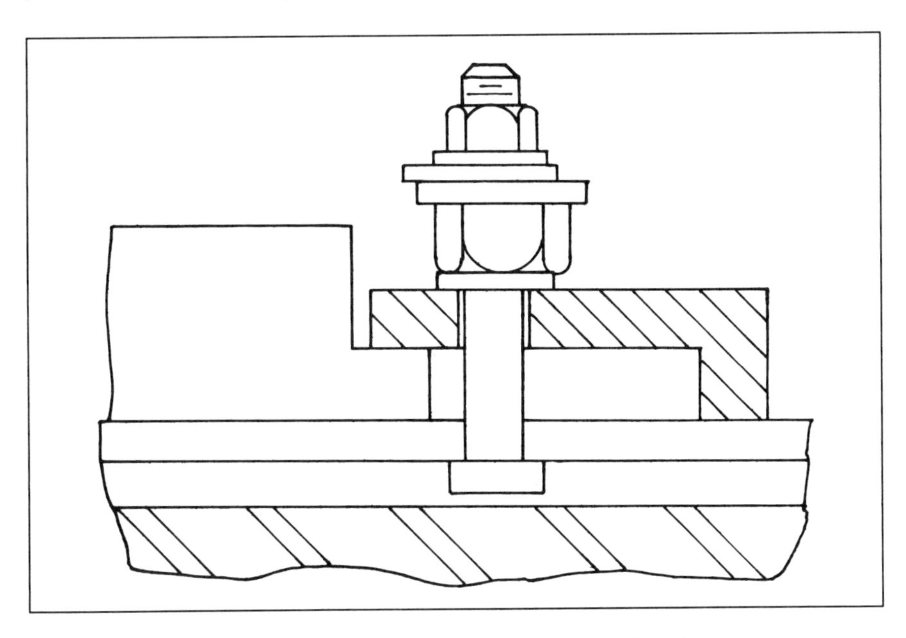

Figure 13-9. *Always use the shortest bolt that will permit proper thread engagement. Longer bolts are costly and can create a pinch point for the operator in addition to creating an insecure setup.*

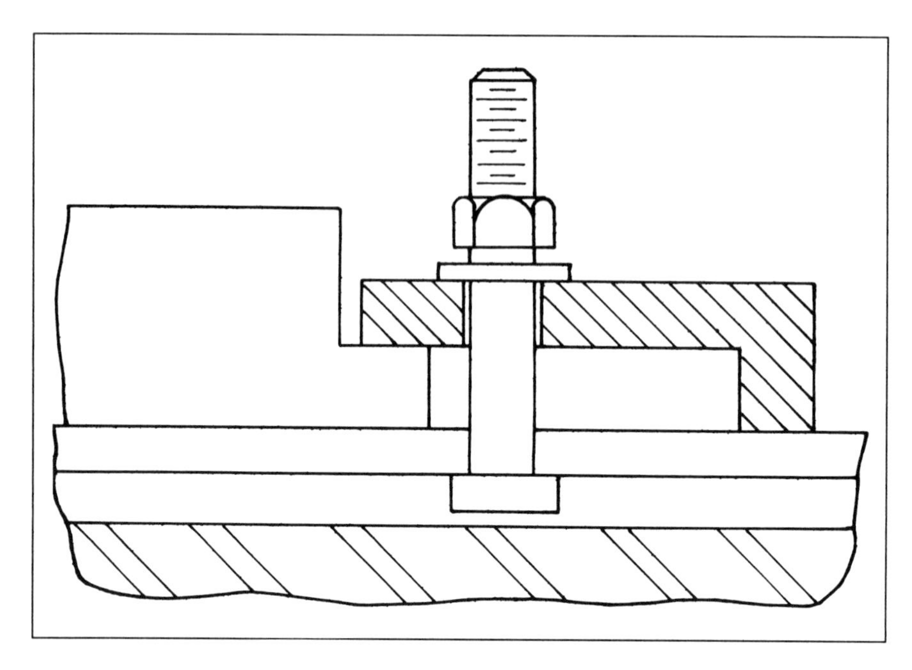

Figure 13-10. A bolt that is too long may not have enough threads to permit the nut to be tightened properly.

as providing a protruding ledge on a parallel attached to the die shoe (Figure 13-12). A recessed ledge (Figure 13-13) is less susceptible to damage during die handling operations. The attachment of the parallel to the die shoe must be at least as strong as the clamping system to provide for emergency stripping loads in the event of a die mishit.

Forged Steel Clamps

A number of styles of forged steel clamps are available from commercial diemaking supply houses. The offset type clamp illustrated in Figure 13-14 engages a drilled hole in the die shoe. The offset design has a short profile to permit ease of scrap shedding. Figure 13-15 illustrates a straight type clamp. An advantage of the clamps illustrated in Figure 13-14 and Figure 13-15 is that the hole is easily and accurately drilled on location with a light-duty radial-arm drill press. Horizontal-spindle milling machines are ideal for cutting clamping flats in parallels efficiently. Shops lacking such mills usually have a suitable drill press for drilling the tie-down holes.

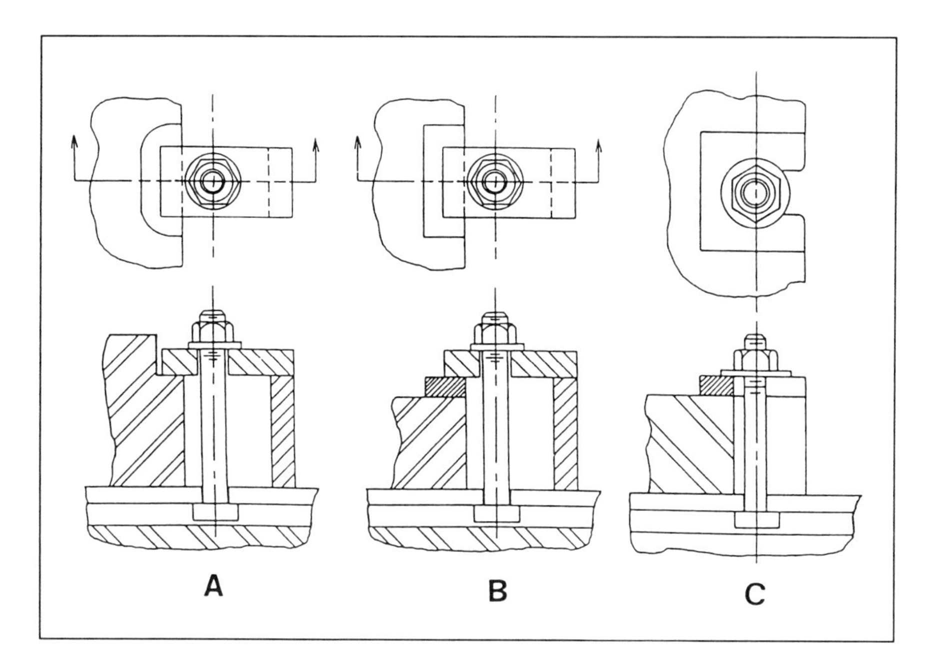

Figure 13-11. Three simple ways to adapt to a constant clamp height: (A) pockets milled to a constant clamp height; (B) spacers attached to the edge of the die shoe; (C) attaching a horseshoe-shaped spacer.

A forged gooseneck clamp engaging a ledge on a die shoe or parallel is illustrated in Figure 13-16. The forged commercially available clamps shown are used with swivel washers, and either swivel head capscrews or swivel nuts.

An obvious improvement to speed diesetting is to place a spring on the bolt to hold the toe-clamp up in position. A washer may be needed under the spring to keep it from catching in the T-slot.

Home-made One-piece Clamp

The clamp illustrated in Figure 13-17 rests near the correct clamping height before it is slid onto the die clamping ledge. This style clamp was developed in a high-speed pressroom in which circuit breaker parts are stamped. Originally, a one-piece toe-clamp with a spring was used. The simplified design has a lead on the portion of the clamp engaging the die clamping ledge. The palm of the hand is placed on the large chamfer at the rear of the clamp. It is simply pushed into place. Figure 13-18 shows the clamp in position.

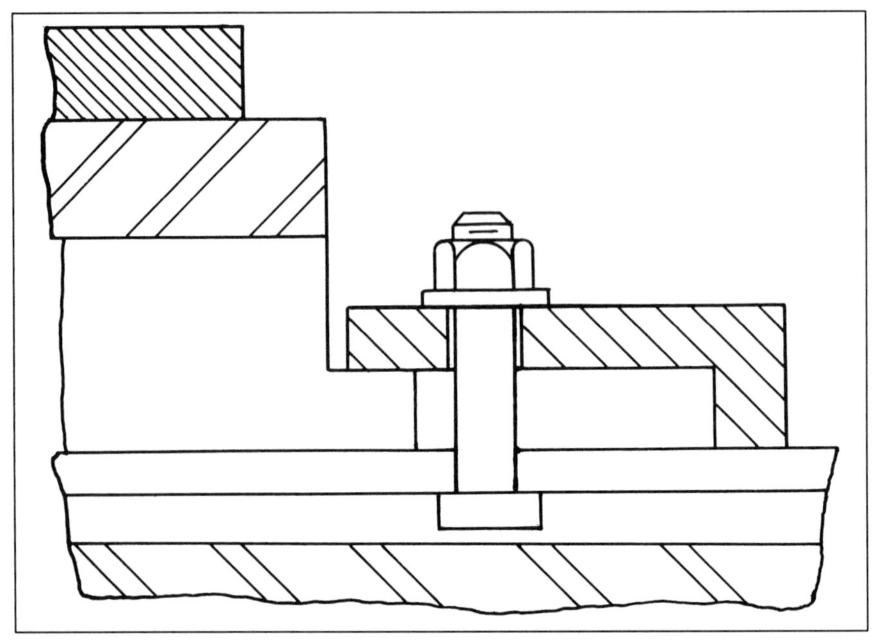

Figure 13-12. *Clamping to a standard height ledge on a parallel permanently attached to the die shoe.*

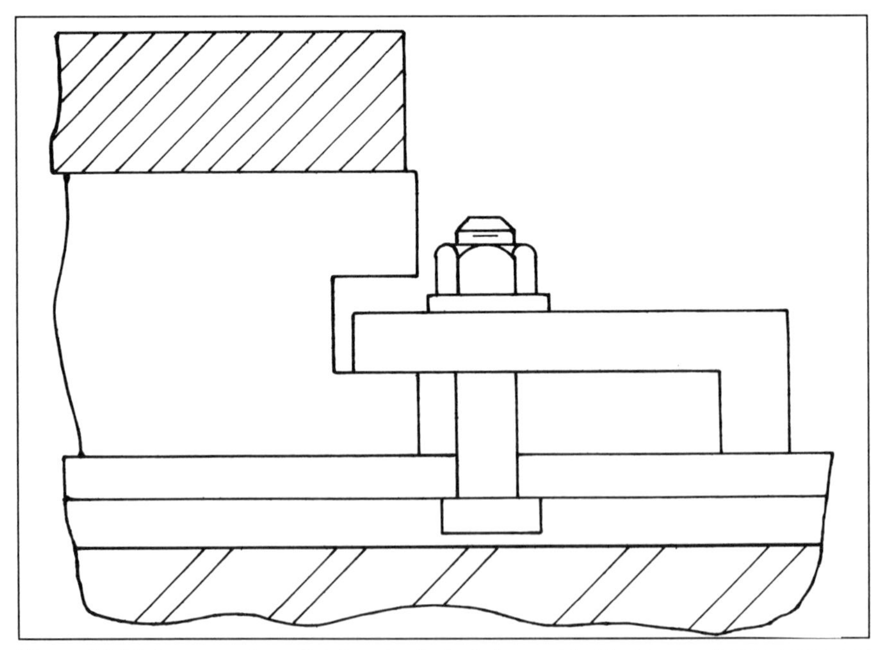

Figure 13-13. *A recessed parallel clamping ledge is less subject to damage than the exposed style in Figure 13-12.*

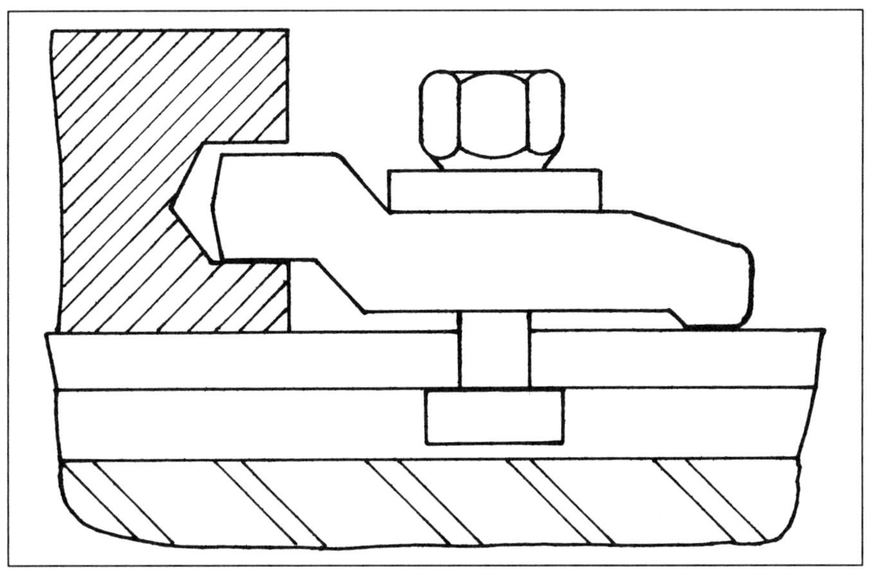

Figure 13-14. *A commercially available forged-steel offset type clamp designed to engage a drilled hole in the die shoe.*

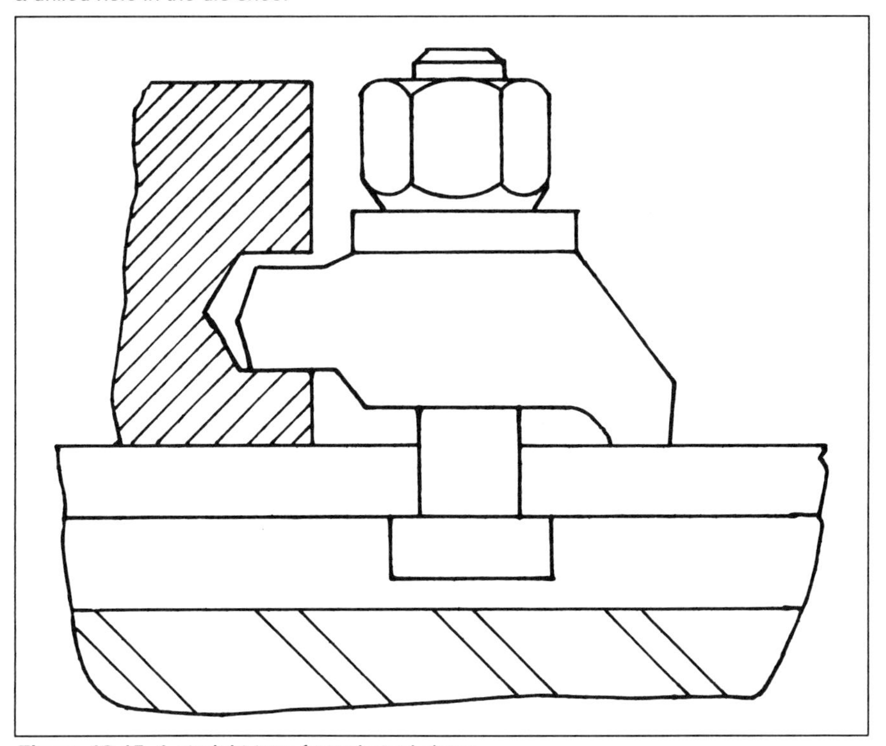

Figure 13-15. *A straight-type forged-steel clamp.*

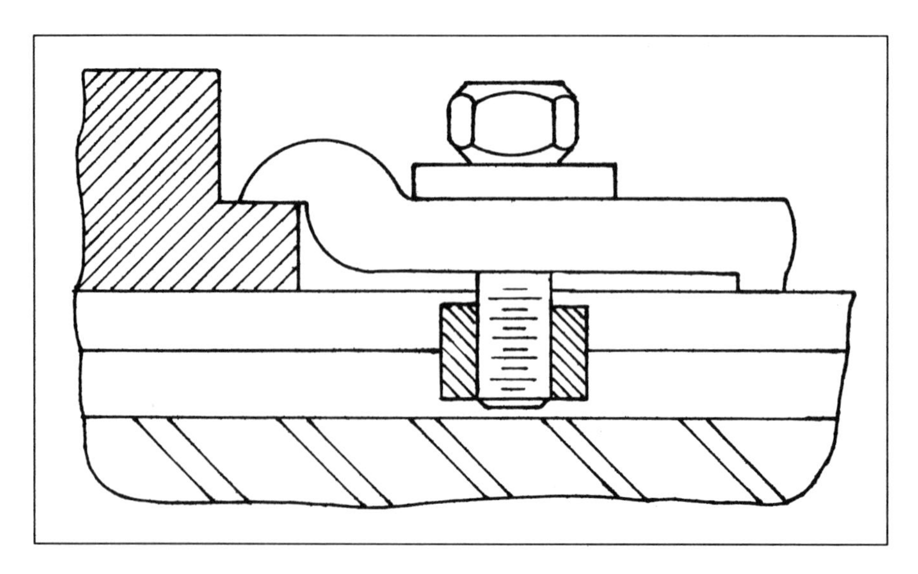

Figure 13-16. A forged-steel gooseneck clamp.

For light-duty applications, the clamp can be machined from SAE 1018 steel and gas carburized for wear resistance. There is substantial stress concentration at the place where the edge of the clamp is attached to the body near the bolt. This is a disadvantage of the design. The radius at this point should be made as generous as possible. If substantial clamping forces are anticipated, the clamp should be made of alloy steel and heat-treated for maximum toughness.

Benefits of Good Clamping Practices

Good clamping methods help ensure consistent setups, which in turn reduce stamping process variability. Poor clamping methods can cause product inconsistency as well as endanger personnel and equipment. Proper die fastening methods with special attention given to safety are an absolute necessity. Every press room should be equipped with good clamping equipment so the diesetters will not resort to unsafe fastening methods to get their job done.

Threads should be protected from damage. Racks near the press or on a die cart should be designed so that the bolts can be stored by hanging them by the heads to help protect the threads from damage that might occur if they were stored in a bin.

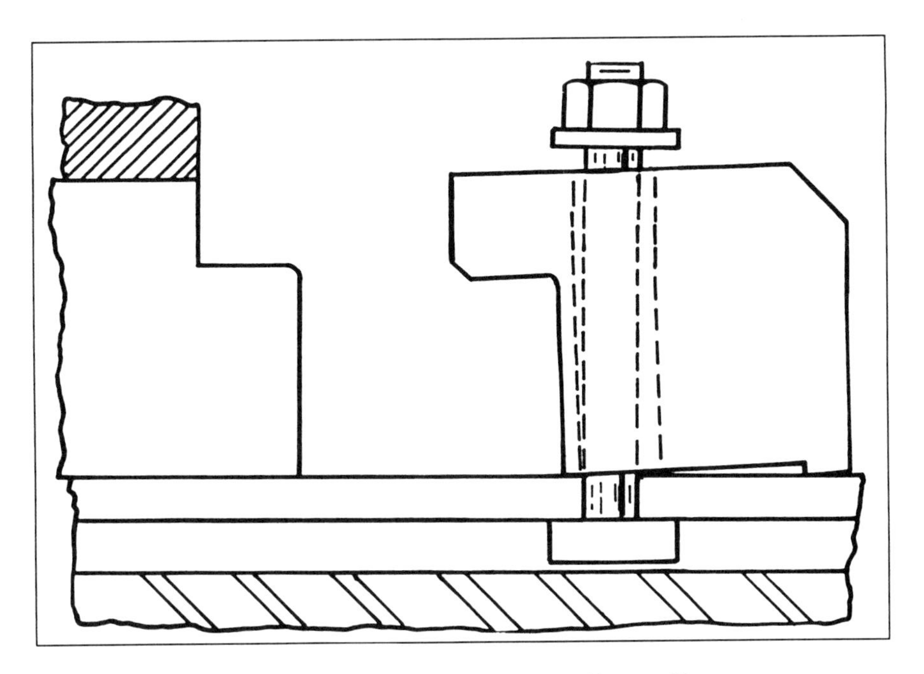

Figure 13-17. *Small one-piece toe-clamp is easily shoved into position.*

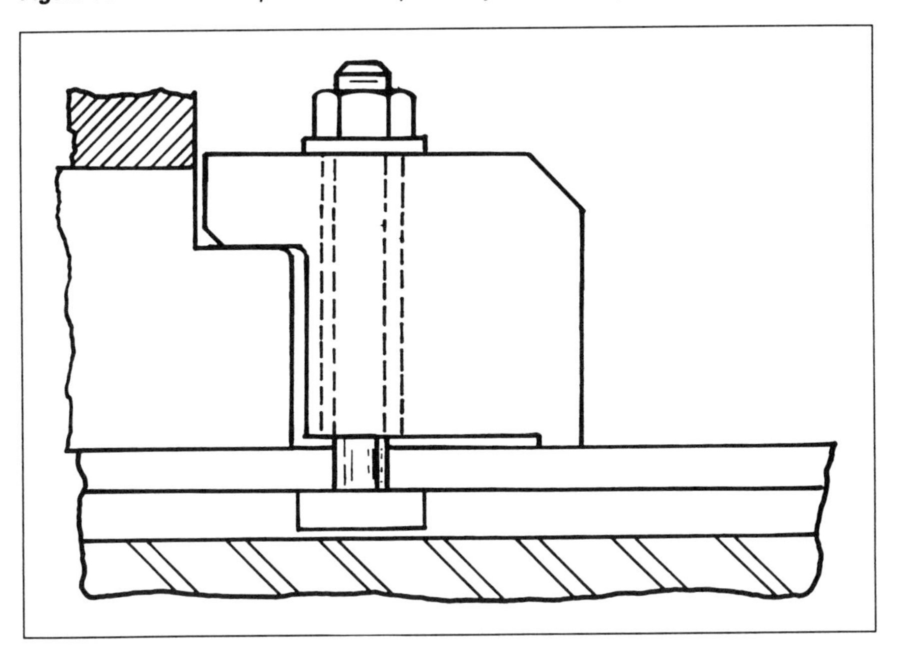

Figure 13-18. *The clamp illustrated in Figure 13-17 in position. A limiting factor is the allowable stress concentration at the radius under the clamping lip.*

Special diesetting wrenches with long lightweight tubular handles are commercially available. The use of "cheater bars" on automotive wrenches and adjustable wrenches should never be used for diesetting. Heavy-duty ratchet wrenches are suitable, provided they have sufficiently long handles to permit proper torquing of the bolts without excessive effort. Using pneumatic impact wrenches with proper sockets is a good way to increase the diesetting rate and reduce fatigue. Diesetters should check the final tightening with a torque wrench periodically to know the tool is operating properly. In critical applications, final hand-tightening is advisable.

SELECTING THE CORRECT CLAMP

Partial Turn Mechanical Clamps

Once a diesetting bolt is snugged up by hand, only a half to one and one-half turns are needed to tighten the fastener to the proper torque value. The exact amount is determined by several factors, including the thread pitch and length of the bolt. Any bow in the die shoe or subplate must also be drawn up. Finally, if spring washers are used, they must be compressed until all clearance is drawn up.

Spring Washers

For decades, Ford Motor Company and many other manufacturers engaged in heavy stamping had excellent standardized die bolting provisions. Features include U-shaped cutouts in the die shoes, and uniform clamping heights. Grade eight or equivalent T-slot bolts and nuts are used. The U-shaped slots are bridged with thick carburized steel straps.

Figure 13-19 shows a T-slot bolt, carburized steel strap, and heavy spring washer in place after hand-tightening. The large square spring washer serves several purposes. First, it helps prevent the nut from working loose under high-impact loads. The pressure of the nut is distributed over a large area of the heavy strap. The spring washer also serves as a visual indicator. Should the bolt loosen in service, a gap will be visible under the washer. Figure 13-20 illustrates the bolting assembly after it is properly tightened.

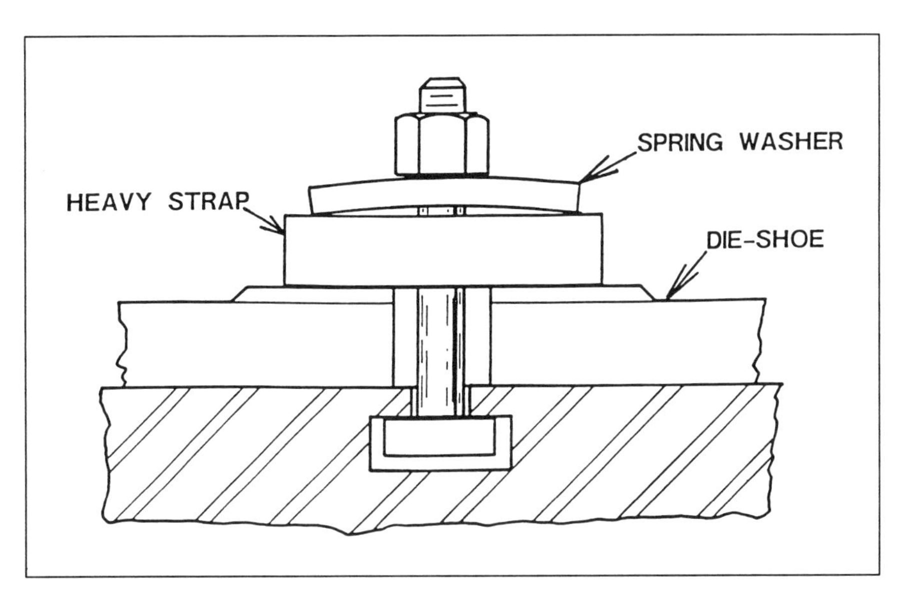

Figure 13-19. *A T-slot bolt, large carburized steel strap, and heavy-duty spring washer in place after hand-tightening. (Ford Motor Company)*

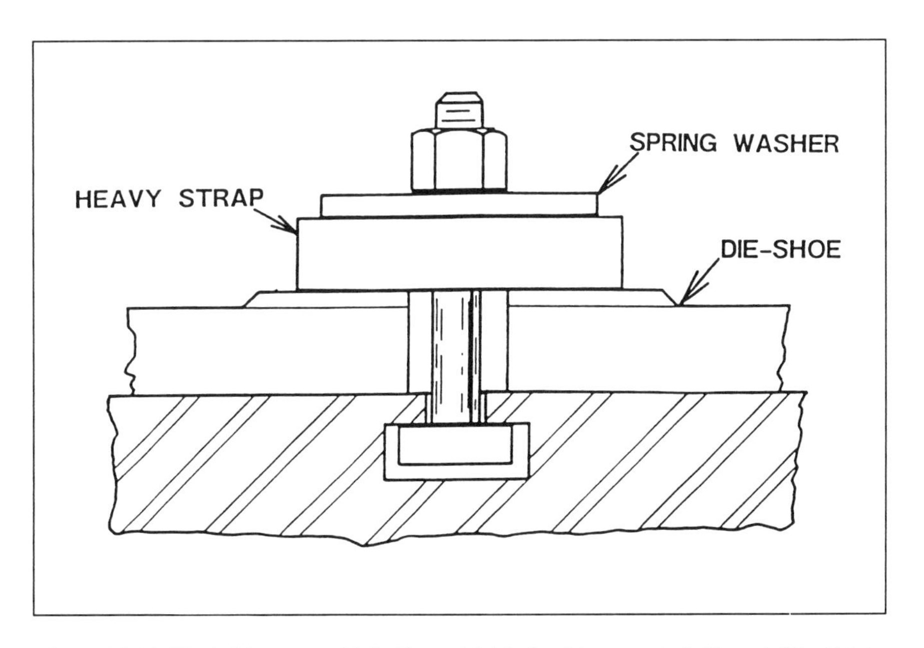

Figure 13-20. *The bolting assembly in Figure 13-19 after it is properly tightened. (Ford Motor Company)*

Example of a Half-turn Mechanical Clamp

Some die fastening assemblies are marketed as "half-turn clamps" as that is the approximate amount of rotation required to tighten them. However, if they employ a standard screw and nut, the mechanical effort required to tighten them will be the same as an ordinary threaded fastener of the same size. Figure 13-21 illustrates an Optima™ brand half-turn clamp.

The total movement of the toggle is approximately 0.008 inch (0.2 mm). This is sufficient to properly tension the clamp provided

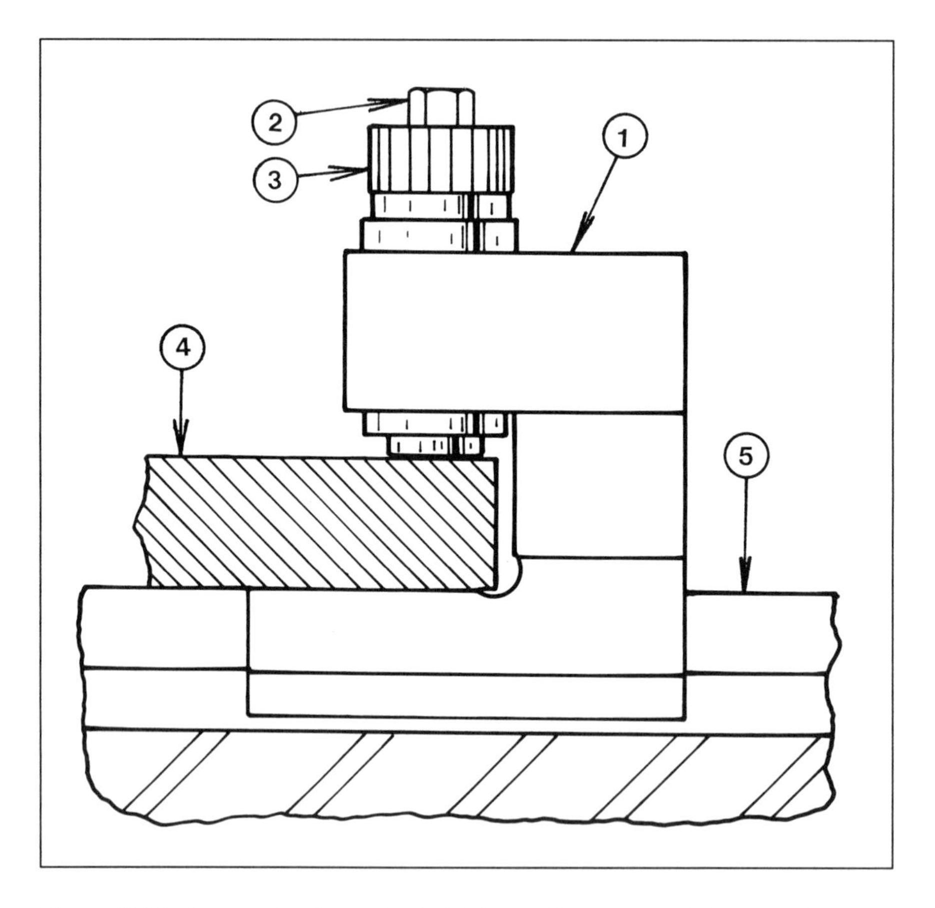

Figure 13-21. *An example of a mechanical half-turn die clamp employing a toggle locking mechanism. The clamp body (1) is slid into place in the bolster T-slot (5) and engages the die shoe or subplate (4). A serrated knob (3) is hand-tightened until the plunger contacts the plate to be clamped. Finally, the over-center toggle mechanism is activated by turning the small screw (2) approximately one-half turn. Since the over-center toggle mechanism provides substantial mechanical advantage, relatively little force is required. (Optima USA)*

the clamp and plate are in intimate contact with the bolster. Should the plate be warped, the serrated nut must be pretightened with a special wrench. When this occurs, the advantage of the clamp as a quick die change device is essentially lost.

Power Actuated Clamps

Automatic die change requires some type of power-actuated die clamping system. The clamps can be actuated by electrical energy, compressed air, or hydraulic pressure.

Some clamps are actuated by electrical motors, but these are not common. The required gear reduction and electrical control systems tend to be complex and expensive.

Clamps powered by compressed air and hydraulic pressure have been used for many years. Many good designs feature over-center toggle or eccentric mechanisms to provide a large clamping force in the locked position. Hydraulic clamps are popular for many new installations as well as for retrofitting to older presses.

Safety Considerations

Automatic systems should be of fail-safe design where the die cannot become detached from or shift position on the ram or bolster during press operation. This could occur because of a failure of the clamping power source or by an unclamp command while the press is in motion. To avoid this, good designs incorporate safety features such as:

- Hydraulic power sources across diagonal corners of the machine, much like the dual-brake system on automobiles;
- Pressure switches to detect a loss of pressure;
- Automatic machine shutdown in the event of the loss of pressure;
- Over-center toggle locking mechanisms that hold in the event of pressure loss;
- Wedge and ramp locking mechanisms that require hydraulic pressure for both activation and release;
- Mechanical locking mechanisms that are either hand or remote operated;
- Hydraulic clamps with check valves to prevent the release of fluid because of a pressure failure. These require a second hydraulic line to release the pilot-operated check valve;

- Limit switches to detect proper clamp position; and
- Pressworking lubricants that won't deteriorate seals, or damage position limit switches.

REPRESENTATIVE DESIGNS

Hollow Piston Cylinder Clamp.

Figure 13-22 illustrates a clamp that is essentially a hydraulically powered nut. This design is popular for quick die change retrofit applications. A major advantage is the clamp can be adapted to a wide variety of clamping heights in the same manner as a nut.

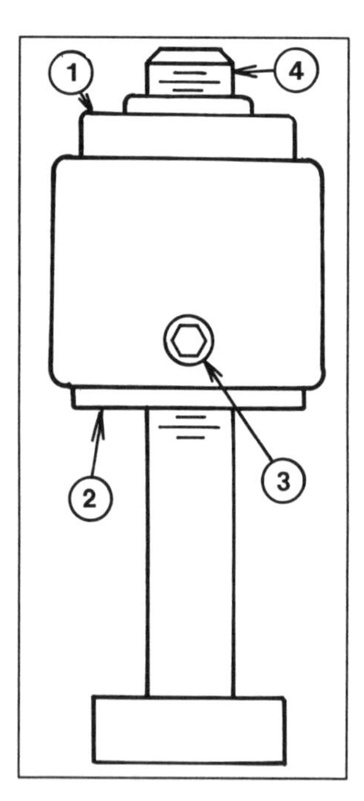

***Figure 13-22.** A hollow piston-type hydraulic clamp for use with a standard T-slot bolt. The clamp (1) screws onto a standard T-slot bolt (4). The piston (2) applies pressure to the die clamping surface (not shown) just as a conventional nut does when tightened. Hydraulic pressure is supplied by a hose which is attached to the hydraulic port (3).*

Hydraulic Ledge Clamps

Figure 13-23 illustrates an end-view of a hydraulic ledge clamp. The clamp body and user-supplied spacer block are fastened directly to the press ram or bolster by socket-head capscrews.

Up to six or more individual spring-return pistons are available in this design. The pistons are supplied with hydraulic pressure by means of internal drilled passages. For safety, this design can be supplied as a split system supplied by two individual pressure sources, or built with hydraulically controlled check valves. Pressure is applied to the hydraulic ports.

The clamping surface on the die shoe or subplate must be of a standard height for all dies. This is a basic requirement for virtually all quick die change systems.

Sliding Clamp for Use in a Conventional T-slot

The sliding clamp is popular for retrofitting existing presses to hydraulic clamping. It is illustrated and explained in Figure 13-24.

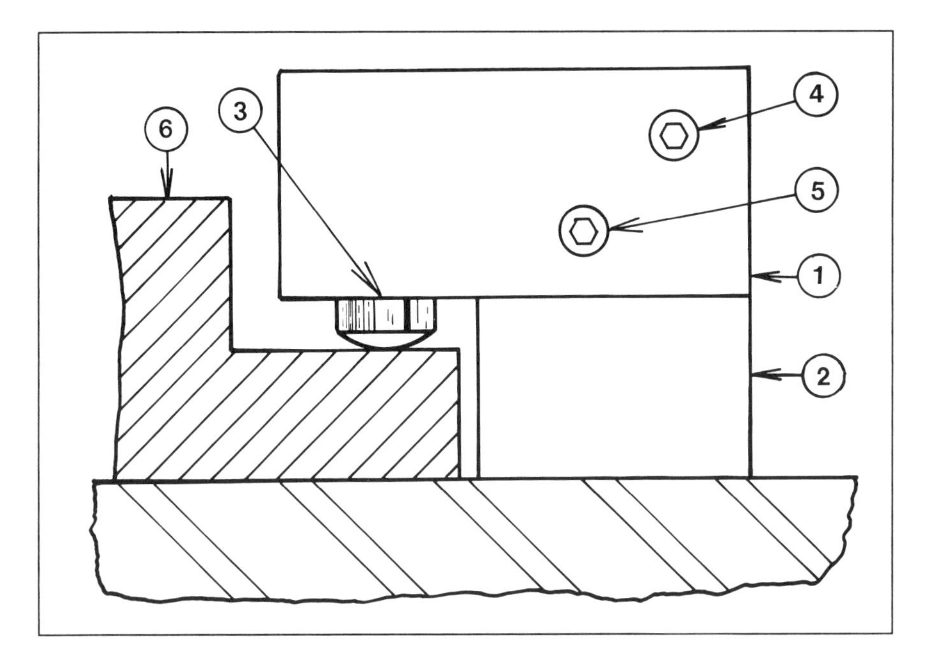

Figure 13-23. *A hydraulic ledge-type clamp. The clamp body (1) and user-supplied spacer block (2) are fastened directly to the press ram or bolster by socket-head capscrews (not shown). Up to six or more individual spring-return pistons (3) are available in this design. The pistons are supplied with hydraulic pressure by means of internal drilled passages. For safety, this design can be supplied as a split system supplied by two individual pressure sources, or built with hydraulically controlled check valves. Pressure is applied to the hydraulic ports (4 and 5). The clamping surface on the die shoe or subplate (6) must be of a standard height for all dies. This is a basic requirement for virtually all quick die change systems.*

Pull-in Type Clamp

Figure 13-25 illustrates a double-acting pull-in type clamp installed in a press bolster. A clamp of this type features up-down motion for clamping and unclamping in a slot cut in a die shoe or subplate. It is shown in the unclamped position with a T-slot cut in a die shoe. In this position, the die shoe can slide in or out of either side of the press.

The cylinder body is fastened into a bored and countersunk hole in the bolster. Two inductive proximity switches inside the cylinder sense proper piston position at either end of the travel. These switches transmit an electrical signal through a connector on the bottom of the cylinder. This is a safety feature and can be

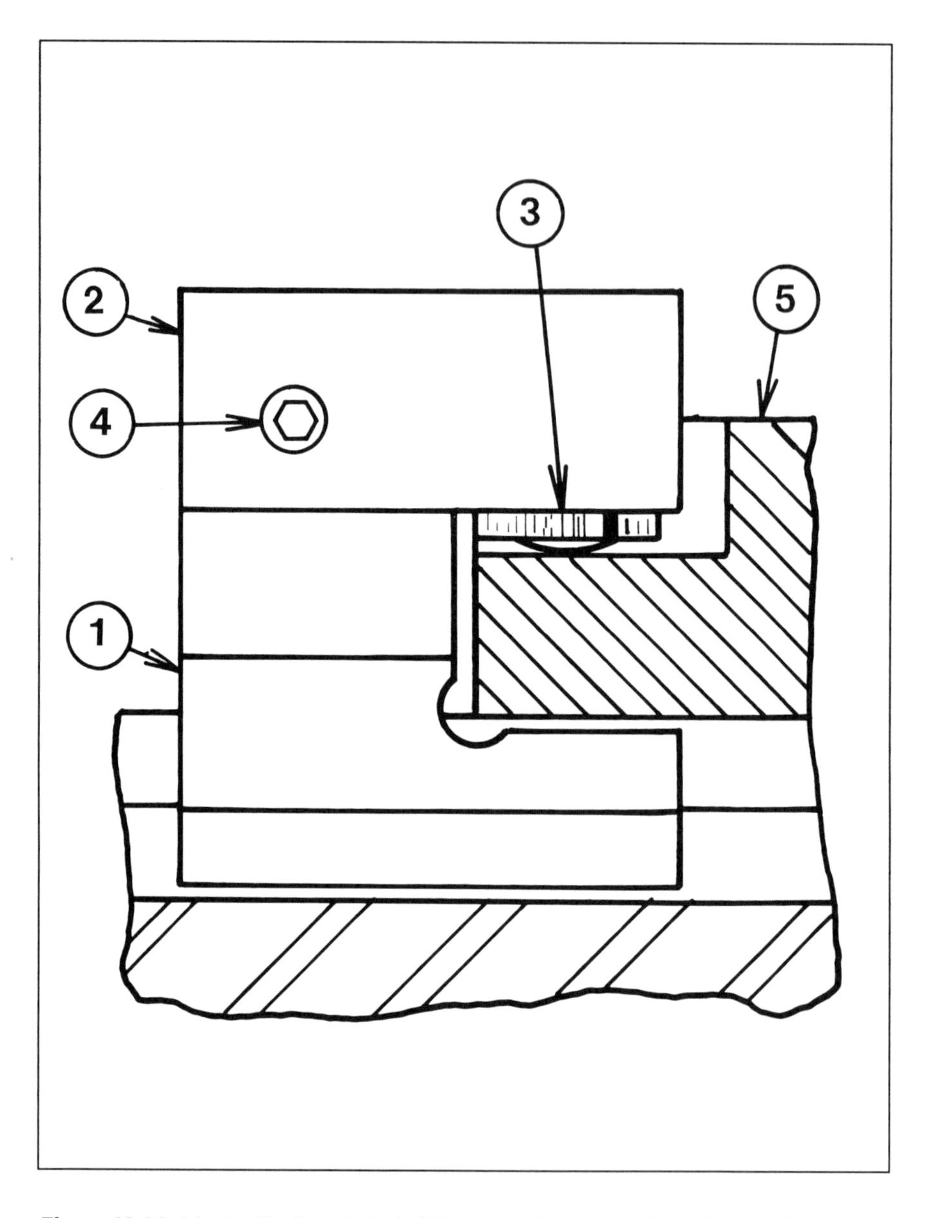

Figure 13-24. *A hydraulically actuated sliding clamp for use in existing T-slots. 1 is a T-slot adapter for several standard types of T-slots. The clamping block (2) is attached to the T-slot adaptor with screws. The hydraulic piston (3) is of the single-acting spring-return type. Hydraulic pressure is applied through a port (4). An optional design employs a second port and a hydraulically controlled check valve to prevent the release of holding force in the event of a line failure. The clamping surface on the die shoe or subplate (5) must be of a standard height to permit die interchangeability using this type of clamp.*

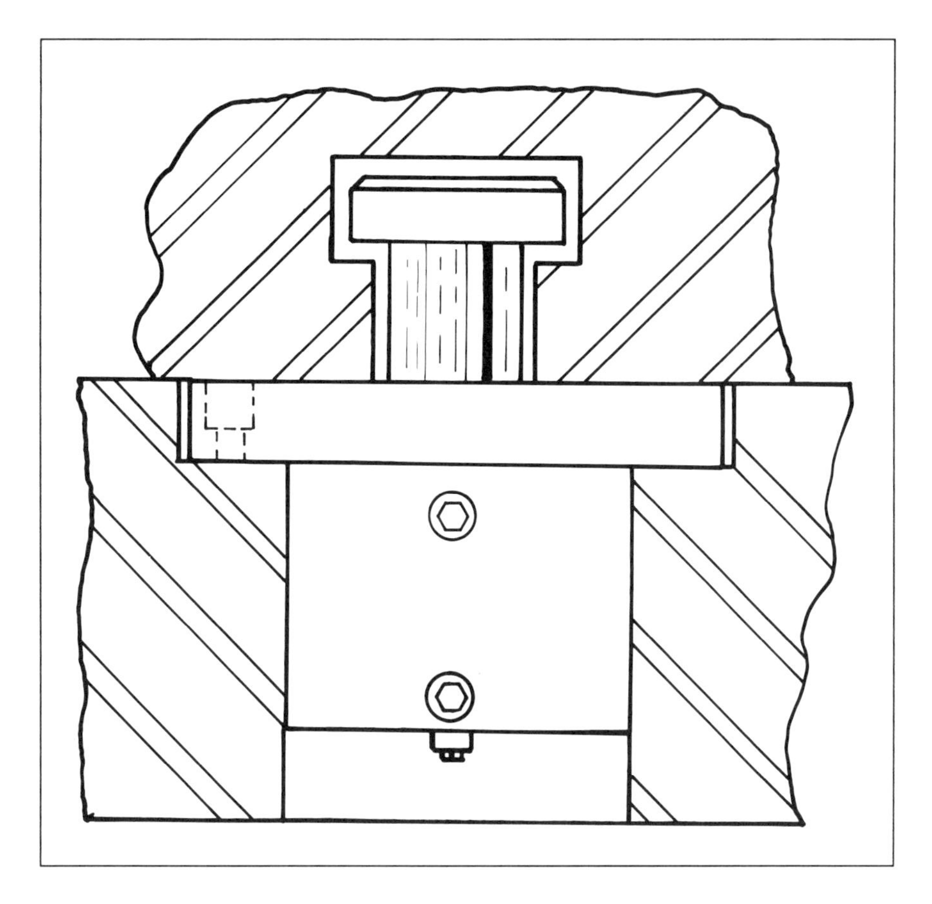

Figure 13-25. Double-acting hydraulic clamp installed in a press bolster.

interlocked with the press controls to permit proper sequencing of automatic die movement.

Specialized Pull-in Type Clamps

Other types of clamp movement can be achieved in the same basic type of clamp body. By adding a second hydraulic actuator and additional proximity switches, the head can be made to lift and swing 90 degrees or lift and swing 90 degrees and then sink below the bolster surface. These features permit the clamp to engage a slot or, in the case of the swing-sink clamp, engage an elongated hole in a die shoe or subplate. The main feature of this type of clamp is the ability to engage a blind slot milled in a die shoe. Using this type of clamp is not advised unless absolutely necessary. Should it fail to completely release, it is virtually inaccessible for repair.

OPERATING DIES ON PARALLELS

Setting dies on parallels or risers is common practice in many shops. Parallels are often used under the die to accommodate a standardized pass height, as well as provide clearance for getting rid of scrap. Often parallels are placed on top of the die to build the die up to the required shut height. It is important that the parallels are spaced to avoid excessive die shoe deflection. Standard engineering formulas and tables are available to calculate parallel spacing requirements for the permissible value of shoe deflection. Commonly used die-parallel materials include:

- Cold-rolled steel bar;
- Hot-rolled steel bar;
- Hot-rolled steel plate; and
- Cast iron.

Cold-rolled Steel

Rectangular cold-rolled steel bar is a popular material for die parallel applications. It is readily available from steel suppliers and can be quickly cut to length with a power saw. The top-to-bottom dimensional accuracy is often close enough for many noncritical applications if the bar stock is selected and matched with care.

The minus dimensional tolerance of cold finished steel bars ranges from 0.004 in (0.1 mm) for stock up to 0.75 inch (19 mm) thick, to 0.014 inch (0.36 mm) for thicknesses over six inches (152 mm). This is not accurate enough for high-quality presswork. The most commonly specified parallelism requirement for the alignment of the press slide to the bed is 0.001 inch (0.025 mm) per foot (305 mm).

Hot-rolled Steel Bar

Generally hot-rolled material is less expensive than cold-finished bar stock. Another advantage is that hot-finished bar stock has fewer residual stresses than cold-finished material, reducing warping during machining.

Normally, only the top and bottom of the parallels require machining. If a large number of parallels are to be sized to a common dimension, rotary-table grinding is often the most economical method.

Hot Finished Steel Plate

This material is excellent for the construction of large irregularly shaped parallels and die risers. The usual method of fabrication is to flame cut the required shape with an automated plate burning machine. The top and bottom dimensions are finished by milling or grinding. To ensure dimensional stability, the plate should be normalized after being flame cut.

Figure 13-26 illustrates a standard type of flame-cut hot-rolled steel parallel. Shops specializing in heavy plate fabrication can usually supply such parallels on short notice. The parallel has the type of milled center line locating key slot shown in Figure 12-31. The top of the clamping surface should be milled accurately to a standard thickness from the parallel bottom. Two inches (50.8 mm) is a frequently specified dimension.

A useful source of low-cost steel plate for parallel fabrication is obsolete die shoes and subplates. Usually the holes in such scrap steel are not objectionable provided they are small and do not cause excessive interruptions of the cutting flame when burning out the desired shape.

Cast Iron

Die risers or parallels requiring lightening holes, reinforcing webs, and feet can be economically made of cast iron. Common gray iron has good compressive strength. The tensile strength is usually adequate in the large sections commonly used for die parallels. Iron alloys with better tensile properties are available at a higher cost than common gray iron.

A disadvantage of cast iron parallels is that they cannot be burned out with a cutting torch if the press becomes stuck on bottom. As a last resort, cutting the parallels may be required to free up a gap-frame or underdriven press that is severely stuck on bottom. Placing a 0.50 inch (12.7 mm), or thicker steel shim on top of the iron parallels is advisable if such a problem is anticipated. Otherwise, in an emergency, an oxygen lance may be required to cut the cast iron.

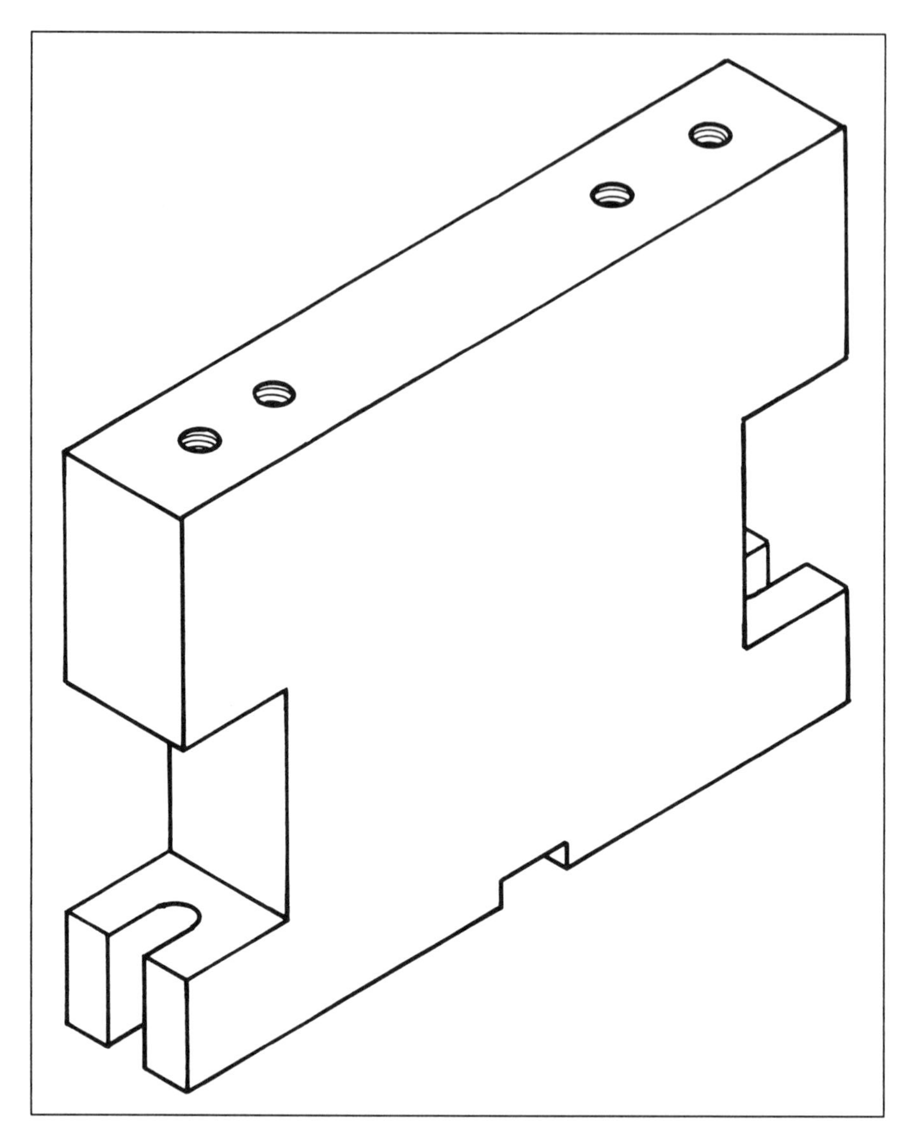

Figure 13-26. *A parallel made of flame-cut hot-rolled steel. The tie down slots can also be flame cut by turning the parallel upright on the flame-cutting table after the basic shape is cut out. (Hydra-Fab, Inc.)*

PROCESS VARIATION PROBLEMS RELATED TO PARALLELS

A common problem in many press shops is that a great many parallels of the same nominal dimension are not of the same exact dimension. The cause of unacceptable variation, when using cold finished steel bar stock for parallels, is the commercial tolerance for the material greatly exceeding the close parallel-size tolerance required for high-quality stamping. It is feasible to maintain the height variation of parallels to ±0.001 inch (0.025 mm) or better.

Selecting and Maintaining Existing Parallels

A short-term solution to the parallel-height variation problem is to sort and select them for uniform height. Figure 13-27 illustrates how parallels can be compared and selected by testing with a straightedge.

Many experienced diesetters carefully feel the parallels for burrs, removing them with a file. Slight differences in height can also be detected by feel when parallels are placed side by side on a flat surface.

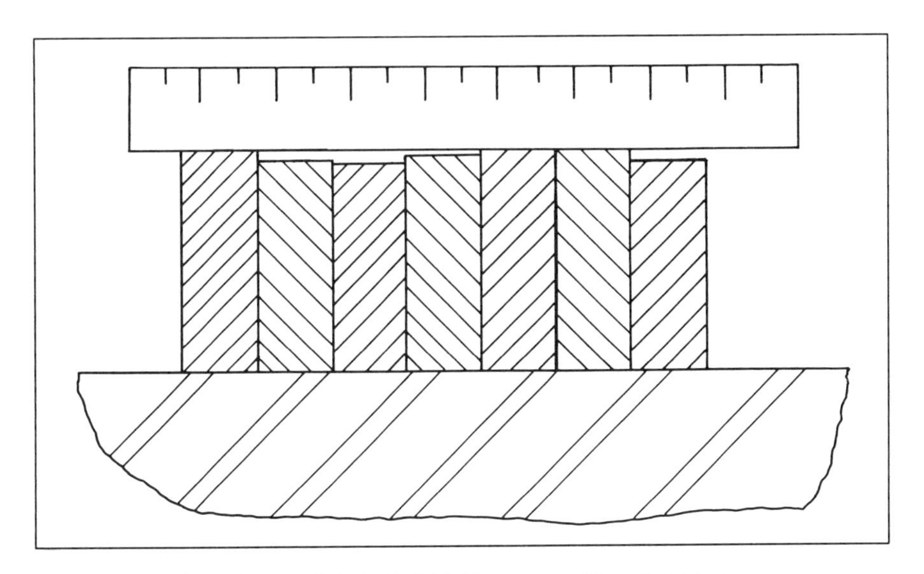

Figure 13-27. *Checking parallels for height differences with a straightedge.*

Resizing Parallels to Standard Dimensions

To eliminate any process variability problems caused by die-parallel height variation, resize parallels to standard on-size and undersized dimensions. Good action planning is needed before the standardization process is begun. It is a poor idea to send random groups of parallels out to be ground for minimum cleanup.

A good action plan is to:

1. Measure and group all parallels according to nominal and actual dimensions.
2. Grind all slightly oversized parallels to the exact nominal dimension.
3. Group all undersized parallels to be ground to a standard undersized dimensions; e.g., 0.040, 0.080 inch (1, 2 mm).
4. Identify the amount of undersize with an easily recognizable stamped designation, color code, or both.
5. Check and standardize any new parallels introduced into the system.
6. Inspect parallels periodically and correct any dimension and straightness problems.

When grinding parallels are made of cold-finished steel, it is important to turn them over several times during the grinding process. The material removed has residual compressive stresses present and the same amount should be removed from each side if the parallel is to remain straight after grinding. Turning and checking the work also affords an opportunity to take an exact thickness measurement of the work in progress.

Identifying Undersized Parallels

It is most important to identify and segregate parallels by both nominal and exact size. A good way to do this is by milling a shallow pocket in the side of the parallel and stamping the amount of undersize, if any, in the pocket as shown in Figure 13-28. Rapid identification can be provided by color coding with spray paint in the pocket and on the ends of the parallel.

By identifying standard size parallels as well as undersized parallels, the diesetter is assured that no oversight in identification has occurred. A suggested color coding scheme is:

- If the amount of undersize is 0.08 inch (2 mm) the color is red;
- If the amount of undersize is 0.04 inch (1 mm) the color is yellow; and
- If there is no undersize, the color is green.

Relationship of Parallels to Clamps

When clamping over parallels, the clamping method should not introduce deflection into the die shoes. Figure 13-29 illustrates a

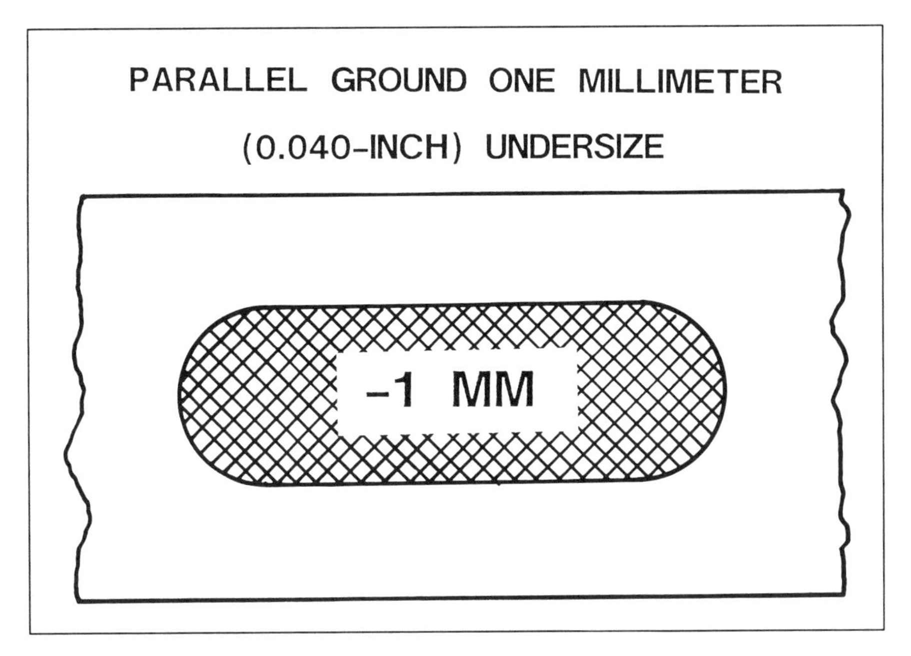

Figure 13-28. A good method to identify undersized parallels with a stamped marking in a milled pocket.

die clamped midway between two of the parallels it is set on. The result is an undesired deflection as shown in Figure 13-30. This deflection causes changes in critical die clearances, which in turn may affect the geometry of the finished part. Maintenance of correct die clearances is mandatory if the stamping process is to be stable.

Figure 13-31 illustrates a simple solution. Placing the clamps on the die shoe at the locations where it is supported by parallels improves clamping security and eliminates a process variable. Both the shoe and parallel are held tightly in compression, with no resulting shoe deflection.

Use Parallels That Are Long Enough

Shoe deflection also can result from using short parallels. Figure 13-32 illustrates a die shoe set on a short parallel or spacer block, which results in die shoe deflection when the clamps are tightened. Again, critical die clearances are changed, adversely affecting the process and the dimensions of the end product.

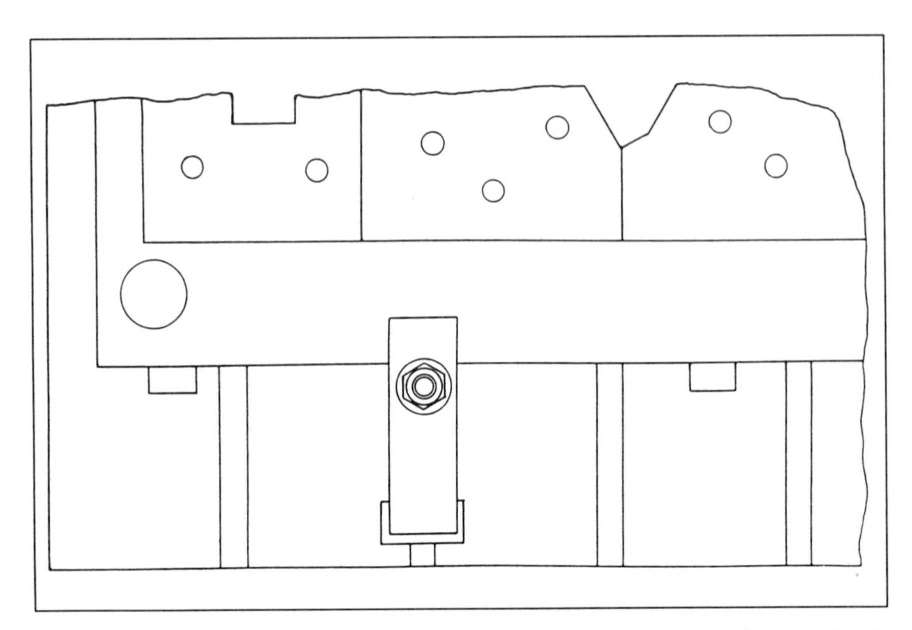

Figure 13-29. *Improper clamping of a die shoe placed on parallels will result in excessive die shoe deflection. Here, a single clamp is installed midway between two parallels.*

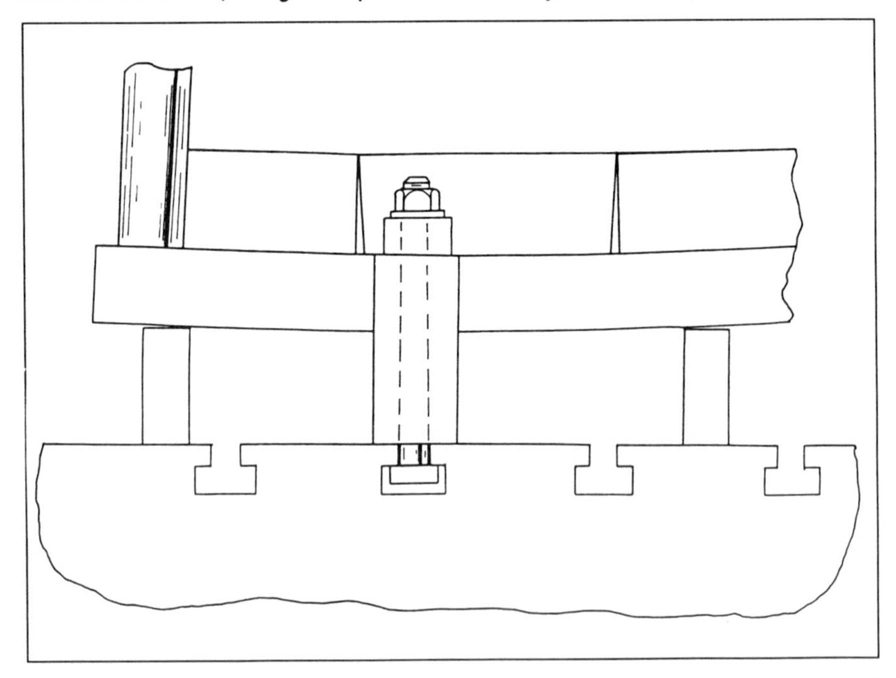

Figure 13-30. *Failing to clamp a die shoe directly over the parallels results in deflection of the die shoe. This can cause changes in critical die clearances.*

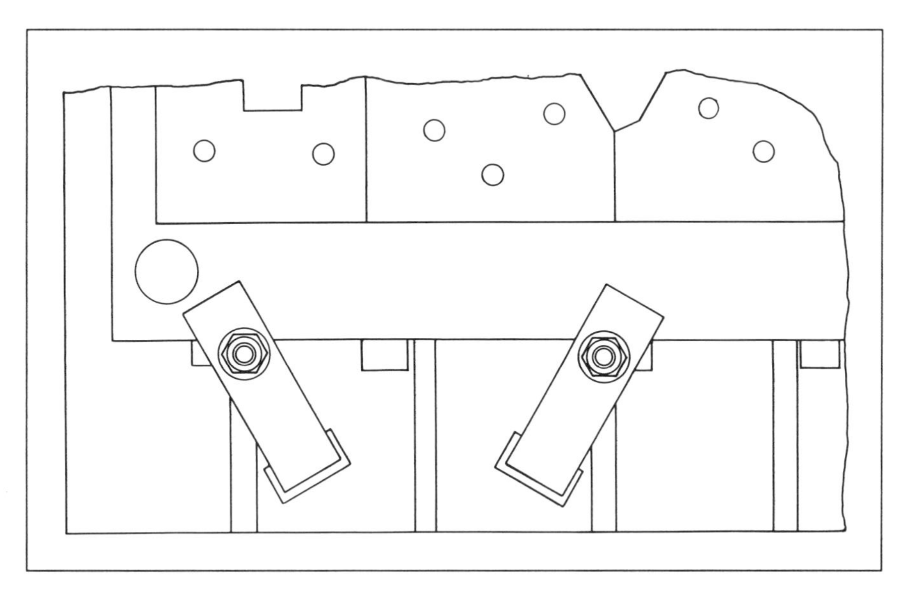

Figure 13-31. *Clamping directly over parallels holds the die shoe and parallel in compression, provides improved clamping security, and reduces the process variability introduced by the die retention method used.*

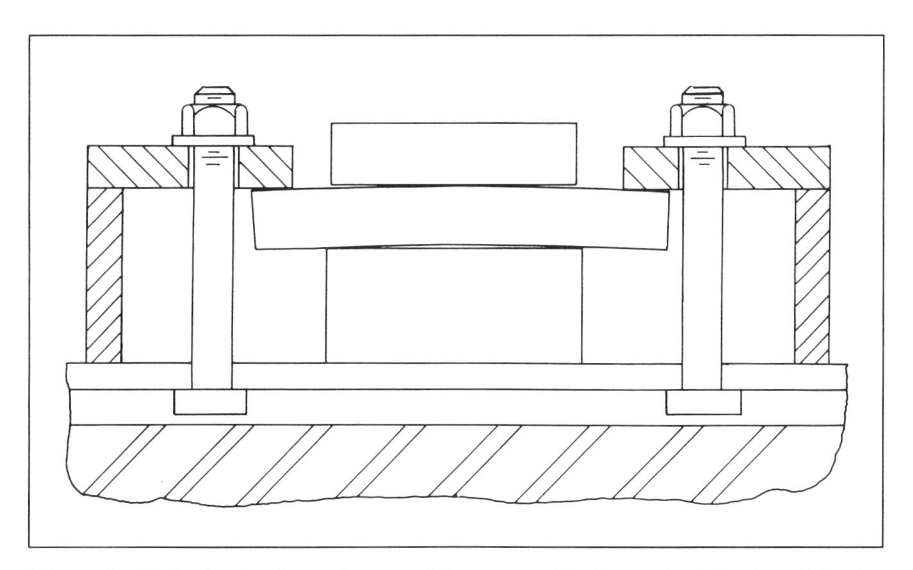

Figure 13-32. *Setting the die on short parallels or spacer blocks results in die shoe deflection when the clamps are tightened.*

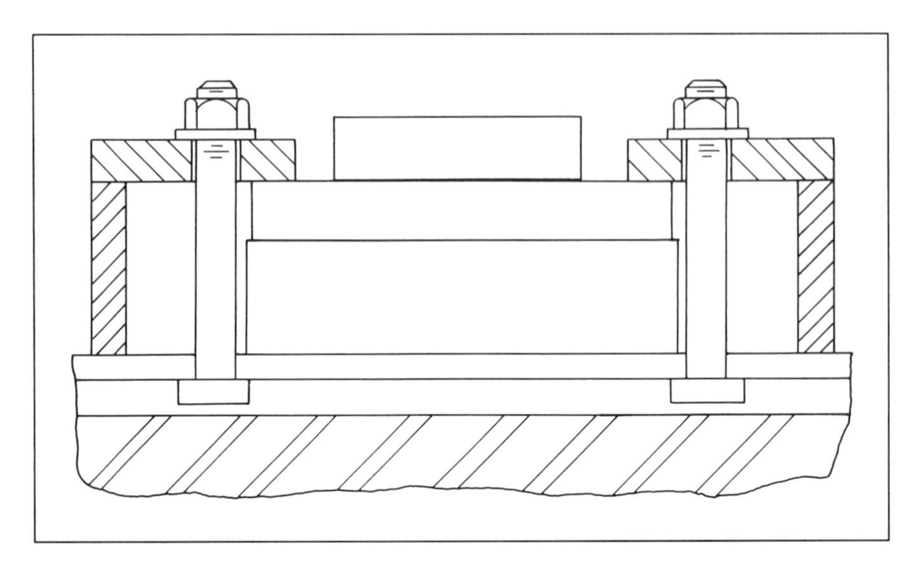

Figure 13-33. *Avoiding the shoe deflection problem shown in Figure 13-32. This is done by using a parallel long enough to extend under the area being clamped, ensuring that the clamp directly holds the die shoe and parallel in compression.*

Figure 13-33 illustrates a simple solution. The parallels must be long enough to ensure the clamp holds the shoe and parallel in compression.

Analysis of the Causes of Variation

Placing the clamps on the die shoe when running dies on parallels can result in an undesired process variation due to many factors. Some are:

- Parallels not always placed in the same location, resulting in varying deflections of the die shoe during different runs;
- Varying parallel-to-parallel height, slightly changing the exact deflection of the die shoe; and
- Difficulty in placing the clamps in exactly the same location every time resulting in slight variations in shoe deflection.

These root causes of process variability can be eliminated by making the parallels captive to the die and clamping directly to the parallels.

Fastening the Parallel to the Die Shoe

A number of requirements must be met in any good system of fastening the parallels to the die shoe. The primary requirement is, of course, safety. The type, number, and size of screws must be sufficient to equal or exceed the strength and impact resistance of the fasteners used to secure the parallel to the press bolster or ram.

Include Diesetting in Die Design Standards

This information should be part of each shop's internal standards for die and diesetting fasteners based upon the nature of the work performed, with special consideration given to the shock and cyclical loading encountered.

If the shock loading encountered is especially severe, very large fastener safety factors must be specified. In some cases, required safety factors may be several hundred times greater than that needed to support the static weight of the die alone.

The choice of attaching the parallels by screws going through the die shoe or by screws threaded into the die shoe is based largely upon individual circumstances. Whenever possible, a standard hole pattern should be adopted to facilitate the reuse of the parallels removed from obsolete dies.

Easy removal of the parallels is important to repair or modify the die. It is critical to be able to remove the parallels quickly in the event the die must run in a different press than the one for which it was adapted.

Parallel Placement in Relationship to Slug Holes

A common mistake made when setting up progressive dies on nonattached parallels is accidentally blocking a slug hole. If it is not economically feasible to follow the recommended practice of equipping the die with permanently attached parallels, the following practice should be followed:

1. Determine the correct placement of each parallel.
2. Scribe and center-punch the locations.
3. Include the parallel placement information in the diesetting instructions.
4. Double check after setting the die by locking out the press in the open position and running a stiff wire through each opening where a slug is cut. Check for obstructions and for proper clearance beneath the shoe.

Figure 13-34 illustrates the use of a stiff wire to probe for possible obstructions when placing parallels under a die shoe. It is

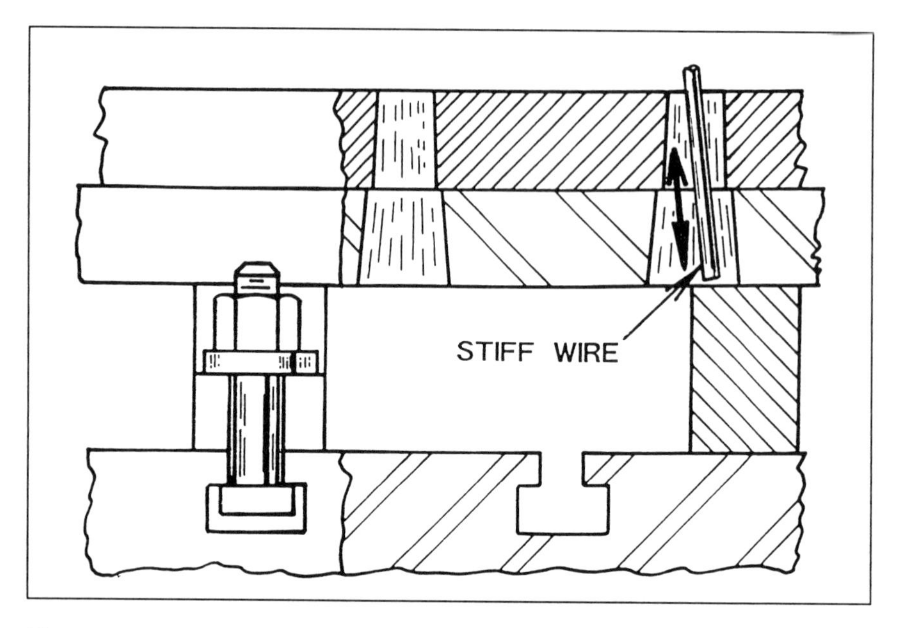

Figure 13-34. Using a stiff wire to probe for possible obstructions when placing parallels under a die shoe. It is important to probe all around the perimeter of the hole.

important to probe all around the hole's perimeter to ensure there is nothing protruding that might start an accumulation of slugs that can damage the die. This procedure also should be followed on the bench whenever parallels are permanently attached to the die; remember, "In pressworking, to assume anything is to blunder."

Milling or Drilling Slug Clearance in Parallels

If a parallel must be placed so that a slug hole is blocked, a clearance must be milled or drilled. The clearance angle should be as steep as possible to ensure positive slug discharge. Heavy pressworking lubricants can contribute to slug discharge problems. Generally, to function dependably, the slug discharge clearance in a parallel should be 30 degrees or less from the vertical plane. Figure 13-35 illustrates a section through a die opening and parallel having a 30-degree clearance angle.

If the parallel is very wide and the slug is small, a drilled hole may be required. Here again, the angle should be as steep as possible. Boring or reaming the hole with a taper or steps in the

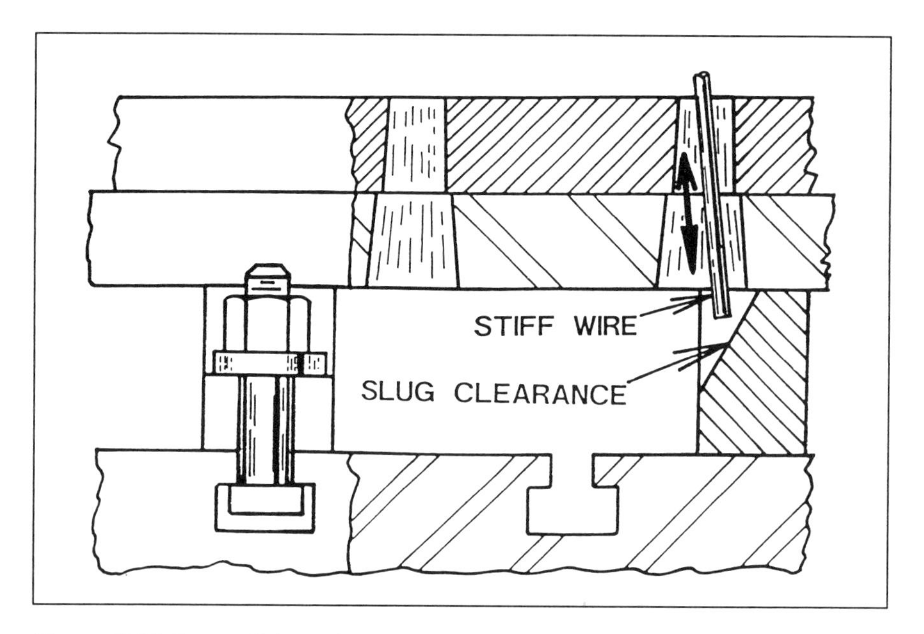

Figure 13-35. *If a parallel must be placed so that a slug hole is blocked, a clearance must be milled or drilled. The clearance angle should be as smooth and steep as possible to ensure against slug buildups.*

diameter that increase toward the discharge end is advisable, if jamming problems are encountered. In severe cases, a timed blast of air can be supplied to blow the slug out.

The work needed to provide slug clearance in parallels generally limits the practice to those permanently attached to the die shoe. If slug-hole blockage is encountered (with a die having unattached parallels), an acceptable solution is to use two short parallels in place of one the entire width of the die. Be especially careful to clamp directly over the parallels to prevent their moving out of position, a frequent cause of die damage caused by slug hole blockage.

Identifying Parallel Locations

Low-cost short-run dies often have no permanently attached parallels. Figure 13-36 illustrates a method for identifying parallel locations to ensure correct placement. After a successful run, the die shoe edge should be scribed and center-punched as shown, before it is removed from the press.

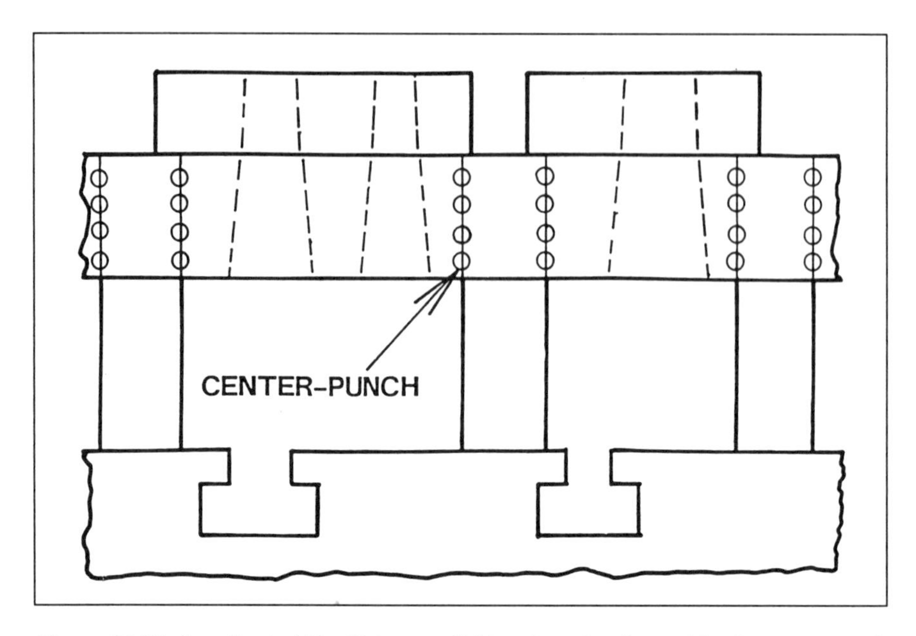

Figure 13-36. A method of identifying parallel locations for dies not having permanently attached parallels. After a successful run, the edge of the die shoe should be scribed and center-punched as shown, to identify correct parallel placement.

The correct location can be further identified by painting the area between the center-punched markings. Even if the paint should weather off during extended outdoor storage, the center-punch markings will remain to identify correct parallel placement.

New Die Tryout Procedure

Scribing and center-punching parallel locations in the press after initial tryout also should be done (as shown in Figure 13-36). This ensures correct parallel location and attachment in the tool room — a way to avoid guesswork and wasted time.

ATTACHING PARALLELS IN RELATIONSHIP TO T-SLOTS

Current part and die design practice often does not take into consideration where die parallels must be placed in relationship to T-slots. The primary design consideration, especially for progressive dies, is efficient stock utilization and die construction economy. If the die designer neglects specifying parallel placement, the tool room must adapt the die to run in the press after it is built.

Figure 13-31 showed a good method of clamping a die in the press using straps and setup blocks. The straps clamp the parallels in compression. The problem of bowing the die shoe by clamping between supporting parallels, as shown in Figures 13-29 and 13-30, is avoided.

Ideally, permanently attached parallels, such as the type flame-cut from steel plate in Figure 13-26, can be used wherever a tie-down location is needed. It is not always possible because of scrap or part discharge interference problems.

Welding Die Components

Figure 13-37 illustrates how permanently attached parallels having slots for tie-down bolts can be mounted on a die shoe. This improvement eliminates the need for diesetting straps and setup blocks (Figure 13-31). The third parallel from the left side is illustrated in Figure 13-26.

The first parallel on the left side had a steel plate with a tie-down attached to the parallel by welding. Safety rules and die design standards may require screws for added security in addition to the weld.

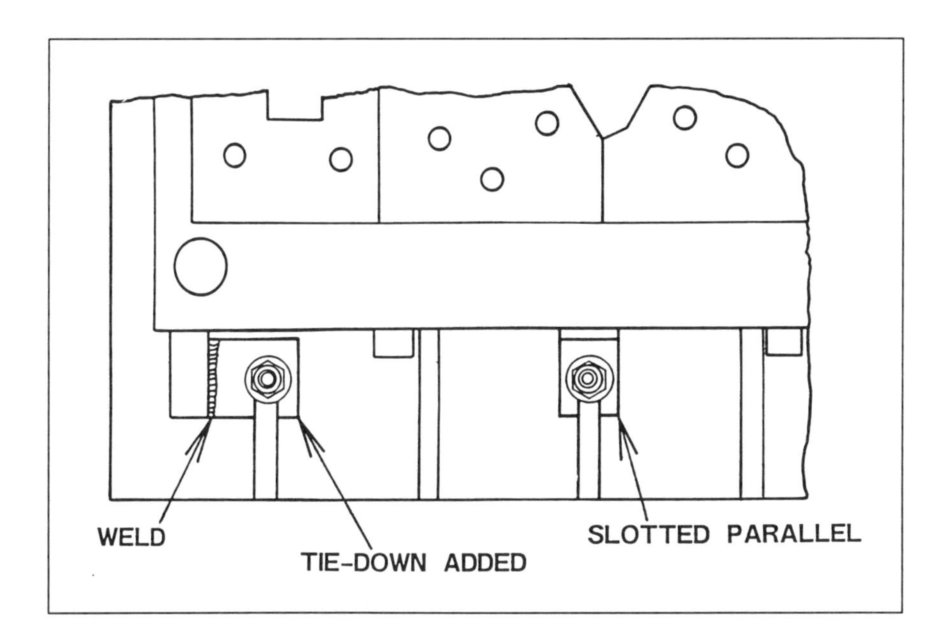

Figure 13-37. *Die tie-down improvements, including permanent attachment of parallels. This permits rapid fastening with T-slot bolts, eliminating the need for straps and setup blocks.*

Any pressworking die fastening method must be *safe* and *secure*. How this is done should be a part of the die design standards and/or safety rules of each shop. In some cases, government and insurance regulations spell out how die fastening is done.

Currently, welding alone is relied on to assemble many critical components of presses, ships, and bridges, as well as dies. Parts of the die to be welded (such as heel blocks and parallel feet) must be correctly prepared to ensure complete penetration. Die welding success factors include proper preparation, material selection, and carefully following good procedures.

Deep-beveled V-grooves must be provided to permit the weld filler metal to completely fuse both parts throughout their thickness. The filler metal must be correct for the application. Preheating to ensure proper fusion in and around the heat-affected zone is usually required. Postheating after welding to normalize or relieve stresses is highly recommended. Normalizing heat-induced stresses helps avoid warpage and distortion of the weldment during the component's service life.

Peening is often required between welding passes for relief of tensile stress. The welder must be trained and certified to follow the correct procedures. Properly done, the completed weld has the strength of the parent metal.

The welded attachment method providing bolt tie-downs can also locate the die. Figures 12-15 through 12-19 showed bumper pins used to locate dies having attached subplates. The same advantage can be realized by welding "V" and flat locators to existing parallels. Eliminating the need for subplates can save both weight and cost.

Other Improved Tie-down Methods

T-slot bolts used in conjunction with constant-height clamping slots or ledges are an essential part of most simple quick die changing systems. To be used effectively, the T-slot bolt, together with either a nut and washer or toe clamp, must be slid into place quickly and tightened rapidly.

Some dies (Figure 13-38) may have drilled holes rather than tie-down slots. T-slot bolts cannot be put in place easily. In such cases, T-slot nuts and cap screws are normally used. If T-slot bolts are required, they must be in place before the die is set.

When setting the die, getting the T-slot bolts to line up with the holes in the lower die can be time consuming — often dangerous. The diesetter may be exposed to hazardous trial-and-error work in positioning the bolts.

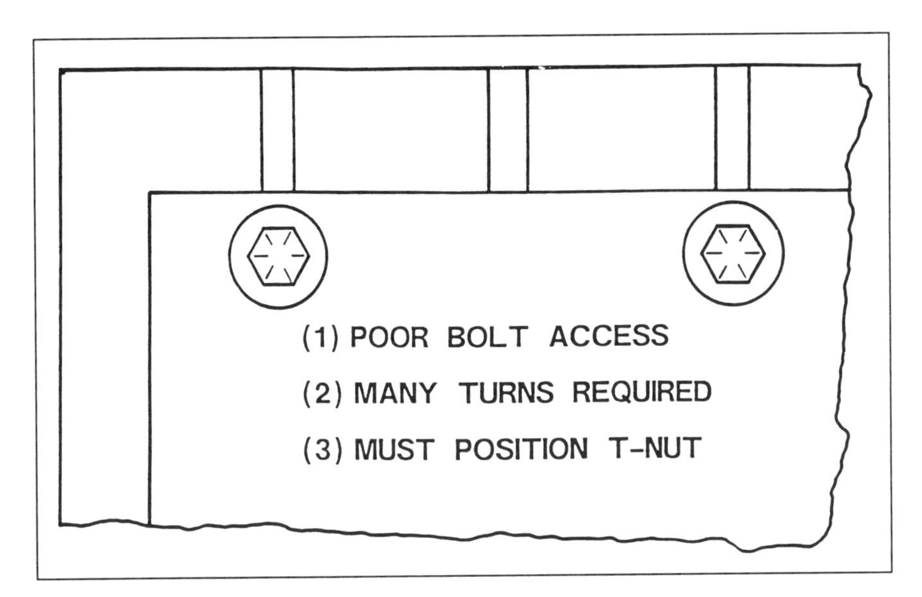

Figure 13-38. *A die shoe or subplate having drilled tie-down holes. It usually must be fastened to the ram or bolster with cap-screws and T-slot nuts; T-slot bolts are preferable.*

Aligning T-slot bolts placed in the ram with drilled holes in the upper die can result in mechanical interference when inching the press closed. A diesetter may attempt to place his or her hand between the ram and the upper die to align the bolts as the press is being inched. This is an extremely unsafe practice.

As illustrated in Figure 13-2, T-slot bolts are much stronger than a T-slot nut and capscrew (Figure 13-3). Figure 13-39 illustrates the advantage of using U-shaped cutouts in the die shoe or subplate for attaching the die to both the bolster and press ram. Providing a constant clamping height (Figure 13-11) can permit the same T-slot bolts to be used with many dies. An added advantage is that only a fraction of a turn is required to tighten the fastener.

Drilled holes are provided rather than U-shaped cutouts so that the holes can be drilled quickly to try out the die. U-shaped cutouts should be provided by milling a tie-down slot from the edge of the plate to the drilled hole before the die is approved for normal production.

Oxyfuel cutting (OFC) of tie-down slots in finished dies is not advisable. The process generates large amounts of heat, which will almost certainly warp the plate. The best solution is to specify the

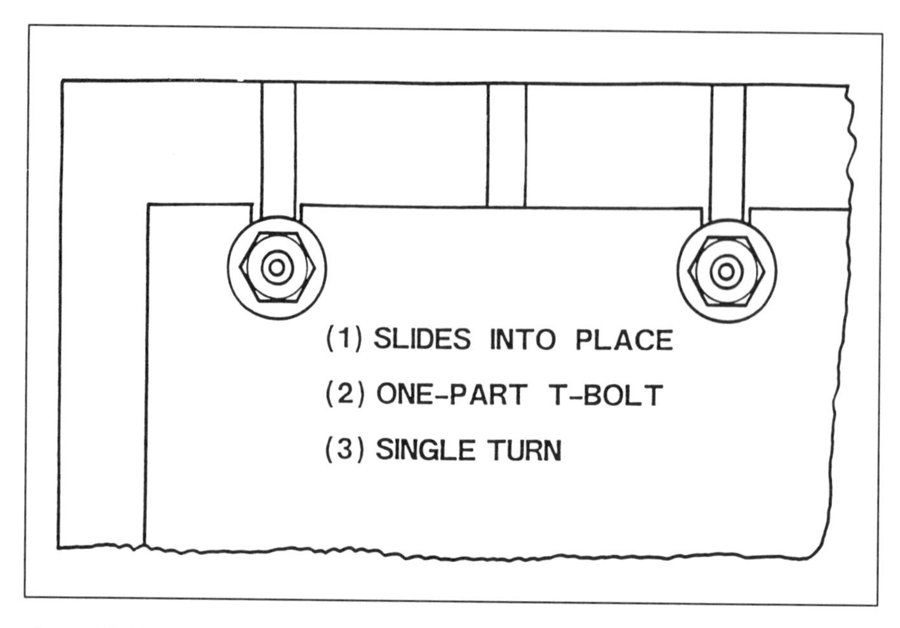

Figure 13-39. Milling tie-down slots for T-slot bolts permits the use of a strong fastener that can be tightened with a fraction of a turn.

location of the slots in the die design. They can be flame-cut at little or no added cost when the die is built.

A major problem in setting some dies side by side in the same press is that some of the T-slot tie-down locations are not accessible. Figure 13-40 illustrates one way in which this problem can occur.

The operation uses a number of die combinations to produce different styles of parts. Round bolster locating pins control die positions. Producing many different styles of parts having different hole patterns and other features is accomplished by changing die details, or one or both of the dies.

The tooling is capable of producing a great variety of product styles with a modest investment in tooling. If an individual die were used for each product style, the tooling cost and die storage space requirements would be several times greater than the split die system.

However, the designer has overlooked a basic problem that the diesetter will have in changing over the dies; the inside T-slot bolts cannot be inserted once the dies are in position. By overlooking

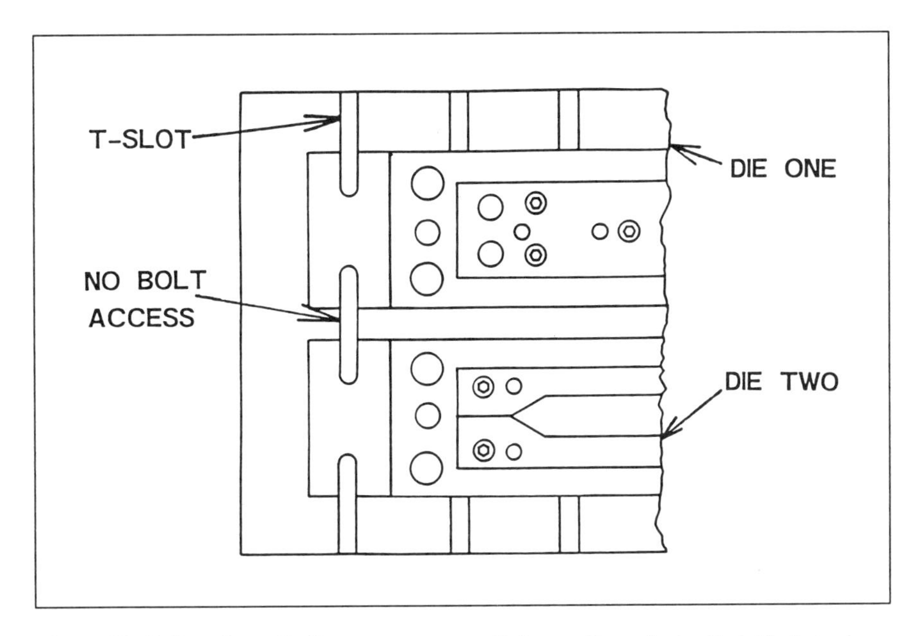

Figure 13-40. *Two dies set in the same press results in a problem placing T-slot bolts to clamp the inner die edges.*

this factor, the designer has created a difficult task for the diesetter (Figure 13-40). If T-slot bolts are used, those on the inside of the upper and lower dies must be in place before the dies are placed in the press. This results in the same dangerous condition when fastening the inside bolts as that illustrated in Figure 13-38.

Figure 13-41 illustrates two simple solutions to the problem of inside bolt placement shown in Figure 13-40. Both the upper and lower dies can be retained by simple one-piece toe clamps and T-bolts. One alternative is to use hydraulic ledge clamps.

Toe clamps used with T-slot bolts are simple and inexpensive. The hydraulic ledge clamps may be more cost effective if many die changes are performed each shift — especially if the access to tighten the conventional bolts is limited by the size of the die or press opening.

A large automotive contract stamper developed a rapid means to accurately locate dies equipped with subplates. A cam-actuated positioner locates the plate to bumper pins on the bolster center line. Their die standards list many configurations to suit different press and die sizes. The basic concept is illustrated in Figure 13-42.

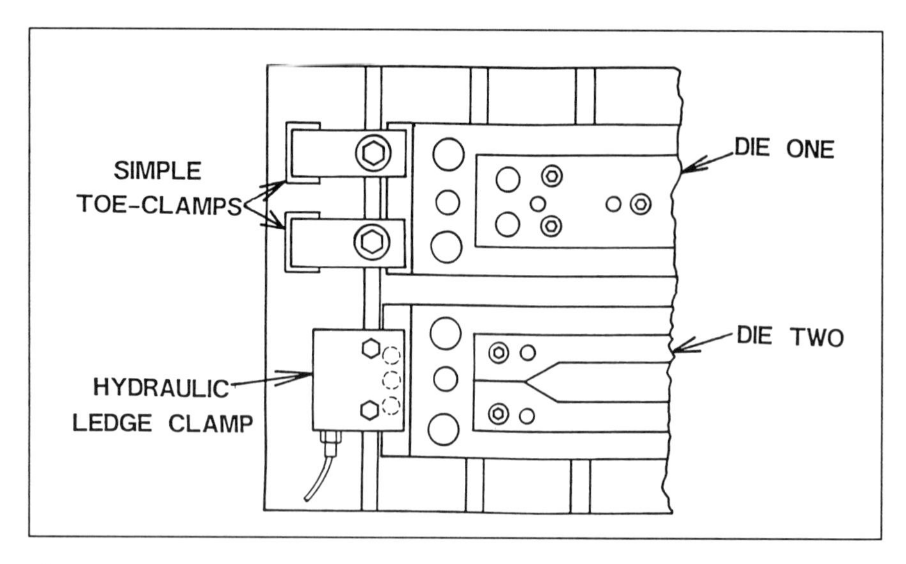

Figure 13-41. *Two ways of solving the problem of inside bolt placement shown in Figure 13-40. Both the upper and lower dies can be retained by either simple one-piece toe clamps and T-bolts, or hydraulic ledge clamps that are actuated when in position.*

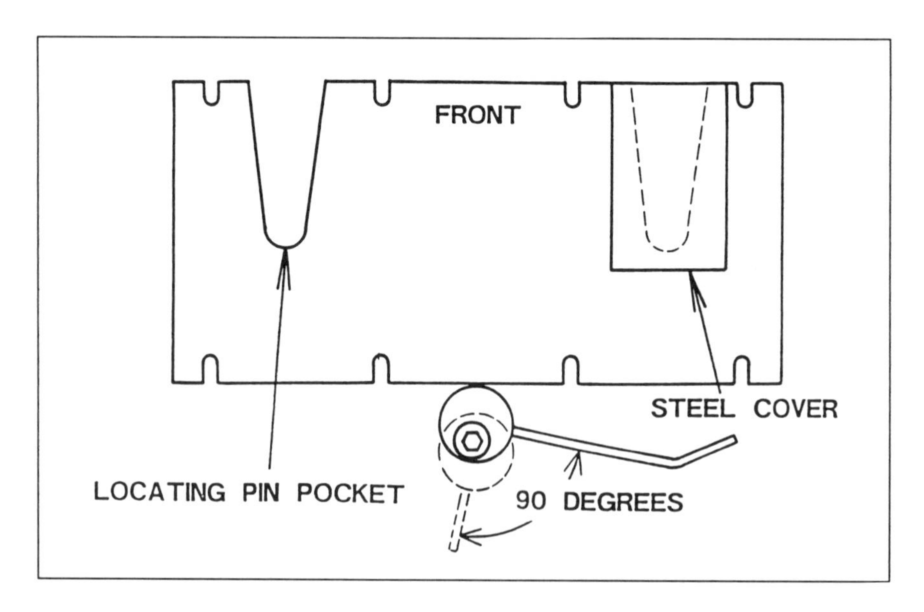

Figure 13-42. *Cam-actuated die subplate locator and eccentric positioning device. The locating pockets mate with bumper pins on the bolster center line. (A.G. Simpson Co., Ltd.)*

COMMON DIE SHUT AND PASS HEIGHT PROCEDURES

Quick Die Change Without Damage

Avoiding the potential for error is one important reason why families of dies are often operated at a common shut height. Setup time and potential for die and press damage are reduced.

Achieving common groups of die shut heights must take into account how much the press must deflect to develop the required amount of tonnage at bottom of stroke. The additional factor of bed and slide deflection must be considered when standardizing die heights for transfer press operations.

Common Pass Height Advantages

Several important advantages are realized in operating dies at a common pass height. These include:

- No need to adjust the coil feeder height for each job;
- More clearance under the die for part and scrap discharge if the maximum pass height can be adopted as the standard; and
- Fixed height stock table and operator platform for each strip-fed job.

Grouping presses and dies according to pass and shut height and assigning each die a primary or home press increases productivity and reduces process variability. A simple press data form should be used to record press data.

FLOATING THE UPPER DIE

Attaching the upper die to the press ram can be avoided if a means to open the die to permit stock feeding and part ejection is built into the die. This method is very popular for setup reduction conversions of small dies. Current United States OSHA regulations require an additional means other than the punch shank to fasten the die to the ram of OBI and OBS presses. A popular alternative to tying the die to the ram is to float the upper die half with mechanical springs. Often they can be placed around the existing guide or leader pins with little modification.

There is no practical limit to the size of the upper die half that can be floated. Large dies generally employ nitrogen cylinders, although elastomer springs find a limited application. Remember, the entire ram to upper die contact area becomes a point of operation and must be guarded.

Figure 13-43 illustrates the placement of self-contained nitrogen cylinders in a die to permit the upper half to float. This opens the die and permits feeding and removing stock and parts without the need to bolt the upper die to the press ram.

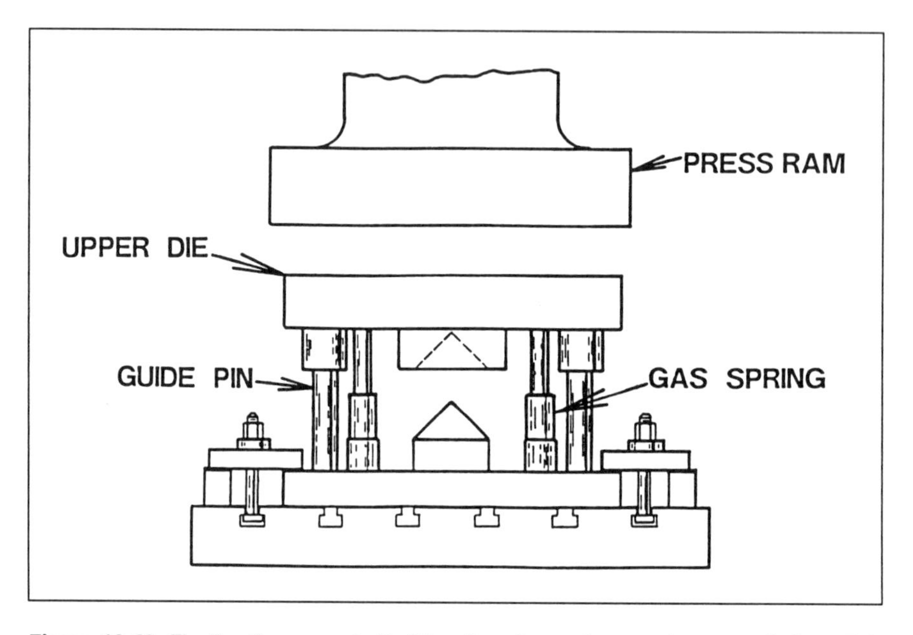

Figure 13-43. Floating the upper half of the die using springs or nitrogen cylinders. This method avoids the need to bolt the upper die to the press ram.

Springs, a low-cost alternative to nitrogen cylinders, may be placed around the guide pins in some cases. Nitrogen cylinders have an advantage of greater lifting force than a spring of the same size. The foot-mounted style of cylinder is popular, but plain cylinders with no mounts are also used. They may be placed in bored holes in the lower die shoe or retained in a holder made of round mechanical tubing fastened to the lower shoe.

SETTING DRAW DIES

If a die cushion is used, the pressure should be set to the correct value before bottoming with a blank in the die. This will speed the work, avoid possible die damage due to folded metal, and reduce scrap.

It is important to maintain a database of correct values for each job for each press. Once the die is bottomed, some fine adjustment may be required to compensate for variations in material properties.

Setting and Adjusting Double-action Press Dies

When bottoming double-action press dies, the blankholder should be adjusted first. If the press has a tonnage meter, the blankholder may be adjusted to a historically correct tonnage value. It is important the outer slide be adjusted before the draw punch fully engages the work. In the event the blankholder is too loose, folded metal may be drawn into the die and cause a slug mark on the draw punch or a reverse. Should severe wrinkling and folded metal problems occur, it is preferable to have the first part fracture than to damage the die. With care, the first blank can usually be made into a good part.

The same sequence is followed for both single and double-action presses. Once the first step of adjusting the blankholder pressure to approximately the correct value is accomplished, the correct depth of draw must be set.

Using a Shut Height Indicator

A shut height indicator can be used to set the proper depth of draw. The procedure is to return to a setting known to produce good parts. There are two exceptions to using this method. First, the die may have been repaired or modified since the last run, resulting in a shut height change. Second, the shut height indicator may have been recalibrated or damaged, producing a reading different from the correct setting for the last run.

The person making the shut height adjustment should never rely on instrument readings alone if shut height adjustments are needed when changing dies. Carefully following procedures is a must.

Visual Observation/Clay

One way to adjust the depth of draw is to observe the features of the drawn part as the slide is lowered a little at a time and the press cycled. A bottoming stamp may be provided to indicate the final adjustment is correct.

Bottoming stamps mark the part in an area unseen in the finished product. The operator or inspector should observe the stamp mark throughout the run to ensure that the slide has not changed adjustment. This could occur should the brake on the slide adjustment motor slip.

For large dies that draw irregular shapes, some diesetters make a close adjustment of the inner slide by placing modeling clay on a *reverse* — an upward projecting forming surface in the lower die cavity.

The inner slide is then adjusted downward a little at a time while the press is cycled in the inch mode. If this practice is permitted, the diesetter must make certain the reverse does not correspond with a milled-out area or eyebolt hole in the draw punch. This oversight has caused very serious press and die damage.

To guarantee that someone will not inadvertently open the press with a die partly fastened in, it is recommended to follow power-lockout procedure with the press on bottom whenever setting and removing any die. This is especially important when reaching through the access holes in the outer slide to manually bolt the inner slide to the punch plate.

The inner and outer slide adjustment screws must not exceed a maximum fixed relationship on many mechanical double-action presses. Safety limit switches installed in the press slides will stop the press if the maximum up-down relationship of the inner and outer sides is exceeded.

These limit switches must be maintained so they function properly. They must *never* be bypassed or jumpered out to set a die or perform maintenance work. Catastrophic press damage is almost certain to occur.

Floating Draw Punch

Figure 13-44 illustrates self-contained nitrogen gas cylinders used to float a draw-die punch in a double action press. For large automotive body panel dies with punches weighing up to eight tons (7.87 metric tons), six 3-ton (27 kN) cylinders are used. Four cylinders would be sufficient. The two extra cylinders permit the production run to continue should any one cylinder fail. Cylinders generally last more than 500,000 strokes between seal replacements in this application.

Cylinders without mounts are placed in holes bored in the blankholder adaptor-plate or "bull-ring," as it is also known. The punch adaptor plate should be inlaid with an air-hardening weld where the cylinder rod makes contact. The use of hardened steel contact blocks is not advised as they may come loose and enter the

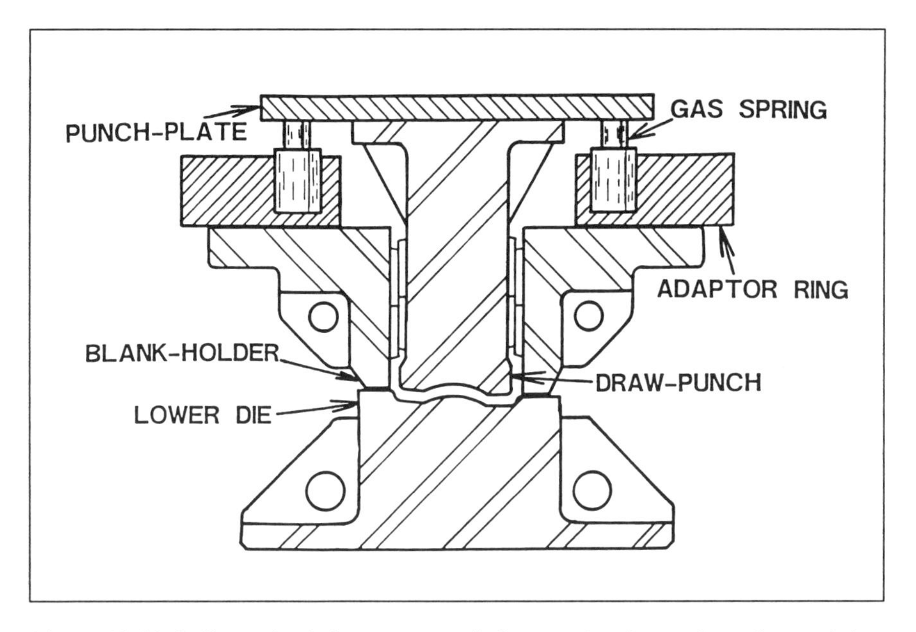

Figure 13-44. *Self-contained nitrogen gas cylinders used to float a draw-die punch in a double action press.*

die. In addition to the obvious advantage of eliminating a diesetter task when setting the die, the floating punch draw-die is affected to a lesser extent by lateral movement of the press slides caused by severe press wear and gear timing problems. The chance of a punch or blankholder wear-plate loosening and falling into the die cavity is practically eliminated.

Some toggle presses have a lower or third action actuated in the upward direction. Both the inner and outer slides dwell on bottom while this occurs.

The inner slide dwell requires a toggle mechanism similar to the outer slide. The extra mechanical complexity adds to the machine's cost and maintenance requirements. The inner slide dwell time also slows the press stroking speed.

Modern practice is to avoid part designs requiring a triple-action press. The diesetter must be cautioned to adjust the third action only after the inner and outer slides are properly adjusted. Should the inner slide be adjusted downward, a corresponding adjustment of the third action is required to avoid severe mechanical interference.

ADJUSTING KNOCKOUT BARS

The purpose of knockout bars in the ram is to strip or knockout the part at the top of the press stroke. The part is then removed from the press by an air blast or shuttle unloader. Usually knockout bars have captive pins extending through the upper platen of the press. The bars are usually supported by springs so the dead weight of the bars and attached pins does not result in the part being ejected prematurely.

The knockout pins in the press engage a plate recessed into the upper die shoe. In the case of die sets with shanks, a single knockout pin is fitted in the center of the shank.

Actuating Knockout Bars

Adjustable-length rods attached to the press frame or crown contacting the knockout bars at the top of the press stroke is the usual method of actuating the knockout bars. The stationary rods are adjustable in length by means of a threaded extension locked in place with a jam-nut. Correctly adjusted, the system provides for positive knockout action.

Errors to Avoid

If the adjustable-length rods do not sufficiently engage the knockout bars, the parts may not be dependably ejected from the upper die. This can result in multiple parts being retained in the upper die, causing serious die damage.

Should the adjustable-length rods be set too long, the rods, knockout bars, and die may be damaged. It is very important that the diesetter ensure the jam-nuts on the fixed rods are tightened properly. Otherwise, the adjusting screws may creep downward resulting in excessive knockout forces.

Should the ram adjustment be raised for any reason, it is absolutely necessary to first shorten the adjustable-length rods to avoid damage.

CONCLUSION

The diesetting methods listed are far from complete. Space alone limits how many principles and techniques can be presented. As new methods are observed, drawings and slides documenting the system are prepared for reference and training purposes.

Diesetting fasteners employing common all-thread rods are omitted for safety concerns stated in the text. Other omissions, because safety may be compromised, include one-sixth turn threaded fasteners, slotted washers, and any bolting system fabricated by welding.

BIBLIOGRAPHY

1. David A. Smith, *How to Solve Die Impact and Noise Problems With Automotive Pull Rod Shock Absorbers*, Society of Manufacturing Engineers, Dearborn, Michigan, 1990.
2. J. E. Shingley and C. R. Mischke, *Mechanical Engineering Design*, Fifth Edition, McGraw-Hill, New York, 1989.
3. Ford Motor Company, *Die Design Standards for North American Operations*, Body and Assembly Division, Dearborn, Michigan.
4. David A. Smith, *Adjusting Dies to a Common Shut Height*, Society of Manufacturing Engineers, Dearborn, Michigan; and Fabricators & Manufacturers Association, International, Rockford, Illinois, 1989.
5. David A. Smith, *Quick Die Change*, "Operating Dies at a Common Shut Height," Society of Manufacturing Engineers, Dearborn, Michigan, 1991.
6. David A. Smith, Interview with Wayne Avers, Engineering Manager of Admiral Tool and Manufacturing, *Quick Die Change Video Course Facilitator's Guide*, pp. 464-466, Society of Manufacturing Engineers, Dearborn, Michigan, 1992.
7. David A. Smith, *Quick Die Change*, "Grouping Presses and Dies for Quick Die Change," Society of Manufacturing Engineers, Dearborn, Michigan, 1991.

CHAPTER 14

DIE MATERIALS AND LUBRICANTS

MATERIALS USED TO CONSTRUCT DIES

Tool steels are used to construct the die components most subject to wear. These steels are designed to develop high hardness levels and abrasion resistance when heat treated.

Plain carbon and low-alloy steels, easy to machine and weld, are used for machine parts, keys, bolts, retainers, and support tooling. Cast-steel dies are used for large drawing and forming dies where maximum impact toughness is required. At carbon levels of 0.35% and higher, cast-alloy-steel dies can be effectively flame-hardened at points of wear.

Cast irons are used for shoes, plates, dies, adapters, and other large components. Cast irons are alloyed and hardened for large sheet metal drawing and forming dies. The ductile (nodular) irons retain the casting advantages of cast iron, while having toughness, stiffness, and strength levels approaching those of steel.

In addition to ferrous die materials, a variety of die components are made of nonferrous metals such as zinc and copper alloys. Elastomer products find widespread application as die pads, rubber springs, and automation components. Even wood and wood fiber products are used for inexpensive dies.

Characteristics of Tool and Die Steels

The steels listed in Table 14-1 are used in the majority of the metal-stamping operations. The list contains 27 steels that are readily available from almost all tool steel sources. Some of these steels have slight variations for improved performance under certain conditions.

The steels are identified with a letter and number code. The letter represents the group of the steel involved. The number indicates a separation of one grade or type from another.

Water-hardening tool steels are designated by the letter "W," mainly W1 and W2. Both are readily available and low in cost. W2 contains vanadium and is more uniform when heat treated, as well

Table 14-1

Tool and Die Steels

AISI Steel Type	Nominal Composition, Percent							
	C	Mn	Si	W	Cr	Mo	V	Other
W1	1.05	0.25	0.20	—	0.20	—	0.05	—
W2	1.00	0.25	0.20	—	—	—	0.20	—
O1	0.90	1.25	0.30	0.50	0.50	—	—	—
O2	0.90	1.60	—	—	—	—	—	—
O6	1.45	0.80	1.10	—		0.25	—	—
A2	1.00	0.60	0.30	—	5.25	1.10	0.25	—
A8	0.55	—	1.00	1.25	5.00	1.25	—	—
A9	0.50	—	1.00	—	5.20	1.40	1.00	1.40 Ni
D2	1.55	0.30	0.40	—	11.50	0.80	0.90	—
D3	2.10	0.40	0.90	0.80	11.70	—	—	—
D4	2.25	0.35	0.50	—	11.50	0.80	0.20	—
D7	2.30	0.40	0.40	—	12.50	1.10	4.00	—
S1	0.50	—	0.50	2.25	1.30	—	0.25	—
S5	0.60	0.85	1.90	—	0.20	0.50	0.20	—
S7	0.50	0.75	0.30	—	3.25	1.40	—	—
T1	0.75	—	—	18.00	4.00	—	1.10	—
T15	1.55	—	—	12.25	4.00	—	5.00	5.00 Co
M1	0.80	—	—	1.50	3.75	8.50	1.05	—
M2	0.85	—	—	6.25	4.15	5.00	1.90	—
M4	1.30	—	—	5.50	4.00	4.50	4.00	—
M7	1.00	—	—	1.70	3.75	8.75	2.00	—
M42	1.08	—	—	1.50	3.75	9.50	1.10	8.00 Co
H12	0.35	—	1.00	1.25	5.00	1.35	0.30	—
H13	0.38	—	1.00	—	5.20	1.25	1.00	—
H21	0.30	0.25	0.30	9.00	3.25	—	0.25	—
H26	0.53	—	—	18.00	4.00	—	1.00	—
L6	0.75	0.70	—	—	0.80	0.30	—	1.50 Ni

W — Water Hardening
O — Oil Hardening
A — Air Hardening
H — Hot Working

D — High Carbon, High Chromium Die Steels
S — Shock Resisting
T — Tungsten-base, High Speed
M — Molybdenum-base, High Speed
L — Special-purpose, Low-alloy

as having a finer grain size with a higher toughness. Both are shallow hardening. In large sections, this results in a hard case and a softer internal core that has high toughness. They are quenched in water or brine and are subject to substantial size changes when heat treated.

The letter "O" identifies oil-hardening tool steels. Among these, O1 and O2, known as manganese oil-hardening tool steel, are very popular. Readily available and low in cost, these steels have less movement than the water-hardening steels and are of equal toughness if the water-hardening steels are hardened throughout. Type O6, which contains free graphite, has excellent machinability.

Air-hardening die steels are identified by the letter "A," with A2 being the most popular. This steel has a minimum movement in hardening and has higher toughness than the oil-hardening die steels, while delivering equal or greater wear resistance. It has a slightly higher hardening temperature than the manganese types. The availability of the popular A2 steel is excellent. Type A8 is the toughest steel in this group, but its low carbon content makes it less resistant to wear than A2.

The principal steels of wide application for long-run dies are high-carbon high-chromium die steels, identified by the letter "D." Grade D2 containing 1.50% carbon is of moderate toughness and intermediate wear resistance. Steels D3, D4, and D7, contain additional carbon and offer very high wear resistance, but somewhat lower toughness. Selection between these is based on the length of run desired, and the machining and grinding problems. D2 and D4 contain molybdenum. They are air-hardening and have minimum movement in hardening.

Shock-resisting tool steels, identified by the letter "S," contain less carbon and have higher toughness than other types. They are used where heavy cutting or forming operations are required and where breakage is a serious problem with higher-carbon materials that might have longer life through higher wear resistance alone. Choice among the grades is a matter of experience. All steels are readily available, with steel S5 being especially widely employed. This grade is an oil-hardening type of silicomanganese steel and is more economical than steel S1, which has equivalent toughness properties with greater wear resistance.

In tungsten and molybdenum high-speed steels ("T" and "M"), T1 and M2 are equivalent in performance, representing standard high-speed steels which have excellent properties for cold-working dies. They have higher toughness than many of the other die steels and offer excellent wear resistance. They are expensive, but readily available. T15 and M4 are hardened by the standard method rather than carburizing because they already combine large amounts of carbon and vanadium.

Type M1 is occasionally used in place of T1 and M2, but it is more susceptible to decarburization. Steel T15 is the most wear-resistant of all the steels in the list, and steel M4 is slightly greater in wear resistance than a steel such as D4. These steels are more difficult to machine and grind than the other high-speed steels, but the improved performance justifies the extra machining expense.

Of the many low-alloy steels ("L") effective as die materials, steel L6 is the only chromium-nickel steel. In large sizes it is water-quenched and has a hard case and a soft core, with an attendant high overall toughness. In small sizes it may be oil-quenched.

Die casting dies, extrusion dies, hot-forming dies, and hot-drawing mandrels are typical hot-work applications. They use hot-working ("H") steels.

HEAT TREATMENT OF DIE STEELS

Simplified Theory of Hardening Steel

Iron has two distinct and different atomic arrangements—one existing at room temperature (and again near the melting point) and one above the transformation temperature. Without this phenomenon it would be impossible to harden iron-based alloys by heat treatment.

Briefly, what happens in the heat treatment of die steels is represented graphically by Figure 14-1. Starting in the annealed machinable condition A, the steel is soft, consisting internally of an aggregate of ferrite and carbide. Upon heating above the transformation temperature, 1414 to 1700°F (768 to 927°C), the crystal structure of ferrite changes, becomes austenite B, and dissolves a large portion of the carbide. This new structure, austenite, is always a prerequisite for hardening. By quenching it and cooling it rapidly to room temperature, the carbon is retained in solution and the structure known as martensite C results. This is the hard-matrix structure in steels.

The rapid cooling results in high internal stresses because the transformation from austenite to martensite involves some volumetric expansion against the natural stiffness of the steel. To overcome this problem, the steel is reheated in an operation known as tempering, or drawing, to an intermediate temperature D to

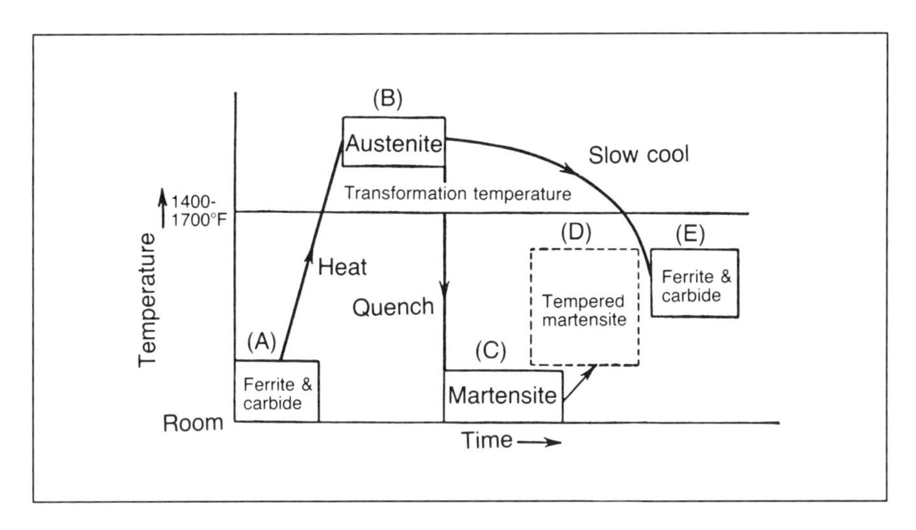

Figure 14-1. Simplified chart of the hardening of steel.

soften it slightly and relieve those residual stresses, which otherwise would embrittle the steel.

If quenching is not rapid enough, there will not be high hardness because the austenite reverts to ferrite and carbide E. The rate of quenching required to produce martensite depends primarily on the alloy content. Low-alloy steels require rapid cooling in water or oil, while highly alloyed steel usually can be air quenched at a much slower rate.

Throughout all these heat-treating reactions, most die steels retain excess or undissolved carbides that take no direct part in the hardening. The high carbon high-chromium steels, for example, have large quantities of excess iron-chromium carbide, which give them the high degree of abrasion resistance possessed by this class of steel.

Each type of die steel must be handled slightly differently for optimum results. Different temperatures, different heating and cooling rates, and variable tempering procedures must be used.

In general, it may be said that the harder a given die, the longer it will wear; the softer a die, the tougher it becomes. Assuming the proper die steel is being used, dies wearing out too quickly should be made harder for improved life; dies that break or crack should be made softer. Within limits, heat treatment can be used to adjust these variables. An oil-hardening steel may work best on one application at Rockwell C62 and on another involving higher stresses and shock at Rockwell C58. Adjustments of the drawing temperature easily produce the hardness desired.

Double drawing and, in some instances, triple drawing are desirable for tools in severe applications. This is because steels retain austenite when quenched. The first temper affects the martensite formed in the quench and conditions the austenite so that it transforms upon air cooling from the draw. Double drawing is necessary to affect the martensite that forms after the first draw. Triple drawing removes the austenite and promotes toughness.

Cryogenic Treatment

Cooling to very low temperatures as part of the drawing process will maintain hardness while actually improving the toughness and fatigue strength of hardened tool steel. This improvement occurs due to the continued transformation of retained austenite into the more desirable martensite, at temperatures ranging from -120 to -300°F (-84 to -184°C). Size change is negligible, but can occur if the material is not properly heat treated. Cryogenic treatment will not correct poor heat treatment practices and multiple draws or tempers cannot be eliminated knowing that the cold process will convert austenite to martensite. Cryogenic treatment must begin almost immediately after heat treatment to be effective. It is often performed between the last two draws.

Heat Treatment

Furnaces for heat-treating tool-steels are often equipped with gas generators that provide an atmosphere containing a controllable amount of carbon. Without a protective atmosphere, surface decarburization to a considerable depth can occur because of the oxidizing effects of free oxygen, water vapor, or carbon monoxide. Typical decarburization rates are 0.010 to 0.030 inch (0.25 to 0.76 mm) per hour. The loss of surface carbon can produce poor wear resistance. It is customary to make tools sufficiently oversized so that they can be ground to remove the surface affected by chemical change during heat treatment.

If the atmosphere used to heat dies for hardening is strongly reducing, it is possible to carburize the surface even though the carbon content is already quite high. A carburized case may be beneficial to tool life, if not carried to excess.

Shops doing only a small amount of heat treating often wrap the die details to be heat treated in a stainless steel foil especially produced for the purpose. This procedure effectively avoids decarburization without the expense of maintaining an atmosphere furnace.

Molten salt baths provide decarburization protection for die steels. Careful maintenance involving desludging and rectification is necessary to maintain a neutral condition.

Control of Dimensional Change

Dimensional changes occur during the hardening operation on dies because the hardened steel occupies a greater volume than the annealed steel from which it came. Unfortunately, the dimensional changes that result from the volume of change are usually not the same in all three directions, making it impossible to predict the changes accurately. The size and shape of the individual workpiece are the most important factors influencing this variable.

The dimensional changes resulting from hardening may vary from nil up to approximately three parts per thousand. Dimensional changes of this amount are of no concern when enough grind stock is provided to allow for scale and decarburization. If processes are used that produce little or no scale or decarburization, grind stock may not be required, but dimensional changes must be considered.

Air-hardening tool steels usually produce dimensional changes that are less than one part per thousand and, therefore, widely used where precision is needed. It is also possible to control the size changes in tools made of type D2 high-carbon high-chromium steel to so-called "zero size change" by an austenite-martensite balance control obtained by multiple tempering. On a series of identical tools made of air-hardening steel, it is customary to allow for expected dimensional changes on the basis of the experience from the first part treated.

Tools cannot be designed on dimension alone. The designer must consider the type of steel that will be used and whether the steel will be hardened by quenching in water, oil, or air. Generally, liquid-hardened tools must be conservatively designed, while air-hardened tools can tolerate some features that would be improper for liquid quenching. The design of a tool must also take into account the equipment available for heat treatment.

Repairing Dies by Welding

Basically three welding methods are used for die repair:

- Shielded metal arc welding (SMAW), or stick electrode welding process, is the most popular because of its versatility and wide range of filler metals available.

- Gas metal arc welding (GMAW), also called MIG or wire welding. The filler metals are available as either solid or tubular wire, which offer the widest range of alloy selection.
- Gas tungsten arc welding (GTAW), also called TIG, or heliarc welding, offers optimum operator control, although the selection of bare metal filler rods is somewhat limited.

Die Welding

In die welding, wear resistant alloys are applied to the surface of dies to increase service life, avoid downtime, or to rebuild or repair dies that have been damaged. Welding is also used to correct machining errors and to increase the wear resistance of die surfaces. Varying degrees of hardness, toughness, and wear resistance are available in welding alloys, letting you choose the best material for each application.

Tool steel welding materials are normally heat treatable alloys. Alloys for SMAW (stick) welding have a wire core covered with a flux coating that may also contain alloy constituents, including rare earth elements. Alloying elements in the flux coating are passed across the arc and form a homogeneous weld deposit.

The tool steel groups most commonly repaired by welding are water, oil, and air-hardening steel. Welding difficulty is dictated by the alloy content of the base metal. Usually steels with higher carbon contents require higher preheats before welding, more care during welding, and greater care in tempering after welding.

Tool steel welding requires the same precautions necessary with any other type of welding process to prevent cracking the base metal during heating and cooling. All base metals with a carbon content higher than 0.35% should be preheated and postheated to decrease brittleness in the base metal near the heat-affected zone. High-carbon alloy steels, such as air-hardening tool steels, are more difficult to weld because of the likelihood of cracking, requiring greater care and procedural control. Hot extrusion tools are notable examples of successful use of this type of repair.

In all die welding it is a normal practice to mechanically relieve the tensile stress that occurs upon cooling. The weld metal is upset by careful peening with a pneumatic hand tool between each pass. The finished weld must be as stress free as possible, although slight compressive stresses are not considered harmful.

Welding cast iron requires a different method of application of the weld than steel does because cast iron melts at a lower temperature. Nickel materials with lower melting temperatures are recommended for underlayment on all cast irons. Generally, after careful preparation involving grinding out the area to be repaired and preheating, a deposit of nickel alloy weld is made, followed by the amount of desired welding material required to obtain the needed surface properties. In all welding operations on cast iron, careful preheating and postheating is required.

Welded trim edges may be provided on iron castings. Generally, two types of welds are deposited. A layer of unique nickel alloy weld is deposited first to act as a buffer for the hard weld material that forms the cutting edge. Materials commonly used include AISI-S.7, H12, and H19. The buffer layer serves to prevent excessive carbon and other elements from the gray iron mixing with the air hard weld and changing its properties. The nickel alloy layer also provides a good bond with the casting. The final cutting edge is built up with an air-hardening weld. After cooling to room temperature, the weld is carefully tempered.

Large broken iron die castings and press parts can be repaired by brazing. Preparation by machining or grinding is required to provide a large area for adhesion of the braze material. Again, careful procedural control, including peening, preheating, and postheating, are necessary for success.

Die Surface Coatings and Treatments

A thin layer of chromium is often applied to forming and drawing dies in order to increase wear resistance and reduce galling. This chromium-plated surface has a very low coefficient of friction with excellent characteristics that resist galling.

The usual practice is to apply a layer of chromium 0.0005 to 0.001 inch (0.013 to 0.025 mm) thick to a very finely ground or polished surface. The time taken to polish the surface is well spent. Otherwise any defects or irregularities on the surface prior to chromium plating will show through and actually be aggravated during the plating operation.

Chromium plating is also used to repair worn dies. It is possible to build up a layer of as much as 0.010 inch (0.25 mm) or more of chromium on a worn surface and increase the total production life of the die.

The use of gas nitriding to produce a hard, wear-resisting case on steels has been practiced commercially for many years. This procedure can be used also on some tool steels to improve wear resistance. Gas nitriding is only advantageous on tool steels that do not temper back excessively at the nitriding temperature, typically 975°F (524°C). This limits gas nitriding largely to the hot-work steels and the high-carbon high-chromium grades. Gas nitriding should not be used on high-speed steels because it forms an exceptionally brittle case.

Gas nitriding of tool steels is usually carried out for a period of from 10 to 72 hr., producing a case depth of from 0.002 to 0.018 inch (0.05 to 0.46 mm).

Ion nitriding applications range from improving the wear resistance of small tool steel die sections to large iron alloy drawing punches weighing 10 or more tons. Unlike the older gas nitriding process, a glow discharge or ion processing takes place when a DC voltage is applied between the furnace, which becomes the anode and the workpiece, which then becomes the cathode, where both are placed in a low-pressure nitrogen gas atmosphere. The nitrogen gas in the furnace is ionized and emits electrons (with a negative charge), and ions of nitrogen (positively charged) that move toward the cathode (i.e., the workpiece), and are accelerated with high speed by the sharp cathode drop just in front of the cathode surface, bombarding the workpiece.

Ion nitriding furnace sizes are available that are large enough to process the largest dies commonly used in the automotive, appliance and aerospace industry. Although often more costly, the ion nitriding process has supplanted chromium plating in many large die applications.

Titanium nitride (TiN) and titanium carbide (TiC) coatings improve the life of tools by acting as a chemical and thermal barrier to diffusion and fusion. The coatings are very thin, typically 0.0001 to 0.0003 inch (0.0025 to 0.0076 mm) in thickness, and quite hard. Although the thin coatings are very brittle, they tend to assume the ductility and deformation characteristics of the substrate material. The coatings have a lubricating quality that lowers the coefficient of friction between the tool and the workpiece. By depositing TiN or TiC onto a steel or carbide tool, the improvement in lubrication makes the tool resist galling.

The physical vapor deposition of titanium nitride (PVD) coating process is carried out in a high vacuum at temperatures between 400 to 900°F (204 to 482°C). This range of temperatures does not exceed those used to draw hardened high-speed tool steel. Because there is very little distortion or size change on the workpiece, this coating process is frequently used to coat finished punches and buttons whenever rapid wear or galling is a problem. The plasma source coats the workpiece in a straight line-of-sight process. Special rotating fixtures with water cooling may be required to ensure that all surfaces are evenly coated and that small sections are not overheated. A TiN coating deposited by the PVD process is easily recognized by its gold color.

With chemical vapor deposition of titanium carbide and nitride (CVD), the process is done at much higher temperatures, 1740 to 1920°F (949 to 1049°C) than the PVD process. For this reason, it is normal practice to follow the coating procedure with conventional heat treatment of the tool steel substrate.

Chemical vapor deposition of TiC is limited to tool steel and solid carbide die materials because the substrate surface must act as a catalyst. It is superior to PVD coatings when extreme abrasive wear is a problem. The coating is deposited from a vapor, so it is possible to get a uniform coating of blind slots and blind holes.

A CVD coating is dull gray in color. When a CVD-coated tool is polished, the resultant tool is silver, quite indistinguishable from the base metal.

Thermal diffusion (TD) is performed by immersing parts in a fused salt bath at temperatures of 1600 to 1900°F (871 to 1039°C) for a period of one to eight hours. Carbide constituents dispersed in the salt bath combine with carbon atoms contained in the tooling substrate, which must contain at least 0.3% carbon or greater. The carbide layer most commonly produced is vanadium carbide, but other carbides can be deposited, depending on the composition of the salt bath. These include niobium carbide, chromium carbide, and in some newer processes, a niobium vanadium combination.

The CVD coating method can apply both TiN and TiC to all tool steels as well as solid carbide tooling. When very high wearability qualities are required, and the distortion caused by the heat treatment following coating that is usually needed is not a problem, CVD may be the best choice.

PVD is a low-temperature process that can be applied to all tool steels, but is generally used to increase the wearability of finished high-speed steel parts, solid carbide, and brazed carbide tooling.

CVD coating requires a post heat treatment to restore the hardness to the steel substrate. In the TD process, popular tool steels, such as D2 and A2, are commonly directly quenched when removed from the fused salt bath. The diffused layer, typically vanadium carbide, is quite thin and has a higher hardness than tungsten carbide. In situations requiring high-volume production runs, tungsten carbide is also treatable.

Powder Metallurgy Process

Tool-steel produced by the powdered metallurgy process is first atomized in a nonreactive gas atmosphere producing very fine particles having uniform properties. This material is placed in large steel canisters that are evacuated and sealed shut. To form ingots, the canisters are placed into a hot isostatic compacting furnace and slowly brought up to the welding temperature of the alloy being produced. A combination of heat and high pressure gas fuses the steel powder into a homogeneous mass.

Chief advantages of this process are uniformity and freedom from imperfections due to segregation of constituents associated with the cast ingot process. Also possible are higher alloy constituencies than are obtained with conventional melt and cast technology.

Wrought Carbon and Low-alloy Steels

Wrought-steel plate, rounds, and shapes are often used in the fabrication of brackets, frames, mechanisms for feeding, ejecting, transfer, and other die auxiliaries when structural strength and weldability are the primary requirements rather than wear resistance. Short-run steel dies are sometimes made with carburized hot-rolled-steel wear surfaces. Where the properties of AISI-SAE 1018 or similar steel (boiler plate) suffice, they have the advantage of being economical and readily available.

Cast Carbon and Low-alloy Steels

Cast-steel shoes, and other die components are often used for large drawing, forming, or trimming dies when a combination of high toughness and strength is required. These steel castings are usually annealed or normalized to provide a homogeneous structure free from casting stresses. Heat treating or flame hardening is often employed to obtain the desired strength, wear resistance, and toughness.

Cast and Ductile Die Irons

High compressive strengths, manufacturing economy, and ease of casting, make gray cast iron useful, especially in large forming and drawing dies. Soft unalloyed gray irons are widely used for plates, jigs, spacers, and other die parts.

Fully pearlitic irons with random uniform flake-graphite structures are excellent for wear resistance. Resistance to wear is significantly improved by flame hardening draw radii or other wear areas. Alloy additions of chromium, molybdenum, and nickel are commonly used to produce uniform pearlite structures and to improve the iron's response to flame hardening.

The ductile (nodular) irons retain the casting advantages of cast iron but, because the free graphite is present in spheroidal shape rather than in flake form, this material develops toughness and strength levels approaching those of steel.

This combination of properties is especially useful in large forming and drawing dies where heavy impact loads or high transverse stresses are encountered. It should be specified for dies where breakage has occurred with gray iron castings.

NONFERROUS AND NONMETALLIC DIE MATERIALS

Dies made of nonferrous materials are used because they are economical for limited production runs, including experimental models, and can offer superior functioning, such as preservation of part finish, relative light weight, and portability for extremely large tools. Other advantages may include corrosion resistance, ease of fabrication, and low lead time requirements. Often, they permit fast, easy corrections when design changes are necessary.

Nonferrous die materials include:

- Aluminum alloys;
- Zinc-based alloys;
- Lead-based alloys;
- Bismuth alloys;
- Cast-beryllium alloys;
- Copper-based alloys;
- Plastics;
- Elastomers; and
- Tungsten carbides.

Dies made with tungsten carbide elements are most economical for large production quantities and for stampings having critical tolerances. Tungsten carbide parts are expensive because of their hardness (which makes them difficult to machine) and the close tolerances to which they are normally held.

Nonferrous Cast Die Materials

Commercially available bronze (e.g., Ampco® metal), cast to the die shape, is used for forming and drawing stainless steel and other difficult-to-work materials without scratching or galling. These alloys are also used for die wear plates and press bushings where a high-load carrying capacity is required.

Zinc-alloy die materials have a higher tensile strength and impact resistance than pure zinc. The alloys can be cast into dies for blanking and drawing a variety of aluminum and steel parts, especially complicated shapes and deeper draws than are possible with plastic or wooden dies. The working surface is dense and smooth, requiring only surface machining and polishing. Dies made of this material are frequently mounted on die sets and used for blanking light gages of aluminum.

Frequently one member of the die set is composed of a zinc alloy, and the other member is made of a softer material such as lead. Drop-hammer operations is an example. Harder punches are required for forming steel sheets where sharp definition is necessary. Worn and obsolete dies made of zinc alloy and lead can be remelted and is nearly 100% reusable. However, concern over the toxicity of lead may limit this application.

Various commercially produced zinc-alloys, such as those called Kirksite® alloys, are available. A typical composition is 3.5 to 4.5% aluminum, 2.5 to 3.5% copper, and 0.02 to 0.10% magnesium. The remainder is 99.99% pure zinc. In order to minimize health hazards from grinding dusts generated when working these materials, the impurity levels of lead and cadmium must be kept very low.

Lead punches, composed of 6 to 7% antimony, 0.04% impurities, and the remainder lead, have been used with zinc alloy dies. Health concerns over the hazards of lead toxicity must be considered when this material is used.

Cast alloys of beryllium, cobalt, and copper have characteristics comparable to those of the commercially available aluminum bronze alloys previously discussed. Beryllium is a very toxic substance. Proper precautions regarding ventilation and industrial hygiene must be taken when working with these alloys.

Alloys of bismuth are used chiefly as matrix material for securing punch and die parts in a die assembly and as-cast punches and dies for short run forming and drawing operations. These alloys are classified as low melting point alloys because the melting points of some are below the boiling point of water.

Carbide Die Materials

Cemented carbides consist of finely divided hard particles of carbide of a refractory metal sintered with one or more metals such as iron, nickel, or cobalt as a binder, forming a body of high hardness and compressive strength. The hard particles normally are carbides of tungsten, although titanium and tantalum are also used.

Thermal expansion is an important physical characteristic of carbide. For most carbide grades, it ranges from one-third to one-half that of steel. This must be considered when carbide is attached to a steel support or body.

Cemented tungsten carbide is widely used for dies intended for high-volume production of difficult-to-stamp materials. For example, motor and transformer lamination cutting dies used for high-volume production are often made of cemented tungsten carbide, which have been known to produce millions of parts before resharpening is needed.

Other widespread applications include cutting, drawing, forming, and ironing dies used in high volume production of parts ranging from razor blades and stainless steel drawn shells to beverage containers.

Restrictions exist on the use of lubricants on some carbide die materials, particularly those containing sulphur. While tungsten carbide is essentially an inert material, the cobalt binder material is attacked by some lubricants. Electrolytic corrosion can also be produced by stray currents from electrical part sensing.

Machinable carbides are machinable ferrous alloys sold under the trade name of Ferro-Tic® and available in several grades. These range from 20-70% (by volume) titanium carbide, tungsten carbide, titanium-tungsten double carbides, or other refractory carbides as the hard phase. These are contained in a heat treatable matrix or binder that is mainly iron.

The grade most used for dies and molds contains neither tungsten nor cobalt, but approximately 45% titanium carbide with a balance of mainly alloy tool steel. The heat treatment needed to harden the matrix is much the same as for conventional tool steels. The manufacturer's recommendations should be closely followed for proper product application, machining, and heat treatment.

Nonmetallic Die Materials

Laminated impregnated wood, hardboard, and plastics have largely replaced hardwood for forming blocks. If they are carefully selected for close grain structure, hard maple and beech are good woods for die applications.

High density panels composed of compressed wood fiber and lignin are used for jigs, dies, fixtures, templates, patterns and molds. One widely known hardboard, Masonite™, is a cellulose semiplastic material available in various thicknesses. It can be readily laminated with cold-setting adhesives. Such wood based materials are suitable for short run dies for prototype. Such dies are popular in the aircraft industry.

Molded rubber female dies and rubber covered punches are used in difficult forming operations, such as the production of deeply fluted lighting reflectors. Many types of rubbers and rubber compounds are used in pressworking including natural rubber, neoprene, and polyurethane.

Specifications for rubbers used in the conventional Guerin, Marform, and Hydroform processes are determined by the performance needed for the process. Some rubber compounds, especially polyurethane, can be cast in place and cured to form the needed shape. This permits economical in-house production of forming die components, part strippers, pressure pads, and nonmarring automation fingers.

Soft, medium, and hard cork layers compressed into sheet form, are sometimes used with, or in place of, rubber pads. Cork deforms only slightly in any direction other than that of the applied load, while rubber flows in all directions.

Like wood, plastic materials are limited for dies by the pressures involved in the process. Selection of plastics is based on economy relative to die life expectancy. Draw radii can be expected to be a primary source of concern because maximum loads and abrasion occur in these areas.

Draw dies having a metal core of either ferrous or zinc-alloy materials, capped with a working face of epoxy, are used in the aircraft, appliance, and automotive industries. Rubber forming dies are made of combinations of cast and laminated epoxy applied to a heavy steel base.

Polyester resins are also used for low-volume production tooling. Cost is the chief advantage this material has over the stronger and more stable epoxy resins.

Combining many of the good properties of both elastomers and plastics, polyurethanes have demonstrated a unique combination of abrasion resistance, tensile strength, and high load bearing capacity not available in conventional elastomers. Also, polyurethanes offer impact resistance and resilience not available in plastics.

Because of their liquid uncured form and their excellent cured properties, these polymers are useful in many tool design applications, including:

- Draw dies;
- Drop-hammer dies;
- Forming and stamping pads;
- Press-brake forming dies;
- Mandrels; and
- Expanding punches.

A major use of this material is for die automation components such as kickers, lifter heads, and rollers where the excellent wear-resistance of the material, together with its nonmarking characteristics, are very useful. Polyurethane is available as a two-part liquid formulation that can be mixed and cast in place to form custom-made die pads and transfer automation jaws.

PRESSWORKING LUBRICANTS AND PROPER APPLICATION

Selecting the proper pressworking lubricant for a specific application is based on several major factors. These include:

- Condition of purchased material;
- Type of tooling;
- Severity of the pressworking operation;
- Desired application technique;
- Cleaning and finishing requirements;
- Disposal cost; and
- Worker acceptance.

Current United States federal law requires lubricant suppliers to furnish a complete safety data sheet to users. This data must be available to all employees using the material.

Guidelines for Choosing Pressworking Lubricants

Presswork operations are becoming more integrated. For example, some dies incorporate an assembly station and tapping operations requiring the lubricant to serve multiple needs.

Pressworking lubricants fall into four categories:

- Fluids;
- Pastes;
- Dry films; and
- Soaps.

Fluids, the most widely used pressworking lubricants, have two major classes, solutions and emulsions.

A solution is a fluid in which all ingredients are mutually soluble. Solutions utilize either oil or water as a base. A typical oil-based solution for difficult operations may consist of a mineral oil base, a wetting agent, a rust inhibitor, and an extreme pressure agent. Mineral oil solutions provide good fluid integrity and are generally safe from biological attack. Oil based solutions can be recycled and clarified.

Water-based solutions may contain surfactants, soluble esters, soluble rust preventives, and, in some cases, extreme pressure agents. These solutions differ greatly from oil-based lubricants. One advantage is that the evaporation of water helps cool tooling used in severe operations. These solutions can be recycled, provided water and other constituents are added as needed to maintain the correct composition.

An emulsion is a fluid system where one immiscible fluid is suspended in another. The mixture with formed droplets is an emulsion. In pressworking lubricants the continuous phase is generally water containing suspended oil or a synthetic solution. The water can contain several additives such as extreme pressure agents, fats, or polymers depending upon the severity of the work. Stable emulsions require proper composition as well as mixing devices and correct application techniques.

Synthetic solutions are pressworking lubricants combining excellent high-temperature properties and good boundary lubrication. The main ingredients for these types of synthetics are synthesized hydrocarbons (polyalphaolefins) and polybutane derivatives.

Synthetic solutions are much like oil-based solutions in their physical characteristics, but have a higher degree of resistance to oxidation. The polyglycols, polyesters, and dibasic acid esters have superior high-temperature stability. Synthetic solutions are more costly than oil-based solutions.

For severe pressworking operations, high film strength lubricants are sometimes needed. Pastes, suspensions, and conversion coatings are often used.

Pastes

Pastes can be made in several ways. They can be formulated with oil or be water-based, containing pigments. Much like pigments in paints, pigments used in pastes are fine particles of insoluble solids. Some commonly used pigments are talc, china clay, and lithopone. Some pigmented pastes are available as emulsion compounds composed of fats and fatty oils and sometimes mineral oil pigment, emulsifier, and soap and water.

They can be used as supplied or diluted with water or oil for easier handling and applications. For more severe work, some pigments are dispersed in an oil base treated with extreme pressure agents. These compounds may be diluted with mineral oil.

Nonpigmented pastes are available in several forms. Emulsion drawing compounds are pastes composed of fats and fatty oils and their fatty acids, that may contain free mineral oil, various emulsifiers, and water. These products are diluted with water before use. Fats, fatty oils, and fatty acids are sometimes used straight, but usually mixed with mineral oil. Mineral oil and greases can be used straight when necessary.

Dry Films

Dry-processed coatings are increasingly used because of their economy, cleanliness, and ease of handling. The coatings, which may consist of dry soap films, wax or wax-fatty compositions, can be applied by hot dipping or spraying the material hot. Also used is cold application in a solvent, in which the vehicle evaporates, leaving a dry coating. For high production, roller coating is preferred for sheet and coil stock.

Phosphate coatings are chemical immersion coatings that provide a measure of lubrication. Graphite coatings are useful under high temperature and heavy-unit-load conditions where it is not feasible to use other lubricants. Graphite is difficult to remove and is used only when strictly necessary.

Suspensions are lubricants consisting of fine particles of various solid lubricants dispersed throughout a carrier fluid. The lubricant is usually insoluble in the carrier or vehicle fluid.

Warm and hot forging of steel and other metals relies on the use of graphite suspended in mineral oil or, more often, water. The structure of graphite, its stability at temperatures over 1000°F (540°C), and positive side effects of air and water vapor on its lubrication, favors the application of graphite with a carrier as a very cost-effective high temperature lubricant.

Soaps

Solid lubricants are used in pressworking, particularly where operations encounter high unit pressures or high temperatures brought on by deformation of metal. The two major types of solid lubrication are soaps and soap combinations or graphite and molybdenum disulfide, either alone or in combination.

There are four types of what are called soap lubricants. The first class are dry powders, which usually are sodium or other metallic type soaps. They are generally furnished in powder form for use in tube bending and wire drawing.

The second class are dried-film compounds. These are usually soluble soaps, often mixed with other soluble solids. An example are those containing borax, waxes, wetting agents, and other chemical ingredients. These dry films are used for drawing and can be applied by spraying or dipping with a 10 to 20% hot solution and then dried prior to the forming or drawing operation.

The third class are sodium or potassium soluble soaps, which are diluted up to 10% with water.

Metallic soaps such as aluminum stearate, calcium stearate which can be used alone or in combination with molybdenum disulfide (MoS2), and/or graphite. This combination, which makes up the fourth class, is used for wire drawing and warm forming.

Advantages of Water-based Pressworking Lubricants

There are many benefits from the use of water-based lubricant solutions. When compared to mineral oil-based lubricants, the advantages include lower initial cost, compatibility with secondary operations, and elimination or reduction of cleaning.

In some instances, portions of the production process require modification to permit the use of water-based lubricants. For example, if the lubricant is recycled, additional clarification and contamination control equipment may have to be installed to maintain lubricant stability and product quality.

When selecting material for use with water-based lubricant solutions, it should be ordered clean, dry, and as free as possible from mill oil and rust preventives.

Paper clad and plastic film protected material finishes can be formed with water-based solutions without damage to the protective coverings by simply applying the lubricant with roller coaters or spray units.

Tests should be conducted to determine that the water-based solution does not react with materials, such as galvanized steel and aluminum, resulting in corrosion or staining. The proper dilution strength should be carefully noted. Operating with the proper concentration can be the difference between success and failure. Surface tests should also be performed on coated stocks.

Material Surfaces

The four most common categories of material surfaces are:

- Normal surfaces;
- Active surfaces;
- Inactive surfaces; and
- Coated surfaces.

Normal surfaces readily retain lubricant. Generally, a special wetting agent is not needed. Bare mild steel is the most common normal surface. Normal surfaces are easy to lubricate if there is not any dirt or contaminants present.

An active material surface is one in which the strength of the bond between the lubricant and metal atomic structure is great. Because the attractive energy is high, surface chemical reactions are encouraged. The surface attraction is also a function of the lubricant's composition. This explains why chemically active additives, wetting agents, and extreme pressure agents are so effective.

An inactive surface is one with low bond strength between the lubricant, including additives and the metal atomic structure. Typical inactive surfaces stainless steel, aluminum, and nickel. To be effective, the lubricant must have a high film strength that provides a mechanical barrier between the tool and part to prevent metal to metal contact.

The most common coated stock is galvanized steel. Material pickup can occur when forming galvanized material. Special water soluble lubricants are available that can keep the galvanized metal particles from building up on the tooling. Dry film and water-based solutions can also be successfully used on galvanized stock.

Discoloration of Zinc Coatings

The term "white rust" describes destruction of the surface of galvanized steel or zinc by oxygen or other chemical elements. The reaction is accelerated by the presence of moisture.

The rate of attack is related to temperature, pH, and composition of moisture and the concentration of dissolved gases within the moisture. The rate of surface breakdown increases with the amount of dissolved oxygen and carbon dioxide. The corrosion cycle starts with the formation of zinc oxide, which in turn is converted to zinc hydroxide and then to basic zinc carbonate in the presence of carbon dioxide.

It is essential that parts be drained of as much water-based lubricant as possible. Parts in process should be used rapidly in order to avoid stagnation of any residual fluid. Conditions of high relative humidity and temperature aggravate the condition.

Progressive dies and transfer press operations require that the lubricant used provides correct lubrication for every operation, and is compatible with the materials used to build them. Some lubricants may attack the self-lubricating graphite composition plugs pressed into guide bushings and wear plates. Damage to pads and seals made of rubber and elastomer products is also a consideration.

Maintaining and operating tooling when it is used with water-based solutions requires special consideration. Most synthetics are alkaline in nature and act as detergent soaps. These solutions remove most conventional greases and machine oils. It may become necessary to protect the tooling and related die components with rust preventives, especially during extended periods of time when the tooling is not in operation.

Some lubricants attack the binding material of tungsten carbide. Electromotive forces (EMF) can be generated between the carbide surface and the machine tool with resulting erosion of the cobalt binder. This same condition can be caused by electrical current, present in the die from sensing probes and other electrical hardware. In such cases, the coolant must not act as an electrolyte.

APPLICATION TECHNIQUES

There are many ways to apply lubricants for pressworking operations. Among the best are recirculating systems, roller coaters, and airless sprays.

The roller coating method is one of the most efficient ways to apply lubricant. The preferred position is between the die and the feeding mechanism. Placing the coater before the coil feed can cause the lubricant to be mechanically worked off the metal surface, often resulting in feeder slippage.

Roller coaters range in complexity from homemade units using paint application rollers to the system illustrated in Figure 14-2. With this type of roller coater either one or both sides of the material can be lubricated. The stock is coated with lubricant as it passes through the applicator rollers. Any excessive lubricant is then squeezed off by wiper rollers and returned to the recirculating reservoir where it is filtered and available for reapplication.

An airless spray is a mechanical method of producing a finely divided spray of lubricant. Pressure is applied by an intensifier, then carried through a high-pressure hose to a nozzle with a tiny orifice where the lubricant is expelled as a fine spray.

A typical airless setup consists of:

- Air-powered intensifier assembly;
- Check valve;
- Lubricant reservoir;
- Nozzles; and
- Valves to activate the nozzles.

The spray pattern can be either round or fan shaped. Airless spray systems are excellent for spot lubrication within the die or lubricating the stock before it enters the die.

A modern airless spray system produces very little mist, minimizing overspray problems. It can be directed precisely at a target area in the die and timed to operate in conjunction with the equipment cycle.

The use of drip application methods is difficult to control. A typical drip lubricator is generally mounted after the stock or roll feed. The drip system is regulated by a petcock, adjusted for the flow desired. This method does not provide for automatic shutoff when the equipment stops, resulting in wasted lubricant, messy parts, and housekeeping problems. If a large stock area is to be lubricated, a drip lubricator is a very unwise choice.

The lower lubricant consumption and increased productivity achieved by eliminating costly drip or manual application can often pay back the cost of automatic application equipment in a short time. In laying out a new press line, it may be advantageous to install a fully automatic coil feeder, stock straightener, and stock lubricator to obtain the maximum productivity from the press and tooling.

Cleaning and Secondary Operation Requirements

The cleaning and finishing processes used often limits the choice of a pressworking lubricant. For example, low temperature cleaning lines generally are not capable of removing heavy oils and extreme pressure agents. For low-temperature cleaning, specially formulated water-soluble lubricants are recommended.

Hot alkaline wash systems can clean heavy residual oils and other difficult-to-remove drawing compounds. The waste disposal costs for alkaline systems can be quite high unless skimmers and clarification equipment are used to lengthen the cleaner's life.

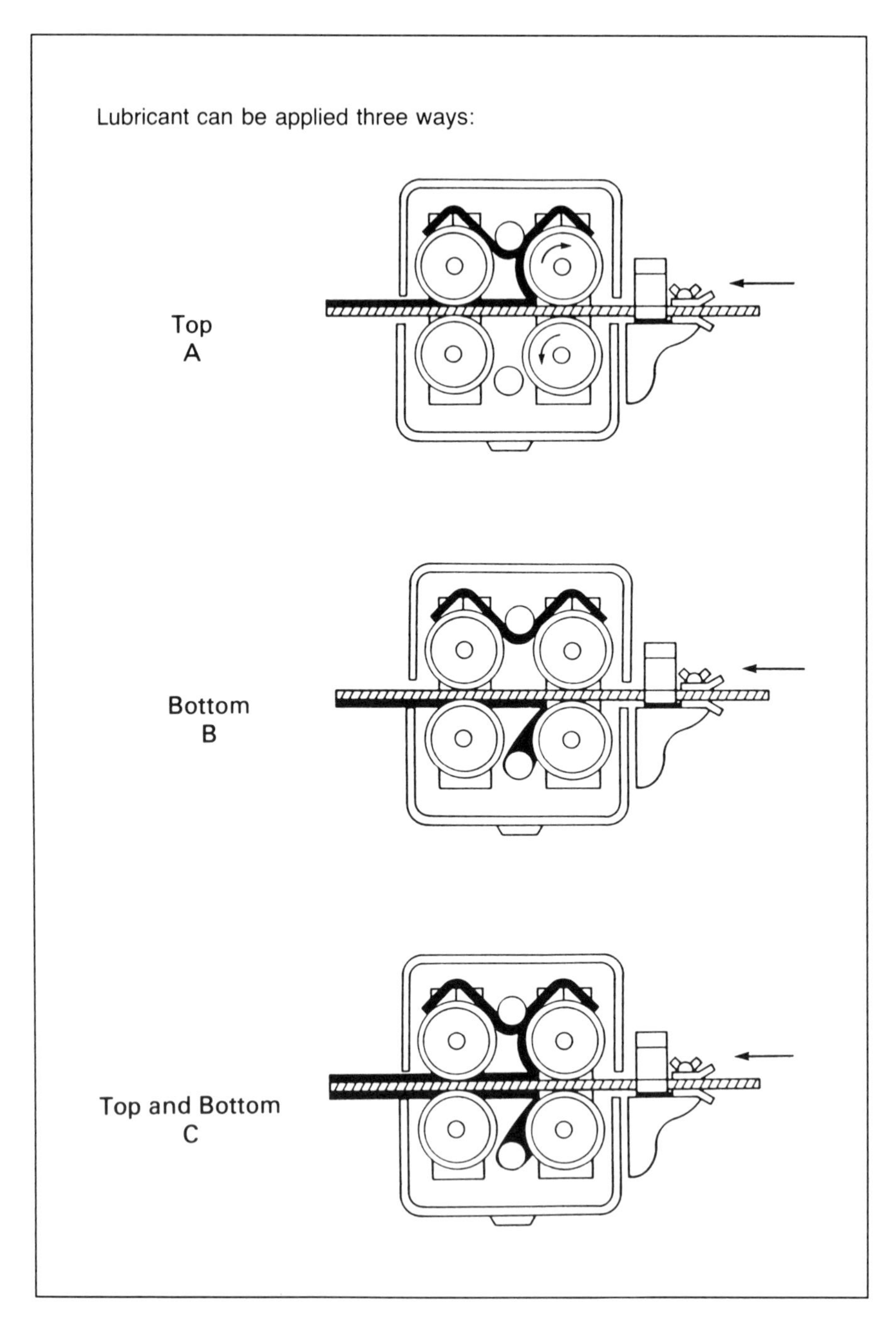

Figure 14-2. Roll coaters permit versatile lubricant application. (Tower Oil and Technology Company)

Other considerations for compatibility with secondary operations include suitability for use with finishing methods such as painting or powder coating. Also, compatibility with assembly or joining is important.

Using some types of water extendable pressworking lubricants (emulsions and chemical solutions) may allow welding without smoke and weld integrity problems. Painting can be performed without prior cleaning in some cases. Heat-treating, annealing, and stress relieving operations may also be affected by the choice of lubricant.

Lubricant Clarification Recycling and Disposal

Recycling pressworking lubricants reduces the amount of new lubricants required as well as disposal costs. Successfully recycling pressworking lubricants depends largely on controlling the contaminants that can affect its useful life. If the lubricant is oil based, some chemical components may precipitate or be deposited on surfaces wetted by the fluid. If solid lubricants are used in suspension, these too deposit.

Stability of emulsions and soluble oils decreases with time and amount of reuse. In the case of water-based solutions, changes in PH may cause precipitation. These changes determine the choice of the recycling equipment and disposal requirements.

Contaminants in lubricants can include particles from the stock that break loose during the pressworking operation. Some typical contaminants are mill scale, aluminum oxide, and galvanized particulate. In addition, contaminants other than those directly resulting from the process may be present. These may include cleaning fluids, oil absorbent from the floor, and lubricants from the pressworking equipment.

If lubricants are to be recycled, appropriate clarification equipment is necessary. A simple gravity-settling tank will remove the debris heavier than the lubricant. Gravity-setting systems, because of their low cost and simplicity, are often used as the first stage in a total system, which might include cyclone separators, filtration devices, and/or a centrifuge. This approach reduces operating costs of the more sophisticated equipment and frequently improves productivity of the system.

Removing tramp oil from the surface of a gravity-settling tank can easily be done by skimming. High-volume recycling systems often incorporate cyclone or centrifuge separators. However, simple filtration is usually employed for low-volume work.

Gravity, pressure, and vacuum filters are commonly used for final clarification of pressworking lubricants prior to reuse. The equipment can be as simple as old-fashioned dairy cans and milk filters to sophisticated suction filters. In all systems, the filter media should have a scheduled replacement based on use that includes proper disposal.

Pressworking lubricants containing water can be affected by any number of microbes that occur in the environment. Lubricant constituents is the food eaten by these microbes. When growth occurs by attacking a constituent, the lubricant can be changed and will no longer perform as intended.

The use of an effective biocide together with proper clarification and filtration is a good procedure to follow. Water-based solutions usually require the addition of water to make up for evaporation. Impurities in the water should be avoided to help ensure long lubricant life. Other constituents may also require replenishment. Most water-based solutions can be tested and controlled with the aid of a chemical titration kit available from the manufacturer of the particular solution.

Clarification, maintaining a correct balance of lubricant constituents, and avoiding microbe growth all promote the recyclability of the lubricant. These measures also help avoid irritation of the operator's skin, which can lead to dermatitis problems.

BIBLIOGRAPHY

1. D. Smith, *Die Design Handbook*, Section 28, "Ferrous Die Materials," Society of Manufacturing Engineers, Dearborn, Michigan, 1990.
2. S. G. Fletcher, "The Selection and Treatment of Die Steels," *The Tool Engineer*, April,1952.
3. R. Denton, "Application of Ion Nitriding," Society of Manufacturing Engineers, Selecting Tooling Materials and Tooling Treatments for Increased Tool Performance Clinic, November 1989.
4. J. Ivaska, Jr., "Lubricants—A Productive Tool in the Metal Stamping Process," SME Technical Paper TE77-499, December, 1977.
5. E. Nachtman, "A Review of Surface-Lubricants Inter-reactions During Metal Forming," SME Technical Paper MS77-338, July, 1977.
6. J. Ivaska, Jr., "How Metal Forming Lubricants Affect the Finishing Process," SME Technical Paper FC83-690, 1983.
7. J. Ivaska, Jr., "Lubricant Implications for Integrated Pressworkers," SME Technical Paper MF87-003, July, 1987.

CHAPTER 15

THE FUTURE OF METAL STAMPING

A LOOK TO THE FUTURE

Engineering materials requiring less complex tooling are displacing steel and other metals. The concern is that a complex stamping, particularly one requiring deep drawing, may not be produced successfully on the dies proposed by the tooling designer. This can introduce uncertainty into the time to market schedule.

Molded plastic components are generally successful when the tooling design is capable of part production. Weight reduction in automotive and aircraft applications is another factor favoring nonmetallic, often more costly, materials.

The design cycle for many stampings is needlessly complex. Reducing the cycle time from design concept to market requires teamwork and an interdisciplinary approach to identify and solve potential stamping problems. Staff members from product designers through tool building and facilities engineering management — developed highly compartmentalized views of their individual roles. Delays result in the overall process of bringing a new product to market.

Developing Teamwork

Training in how the material forming process works is one way to increase cooperation. Training can be done by bringing the persons responsible for launching the product together for hands-on instruction in the basics of pressworking technology. This instruction should include sheet metal formability and tool design. The goal is to create a new appreciation and sense of responsibility for the success of the overall process and infuse it into the plant culture.

Problem solving and direct application should be this training program's strength. Participants must develop a broad perspective how their job function ties in with stamping success from design through the life of the stamping assembled in the end product. Participants from process engineering and tool design to production scheduling found this training provides them the tools needed to effectively perform their jobs. A sample of topics might include:

- How the various stamping processes work;
- Sources of stamping variability;
- Building a database of good versus bad designs;
- Team player concepts with new product introduction;
- Control the interaction between press and die functions;
- Circle grid and waveform signature analysis; and
- Computerized formability analysis.

Factors Favoring Metal Stampings

In spite of competition from nonmetallic materials, metal stampings remain very attractive from a cost-to-strength ratio in high-volume production. These factors favoring metal stamping include:

- Maximum sizes are limited chiefly by press capacity.
- Few limitations upon minimum size; sections as thin as 0.003 inch (0.08 mm) are possible; parts so small that 10,000 may be held in one hand.
- Tolerances are very good. For small stampings made with progressive or compound dies, plus or minus 0.002 inch (0.06 mm) is common, and closer limits are possible on small and thin parts. Precision fineblanked parts typically are stamped to much closer tolerances.
- The weight factor is highly advantageous. Parts formed from sheet metal have favorable weight-to-strength ratios.
- Surface smoothness is excellent. The surface condition usually is not affected by the forming operation. Prepainted metals are in common use as stamping materials. The sheet steel's surface finish can be optimized by special texturizing of the finish mill rolls for maximum paint luster on the finished stamping.
- A wide choice of materials is available, including most metals and alloys in sheet form.

Production and Economic Factors

The lead time required for complex tooling is long compared with some other production methods. Traditional die design and tryout may take many months. Integration of the computer-aided part and die design (CAD) with computer-aided die manufacturing (CAM) can greatly reduce tooling costs and lead time. Output is very high; over 600 large automotive body stampings per hour are commonly produced on automated transfer presses. Small stampings produced on high-speed equipment often have production rates well in excess of 36,000 pieces per hour.

Economic factors include the following:

- Stamping material costs are comparatively low. A favorable cost factor is the minimum scrap loss achieved through careful selection of stock, and strip layout.
- Tool and die costs are high — often higher than tooling for comparable parts to be die-cast. Costs are most favorable where large production is planned.
- Direct labor costs depend on the part size and shape and extent of automation. Usually these costs are low.
- Presses, except for small manual punch presses, typically cost more than standard machining equipment such as lathes and grinders, and require higher machine-hour rates.
- Finishing costs are low. Often no finishing is required. Painting or plating can provide a long-lasting attractive appearance on low-cost materials such as low-carbon steel.
- Inventory costs can be quite low. Quick die change techniques permit parts to be produced as needed with a changeover time typically under 10 minutes.

APPLICATIONS

Automotive

Stamped automotive body panels have been made of increasingly thinner or lighter sheet steel for decades. In the United States, government-mandated body panel thickness reduction was ordered at the onset of World War II. During the war, a few cars were made to this standard as the auto makers rapidly converted to the production of war material. The thinner material proved to be satisfactory for motor vehicle construction, pointing out that cars of that era were over-designed.

Thinner metals are more apt to fail in deep-drawing applications than thicker gages. Over the last several decades, fuel economy concerns and other economic considerations resulted in body part designs employing designs not requiring deep drawing operations. Consequently, thinner metals are feasible for stretch forming modern body panels. Economy of tooling also was a driving factor. Today, a fender or quarter panel once taking seven or eight operations now only requires four or five hits.

Simultaneous engineering in globally competitive car companies is shortening the lead time from concept to production. An important factor is controlling product changes affecting tooling.

Considerable tooling and production economy is achieved by reducing the number of stampings required to build a car and the number of dies required per stamping. The average number of hits per part is now under four. These factors make steel and other metals compete as engineering materials for vehicles produced in lower volumes.

Energy Absorbing Designs

In the United States, fuel economy and occupant safety are important government public policy issues. Crash testing vehicles determines occupant safety in a collision. The occupants of a vehicle are subject to rapid deceleration in a collision. The vehicle's kinetic energy is best absorbed by a body made of thin easily deformed parts. Seat belts and air bags together with an energy-absorbing body structure have dramatically increased the number of passenger miles driven per fatality. Ideally, occupants should survive a collision with little or no injury, even though the vehicle is severely deformed.

Aluminum alloys with sufficient strength for automotive applications are not as readily formed as steel. However, for large body panel applications not requiring severe deformation, such as hoods and deck lids, aluminum is increasingly used. Alloy 2036-T4 is one of several popular materials for such applications. To obtain equal stiffness, the aluminum must be made approximately 44% thicker. The specific gravity of aluminum is about 35% of that of steel. Typically, aluminum stampings weigh approximately half as much as steel for equal stiffness.

Automotive designers use computer-aided analytical tools to predict a panel design can be successfully formed. The material thickness needed to provide sufficient strength as part of the complete assembly well can also be estimated.

Properly engineered automotive body applications can use steel 30% or more thinner than they could several decades ago. The increased corrosion resistance of most cars produced in the last decade is due to the use of galvanized steel. Much of the material is coated by electro-galvanizing one side (which becomes the inside) of the finished panel. Organic coatings containing zinc are also used. The latter are available as precoated material or are selectively applied at the fabrication facility.

Automotive structural and outer body components are often made of either high-strength low-alloy (HSLA) steel or steels that age-harden or bake-harden in the paint drying operation. These steels achieve equal strength with thinner gage material. The driving factor is total vehicle weight reduction. Pressworking processes that cold-work the material also increase strength. Developing the required strength often requires optimum amounts of biaxial stretch during forming.

A possible advantage of plastics over metal in body panel applications is better dent resistance. Side panel damage is mainly due to parking lot door dings. Roof, hood, and deck-lid damage often result from the impact of large hailstones. A third problem is hand imprinting.

Automakers and steel producers have a powerful vested interest in this area. They have production facilities designed for high-volume production of body panels and wide flat sheet steel. Determining what is acceptable to the customer together with standardized testing procedures is a means to determine an acceptable limitation in thickness reduction.

Appliance Applications

Appliances, including heating ventilating and air conditioning (HVAC) equipment, account for many stamped metal parts. Most of these stampings are made to exacting tolerances that provide energy efficient operation and long product service life.

For example, the heat exchangers in HVAC equipment such as air conditioners and high-efficiency gas furnaces have many thin aluminum alloy fins. These fins are stamped to close tolerances. Refrigerant compressor cases are deep-drawn from heavy gage steel stock. The case halves must fit together snugly and hold many internal parts in close alignment. The dependability, value, and service life of most appliances depend on many stampings produced in high volumes to rigid specifications at low cost.

Electrical, Electronic, and Computer Stampings

All metals conduct electricity. Silver is the most conductive of the elements for a given cross section at room temperature. Copper is a close second. Aluminum will conduct more current per pound than either silver or copper, although the cross sectional area is larger.

Electrical stampings range from heavy buss bars and terminal lugs used for power distribution to minute integrated circuit and computer terminals. In each case, electrical conductivity and a long-term trouble-free connection is important.

Decades ago, huge volumes of precision stampings were needed to produce vacuum tubes and telephone relays. Once, the American Telephone and Telegraph Company (AT&T) had the largest amount of gold in private hands. It was found in billions of relay contacts in the telephone system, with each contact attached to a precision stamping. Automatic pressworking equipment assembled, formed, and coined the parts in multiple transfer press operations.

Few uses remain for small electromechanical relays. However, the technology evolved into the tooling needed to produce electronic parts such as integrated circuit connectors and computer cable terminals. Every decrease in the number of stampings to construct electronic components is more than offset by increased complexity and the number of electronic products employing stampings.

An important use of stampings is electrical and electronic enclosures. These products range in size from large housings for power distribution transformers to tiny electromagnetic interference (EMI) shields installed directly on printed circuit boards. Metal enclosures are superior to conductive plastic cases for durable EMI shielding.

Recyclability

Assembled goods containing metallic components are recyclable. Often the value of the reclaimed metals is substantially more than recycling process cost. A large percentage of the primary metals entering the commercial market has been recycled. This is especially true of aluminum, zinc, copper and its alloys, as well as steel.

First, liquids and refrigerants in automobiles and appliances are removed for recycling or safe disposal. Next, salvageable parts are removed for resale. Recyclable plastics, motors, drive components, and large pieces of nonferrous metal are removed. Finally, the remainder is mechanically shredded.

After shredding, steel is magnetically separated from the other substances. Recovery of the remaining metallic scrap by differences in physical properties is the final step. The residue, known as "shredder fluff," must be landfilled. The use of many nonmetallic parts exacerbates environmental problems.

Recycling and refining scrap metals into new material is a practice dating to the dawn of the bronze and iron ages. Metals are easily and economically recycled. The final chapter in the design life of a product should be recycling it into reusable materials efficiently. Design for recyclability is a coming environmental requirement. Metal components are essentially 100% recyclable. This is especially true of steel because of the ease of magnetic separation.

Metal stamping can be a very environmentally friendly process, as well as a safe occupation. Successful stamping is simply a matter of following proven good procedures and engineering principles. Long-term prosperity also requires economic common sense and good human relations policies.

BIBLIOGRAPHY

1. David A. Smith, Editor, *Die Design Handbook,* "Progressive Die Design, Computer-Aided Design and Machining," Society of Manufacturing Engineers, Dearborn, Michigan, 1990.
2. The University of Michigan, *A Key to World Class Manufacture of Automotive Bodies,* The Auto Steel Partnership, 1993.
3. J. Story, *Forming of Aluminum Sheet,* presented at the SME clinic, Practical Sheet Metal Part Design, September, 1992.
4. R. Krupitzer and R. Harris, *Automotive Engineering,* "Dent Resistance Gains for Steel Body Panels," December, 1992.

INDEX

S

T

U

V